W9-BXT-009

The Neural Crest in Development and Evolution

Springer
New York
Berlin
Heidelberg
Barcelona
Hong Kong
London
Milan
Paris
Singapore
Tokyo

Brian K. Hall

The Neural Crest in Development and Evolution

With 65 Illustrations, 8 in Full Color

Springer

Brian K. Hall
Department of Biology
Dalhousie University
Halifax, Nova Scotia B3H 4J1
Canada

Cover Illustration: Migrating cranial neural crest cells in a *Xenopus laevis* embryo, from an original supplied by Douglas DeSimone; see Figure 9.3A.

Library of Congress Cataloging-in-Publication Data
Hall, Brian Keith, 1941–
The neural crest in development and evolution / Brian K. Hall.
p. cm.
Includes bibliographical references and index.
ISBN 0-387-98702-9 (hc. : alk. paper)
1. Neural crest. I. Title.
QL938.N48H347 1999
573.8'6387—dc21 98-51752

Printed on acid-free paper

Production managed by Timothy Taylor; manufacturing supervised by Thomas King.
Photocomposed copy prepared by Carlson Co., Yellow Springs, OH.
Printed and bound by Maple-Vail Book Manufacturing Group, York, PA.
Printed in the United States of America.

9 8 7 6 5 4 3 2 1

ISBN 0-387-98702-9 Springer-Verlag New York Berlin Heidelberg SPIN 10707191

Preface

Knowledge of the development and evolution of the neural crest sheds light on many of the oldest unanswered questions in developmental biology. What is the role of germ layers in early embryogenesis? How does the nervous system develop? How does the vertebrate head arise developmentally and how did it arise evolutionarily? How do growth factors and *Hox* genes direct cell differentiation and embryonic patterning? What goes wrong when development is misdirected by mutations or by exposure of embryos to exogenous agents such as drugs, alcohol, or excess vitamin A?

In 1988, I was instrumental in organizing the publication of a facsimile reprint of the classic monograph by Sven Hörstadius, *The Neural Crest: Its properties and derivatives in the light of experimental research*, which was originally published in 1950. Included with the reprint was my analysis of more recent studies of the neural crest and its derivatives. The explosion of interest in and knowledge of the neural crest over the past decade, however, has prompted me to produce this new treatment. Here, as in my 1988 overview, I take a broad approach to the neural crest, dealing with its discovery, its embryological and evolutionary origins, its cellular derivatives—in both agnathan and jawed vertebrates or gnathostomes—and the broad topics of migration and differentiation in normal development. Cells from the neural crest are also associated with many developmental abnormalities. The book would be incomplete without a discussion of tumors (neurocristopathies) and syndromes and birth defects involving neural crest cells.

The book is organized into three parts. Part I begins with a chapter devoted to the discovery of the neural crest and the impact of that discovery on entrenched notions of germ-layer specificity and the germ-layer theory, a theory that placed a straitjacket around embryology and evolution for almost a century. In Chapter 2, I discuss the embryological origins of the neural crest in vertebrates. Chapter 3 is devoted to the evolutionary origins of the neural crest (is there a protoneural crest in urochordates or cephalochordates?) and to the origin of skeletal tissues from the neural crest in the first craniates. Because the neural crest is not limited to vertebrates, I round out Part I on origins with a discussion of the neural crest and neural crest derivatives in agnathan craniates (hagfishes) and agnathan vertebrates (lampreys). Part II (chapters 4 to 8) presents a survey of our knowledge of the neural crest and its derivatives in the major groups of jawed vertebrates. The first two chapters of Part III analyze the mechanisms that control the migration and differentiation of neural crest cells, the last two tumors and birth defects involving these cells. To avoid interrupting the flow of the text I have placed most references and some supporting statements in endnotes, which serve as an annotated bibliography through which access to the literature may be readily obtained. Boxes are used for items of interest that may be at a tangent to the main argument. The text is extensively illustrated and referenced and there is a detailed index.

It is a pleasure to thank Tom Miyake, Janet Vaglia, Chris Rose, Moya Smith, Lennart Olsson, Jianmin Fang, Ann Graveson, and Steve Smith for discussions on various aspects of the neural crest. Tom Miyake used his considerable artistic skills in adapting some of the figures from the literature and in transforming rough drafts of my figures into their final form. Many thanks Tom. June Hall edited the manuscript with an eye to style and comprehensibility. Many thanks June.

Halifax, Nova Scotia
Canada

Brian K. Hall

Contents

Part I

Discovery and Origins

1

Discovery

Zwischenstrang

In 1868, Wilhelm His (1831–1904), a Swiss embryologist, identified a band of cells sandwiched between the developing neural tube and the future epidermal ectoderm in neurula-stage chick embryos as the source of spinal and cranial ganglia. His called this band *Zwischenstrang*—the intermediate cord. At the time of his discovery, His was professor of anatomy and physiology in his native Basel and an influential member of the city parliament. A man of many accomplishments, His also invented, in 1866, a microtome with a system for measuring section thickness when cutting thin, serial sections of animal and plant material. He discovered that nerve fibers arise from single nerve cells and coined the term *neuroblasts* for those cells, and, in 1895, wrote *Nomina Anatomica*, the treatise that introduced order into anatomical terminology. His also rediscovered the burial place of Johann Sebastian Bach and identified the skull by comparing a clay reconstruction of the skull with paintings of Bach.[1]

His was a vigorous and, if Ernst Haeckel's ferocious response is any guide, very effective opponent of the Haeckelian biogenetic law that ontogeny recapitulates phylogeny. Wilhelm Roux may have been the "father" of experimental embryology, but Wilhelm His founded the causal analysis of development. The first modern scientist to provide a causal explanation for embryonic development, His based his explanation on mechanics (embryos as rubber tubes), a developmental physiology, and predetermined organ-forming germinal regions, each of which contained cells with specified fates. In 1874, His included the intermediate cord, or neural crest as we now know it, as one of the organ-forming germinal regions. Although not an experimentalist, in the view of a preeminent historian of biology: "No one before him described so diligently, lucidly, and exactly the minutiae of the development of the higher animals and especially of man" (Singer 1959, p. 497).

As far as I can determine, the term *neural crest* was first used by Arthur Milnes Marshall (1852–1893) in his 1879 paper on the development of the olfactory organ. By age 27 Marshall was professor of zoology at Owens College in Manchester, England, where he remained until his death in a fall from England's highest peak, Scafell Pike (3210 ft).[2] His scientific reputation rests on his work on cranial nerves, olfactory organs, and head cavities, and to a lesser extent on his later work on corals. In an 1878 paper on the development of the cranial nerves in chick embryos, Marshall used the term *neural ridge* for the cells that give rise to cranial and spinal ganglia. Realizing that this term was less descriptive than was desirable, a year later he replaced neural ridge with neural crest. As told in his own words:

I take this opportunity to make a slight alteration in the nomenclature adopted in my former paper. I have there suggested the term *neural ridge* for the longitudinal ridge of cells which grows out from the reentering angle between the external epiblast and the neural canal, and from which the nerves, whether cranial or spinal arise. Since this ridge appears before closure of the neural canal is effected, there are manifestly *two* neural ridges, one on either side; but I have also applied the same term, *neural ridge*, to the single outgrowth formed by the fusion of the neural ridges of the two sides after complete closure of the neural canal is effected, and after the external epiblast has become completely separated from the neural canal. I propose in future to speak of this single median outgrowth as the *neural crest*, limiting the term *neural ridge* to the former acceptation. (Marshall 1879, p. 305, n. 2)

The neural crest—its embryological and evolutionary origins, the multitude of cells, tissues, and organs that develop either directly from neural crest or under its influence, how those cells migrate and differentiate, and abnormalities that result from neural crest tumors, deficiencies, or defects—is the topic of this book. In this chapter I provide a brief overview of major phases of investigation of the neural crest, discuss the impact for the germ-layer theory of the discovery of the neural crest and of secondary neurulation, and argue for the neural crest as a fourth germ layer in craniates alongside ectoderm, mesoderm, and endoderm. I then devote chapters 2 and 3 to the embryological and evolutionary origins of the neural crest, and Chapter 4 to the neural crest in agnathans; Part II (chapters 5 to 8) to a survey of neural crest derivatives in the different groups of jawed vertebrates; and Part III to mechanisms of neural crest cell migration and differentiation (chapters 9 and 10) and to tumors and birth defects involving neural crest cells (chapters 11 and 12).

A Brief Overview

The neural crest has long held a fascination for developmental and, more recently, evolutionary biologists. Although it was initially associated with the origin of cranial and spinal ganglia and neurons, Julia Platt demonstrated in the 1890s that the visceral cartilages of the craniofacial and branchial arch[3] skeletons and dentine-forming cells of the teeth of the mud puppy, *Necturus*, also arise from the neural crest. Although supported by several workers in the late nineteenth century, her view gained acceptance only slowly; her results on the neural crest origin of the visceral arch skeletons were too controversial, for they ran completely counter to the entrenched germ-layer theory discussed in the following section. Platt was also very outspoken and critical of her male colleagues (justifiably so when she thought they were arguing from weak factual bases), and such attributes did not sit well with those men deciding on academic appointments. Despite a year-long search in the United States and Europe, she failed to obtain an academic position, turning to civic work in Pacific Grove, California. In 1931, aged 74, Julia Platt became that city's first female mayor.[4]

Because a neural crest origin was so contentious, there was a 50-year gap between Platt's papers and the next reports of neural crest contribution to the skeleton by Sven Hörstadius, Sven Sellman, and Gavin de Beer. Nowadays, not only has the skeletogenic capability of the cranial neural crest been documented in representatives of all classes of vertebrates (chapters 4 to 8), but the neural crest and its cells occupy a central position in studies of vertebrate development and evolution, as discussed in chapters 2 and 3.[5]

Standing as a milestone on the road to understanding the neural crest is the monograph by Sven Hörstadius (Box 1.1), *The Neural Crest: Its properties and derivatives in the light of experimental research.* Published in 1950, 82 years after the discovery of the neural crest, it was reprinted in 1969 and again in 1988. The present book updates my analysis of the neural crest that accompanied the 1988 facsimile reprinting. *"Hörstadius,"* as the monograph is known, was based on a series of lectures delivered during 1947 at the University of London at the invitation of Professor (later Sir) Gavin de Beer, then head of the Department of Embryology at University College and later director of the British Museum (Natural History). de Beer had just completed his own extensive experimental study of the neural crest origin of craniofacial cartilages and dentine in the axolotl, *Ambystoma mexicanum*,[6] and had provided suggestive evidence for a neural crest contribution to the splenial, a membrane bone of the head.

Hörstadius had published two papers on the neural crest before delivering his lectures in 1947. Based on work undertaken with Sven Sellman, the papers were devoted to an extensive experimental analysis of the development of the neural-crest-derived cartilaginous skeleton of *Ambystoma* (see Chapter 5). This experimental verification of Platt's observations on *Necturus* ran counter to the dogma enshrined in the germ-layer theory that skeletal tissues develop from mesoderm and no other germ layer. Despite these early studies on the skeletogenic neural crest (but because of the entrenched germ-layer theory; see below), the focus of interest until the 1940s and 1950s was on the neural crest as a source of pigment cells and of neural elements such as spinal ganglia. Amphibian embryos were the embryos of choice.

The 1960s saw a major thrust into investigation of mechanisms of the migration of neural crest cells and a move away from amphibian and toward avian embryos as the experimental organisms of choice. The switch was thrown by the seminal studies of Jim Weston (1963) and Mac Johnston (1966) on migration of trunk and cranial neural crest cells in the embryonic chick, by the discovery and exploitation of the quail nuclear marker by Nicole Le Douarin (1969, 1974, and see Chapter 7), and by an influential review published by Weston in 1970.

Box 1.1
Sven Otto Hörstadius (1898–1996)

Hörstadius was born on 18 February 1898 in Stockholm. His early academic career was spent at Stockholm University, from which he graduated in 1930 and where he was appointed first lecturer and then associate professor of zoology. Early marked for recognition, Hörstadius was awarded the Prix Albert Brachet by the Belgian Academy of Science in 1936. From 1938 to 1942 he also directed the Department of Developmental Physiology and Genetics at the Wenner-Gren Institute of Experimental Biology. In 1942, Hörstadius became professor of zoology at the University of Uppsala, a position he occupied for 22 years, retiring in 1964 as professor emeritus.

A pioneering embryologist, brilliant lecturer, and expert ornithologist, Hörstadius had a reputation for producing some of the best, and among the earliest, close-up photographs of difficult-to-photograph birds. A member of the Royal Swedish Academy of Sciences (and member of Council), the Academia Pontifica (Vatican), the Royal Institution of Great Britain, and the Societé Zoologique de France, Hörstadius held honorary doctorates from the Université de Paris and Cambridge University.

The numerous honors bestowed upon Hörstadius reflect both his standing in European scientific circles and the breadth of his interests and accomplishments.[7] Our knowledge of the most fundamental aspects of echinoderm development derives from his studies (Hörstadius 1928, 1939). For example, he was the first to demonstrate a fundamental feature of life now taken as a given, viz., nuclear control of the species-specific characteristics of organisms. Hörstadius created chimeric sea urchin embryos by enucleating an egg from one species and fertilizing it with sperm from another. The characteristics of the resulting embryos were those of the species providing the nucleus, not the species providing the cytoplasm. Hörstadius' experimental studies culminated in the publication of *Experimental Embryology of Echinoderms* in 1973.[8]

Detailed maps of the fate of neural crest cells appeared during the 1970s. The microenvironment encountered by these cells was revealed as a major determinant of the migration, differentiation, and morphogenesis of neural crest cells in both normal and abnormal (dysmorphic) embryos (chapters 9 to 12). Syndromes involving one or more cell types derived from the neural crest were recognized as separate and identifiable entities and classified as neurocristopathies (Chapter 11). Monoclonal antibodies against individual populations or types of neural crest cells were developed in the 1980s and used to analyze determination, lineage, and multipotentiality (Chapter 10). Mammalian embryos, which are difficult to study, began to yield the secrets of their neural crest cells to skilled and persistent experimental embryologists (Chapter 8).[9]

During the 1980s and 1990s, further subpopulations of neural crest cells were identified. These included subpopulations in the hindbrain associated with rhombomere specification and segmental patterns of expression of *Hox* genes,[10] and a cardiac neural crest that contributes cells to the valves, septa, and major vessels of the heart (Chapter 7). Homeotic transformation, and a code of *Hox* genes that integrates specific regions of vertebrate embryos, were discovered and analyzed in some detail. Mapping of the neural crest became ever more fine-grained as comparative studies were undertaken, some within specific phylogenetic frameworks to test explicit evolutionary hypotheses. Knowledge of the molecular basis of the migration, differentiation, and death (apoptosis) of neural crest cells advanced considerably.

Last but by no means least in this brief overview is the role that studies of the neural crest played in forging a new synthesis between developmental and evolutionary biology. The neural crest is a craniate synapomorphy, a character shared by all craniates. Evolutionary biologists sought the origin of the craniate head in the unique properties of the neural crest, placing the neural crest at center stage in the craniate and vertebrate evolutionary play. Indeed, it has been argued that the vertebrate head is a "new head" added to the anterior of a more ancient invertebrate head (Chapter 3).

Neural Crest and Germ-Layer Theory

The germ-layer theory makes three claims:

- Early embryos are arranged into equivalent layers: ectoderm and endoderm in diploblastic animals; ectoderm, mesoderm, and endoderm in triploblastic animals.
- Embryos form by differentiation from these germ layers.
- Homologous structures in different animals arise from the same germ layers.

The germ-layer theory exerted a profound influence on those claiming a neural crest—i.e., ectodermal—origin for tissues such as mesenchyme and cartilage, traditionally believed (indeed "known") to arise only from mesoderm. Hörstadius (1950, p. 7) commented on the "violent controversy" that followed the assertion of a neural crest origin for mesenchyme by Platt, Brauer, Dohrn, Goronowitsch, Lundborg, Kastschenko, and von Kupffer, and the opposition to such a heretical idea by Buchs, Corning, Holmdahl, Minot, and Rabl—a veritable who's who of comparative morphology. The bulk of the discussion in de Beer's 1947 paper evaluates his own findings of the neural crest origin of visceral cartilages in relation to the germ-layer theory, which had placed evolutionary studies of embryonic development in a straitjacket for almost a century. Hall (1998a, b) contains a detailed discussion of the discovery of germ layers, naming of germ layers, and germ-layer theory. A brief synopsis follows.

Germ-layer theory had its origin in the early nineteenth century. In the course of his pioneering study on the development of the embryonic chick, undertaken for his doctoral thesis, Christian Heinrich Pander (1817) recognized that the blastoderm of the embryonic chick is organized into the three germ layers we now know as ectoderm, mesoderm, and endoderm. Eleven years later, Karl von Baer extended Pander's findings when he demonstrated that all vertebrate embryos are built on a similar three-layered plan. In 1849 Thomas Huxley suggested that the outer and inner layers of vertebrate embryos are homologous with the outer and inner layers of coelenterates, extending the concept of germ layers from vertebrates to invertebrates, from embryos to adults, and from ontogeny to phylogeny.

Pander had referred to an upper "serous," a lower "mucous," and a middle "vessel" layer and coined the terms *Keimblatt* (germ layer) and *Keimhaut* (blastoderm). In 1853, George J. Allman coined the terms *ectoderm* and *endoderm* for the outer and inner layers of the hydroid *Cordylophora*. The middle "vessel" layer, which was rather ill-defined by Pander, was established as a distinctive germ layer by Robert Remak in his encyclopedic treatment of animal development published between 1850 and 1855. Remak was also the first to identify distinctive histological characters of each germ layer. Huxley coined the term *mesoderm* for the middle layer in 1871.

The terms *ectoderm*, *mesoderm*, and *endoderm* were first applied to the germ layers of vertebrate embryos by Ernst Haeckel in 1874 in the context of the Gastræa theory. In developing his classification of the animal kingdom in 1873, the influential English zoologist Sir Edwin Ray Lankester used the terms *ecto-*, *meso-*, and *hypoblast*, terms first applied to embryonic layers by Francis (Frank) Balfour. Lankester expanded the germ-layer concept from ontogeny into systematics, using the observation that not all animals develop from three-layered embryos to divide the animal kingdom into three grades based on the number of germ layers:

- Homoblastica, for single-celled organisms;
- Diploblastica, for sponges and coelenterates; and
- Triploblastica, for the remainder of the Metazoa.

Lankester's scheme stood for 125 years until evidence was assembled that craniates are tetrablastic not triploblastic, the neural crest constituting a fourth germ layer, which is the topic of the final section of this chapter.[11]

Not all embraced the germ-layer theory. Adam Sedgwick, who in 1882 succeeded Balfour as director of the morphology laboratory at Trinity College, Cambridge, rejected germ-layer and cell theories entirely, claiming that one could not even state what the cell theory is: it is a phantom, and when extended to the germ layers it is the "layer phantom" (Sedgwick 1894a, p. 95).[12] His rejection of germ-layer theory raises a fundamental feature of vertebrate development, discussed in the next section, that is either unknown, or underappreciated, by most students of development.

Sedgwick's criticisms arose, in part, from the nature and origin of mesenchyme, peripheral nerve trunks, and the neural crest. The term *mesenchyme* had been coined in 1882 by the brothers Oscar and Richard Hertwig for those cells that leave the mesodermal germ layers to form elements of connective tissue or blood. The term is now used for meshworks of cells irrespective of their germ layer of origin.[13] A century ago, however, an ectodermal neural crest producing head mesenchyme created major problems for the entrenched germ-layer theory. So too did the observation that nerves, muscles, mesenchyme, and connective and vascular tissues all developed from a single layer in the vertebrate head. This was not what was demanded by a rigid germ-layer theory in which:

- ectoderm formed nerves and epidermis;
- mesoderm formed muscle, mesenchyme, and connective and vascular tissues; and
- endoderm formed the alimentary canal.

Multiple Tissues from Single Layers

A diversity of cell and tissue types, however, do arise from single germ layers. As long ago as 1884, von Kölliker argued that epithelial, neuronal, and pigmented cells all arise from ectoderm. Further, under experimental conditions, structures can develop from a germ layer other than the one from which they arose embryonically:

- Somites, which form from mesoderm during embryogenesis, can form from ectoderm if embryos are perturbed.
- The medullary plate, which forms from ectoderm, can form from mesoderm.
- Tail somites can form from medullary plate.
- Nerve cells and cells of the gut, which arise from ecto- and endoderm, respectively, can arise from mesoderm.

de Beer (1947) cited asexual reproduction, regeneration, and adventitious (ectopic) differentiation as further situations in which structures develop from a different germ layer than the one that produced the original structure. Consequently, the germ-layer theory neither speaks to the full developmental potential of individual germ layers nor to determination or cell fate. It speaks only of the norm.[14]

de Beer saw germ-layer theory as double-barreled, though perhaps two-faced would be the better metaphor. On one side, organ systems in different embryos arise from similar layers. On the other, homologous structures in different animals arise from the same germ layer. de Beer carefully pointed out that the germ-layer theory is a morphological concept that does not speak to developmental potencies or cell fate, concluding:

> [T]hat there is no invariable correlation between the germ layers and either the presumptive organ-forming regions or the formed structures. It follows that the germ layers are not determinants of differentiation in development, but embryonic structures which resemble one another closely in different forms although they may contain materials differing in origin and fate. The germ-layer theory in its classical form must therefore be abandoned. (de Beer 1947, p. 377)

In its strictest form, the germ-layer theory regarded as homologous only those structures that develop from equivalent layers. Again, according to de Beer, the "problem" with fitting a neural crest origin of cartilage into the germ-layer theory is largely the result of a misconception of the theory of homology and of a misapplication of homology of adult structures to homology of developmental origin and developmental processes—the "attempt to provide an embryological criterion of homology" (p. 393). This had important consequences. We now know that homologous structures need not arise from the same embryonic area or, indeed, by the same developmental processes.[15]

Despite de Beer's denigration of the germ-layer theory, he was very reluctant to abandon it entirely, as seen in the conclusion to his paper, which provides a nice example of "the dogged attempt of the human mind to cling to a fixed idea" (Oppenheimer 1940, p. 1).

There is just sufficient constancy in the origins and fates of the materials of which the germ layers are composed to endow the ghost of the germ-layer theory with provisional, descriptive, and limited didactic value, in systematizing the description of the results of the chief course of events in the development of many different kinds of animals; provided that it be remembered that such systematization is without bearing on the question of the causal determination of the origin of the structures of an adult organism. (de Beer 1947, p. 394)

A further fine example of the tenacity of the germ-layer theory in relation to the neural crest is the delay in publication until 1958 of a study by J. P. Hill[16] and his graduate student Katherine Watson on the neural crest origin of mandibular and maxillary arch mesenchyme and cranial ganglia in Australian marsupials and in the American opossum. As reported by T. Thomas Flynn in a preface to the Hill/Watson paper, the work on marsupial embryos began in 1911. Read as a paper (presumably by Hill, as Watson was not present) to the meeting of the Anatomical Society of Great Britain and Ireland at University College, London, on 19 November 1920, "its reading brought forth energetic criticism, particularly from those who felt that the germ-layer theory was in danger" (Hill and Watson 1958, p. 493). Hill died in May 1954. When the manuscript of the 1920 talk was found among his papers, its significance was recognized and publication expedited. Katherine Watson commented in a preface to the paper that "Prof. Hill accepted this conclusion [that mesenchyme of the maxillary and mandibular arches is of neural crest origin] with a reluctance which was due in large part, I think, to his adherence to the germ-layer theory." As the late Alfred Sherwood Romer told the story, Hill delayed publication of his student's research for almost 50 years so as not to "commit treason to the germ-layer theory" (Romer 1972, p. 129). As reported by Flynn, the proceedings of the Anatomical Society meetings were not normally published because of financial constraints, an explanation for the delay in publication at variance with that proffered by Romer.

Three Germ Layers in One: Secondary Neurulation and the Tailbud

Of the three claims of the germ-layer theory, the first—that early embryos are arranged into equivalent layers—is always true; the second—that embryos form by differentiation from these germ layers—is not true for caudal development of vertebrate embryos; while the third—that homologous structures in different animals arise from the same germ layers—need not be true, as will be illustrated by a discussion of secondary neurulation, in which cells from the three germ layers and neural crest function as a unit.

In 1928, Holmdahl distinguished two phases of vertebrate embryonic development:

- primary development, when germ layers are laid down; and
- secondary development, when the caudal end of the embryo develops without segregation into three germ layers.

Holmdahl's second phase is predicated on a lack of germ-layer involvement in the formation of the caudal end and tailbuds of vertebrate embryos.

Secondary Neurulation and Tailbuds

The cranial region of vertebrate embryos arises by primary neurulation through germ-layer delamination and migration. Germ-layer exactitude, however, breaks down even in the cranial region; avian forebrain (prosencephalic) mesenchyme, long assumed to arise from prechordal plate (i.e., from mesoderm), arises from endoderm (Seifert, Jacob, and Jacob 1993). The caudal end of chick embryos arises by secondary induction and transformation of epithelial cells into a mesenchymal tailbud. Secondary neurulation is characteristic of all vertebrates studied—lampreys, fish, amphibians, birds, and mammals (including humans)—a finding that is consistent with secondary neurulation being an ancient process. Exceptions, such as the extension of primary neurulation into the tail in *Xenopus*, are derived conditions in highly derived species.[17]

As noted in the previous section, von Kölliker had considered the problem for germ-layer theory of development of the most caudal part of the nervous system from mesoderm 50 years earlier. Using the technique of vital dye labeling published by Vogt in 1925, Bijtel (1931) confirmed von Kölliker's finding when he demonstrated that tail somites in amphibian embryos arise from the medullary plate, i.e., that mesodermal cells arise from ectoderm. Studies using ^{3}H-thymidine-labeled grafts and quail/chick chimeras extended these findings to birds, in which neural, muscular, vascular, and skeletal tissues arise from common tailbud mesenchyme.

Muscle, cartilage, neuroepithelium, and pigment all differentiate from what appears to be homogeneous mesenchyme when tailbud mesenchyme is cultured. Because these tissue derivatives represent multiple germ layers, Griffith, Wiley, and Sanders (1992) argued that the tailbud consists of three unseparated germ layers. Kanki and Ho (1997) showed that pluripotent cells exist within the zebrafish tailbud, which contributes to posterior trunk tissue anterior to the anus as well as to the tailbud. This is a far cry from the way in which germ layers lay down the more rostral portion of the embryo. Neurulation therefore operates differently cranially and caudally. Indeed, the opposite ends of the same embryo develop by fundamentally different developmental mechanisms, primary and secondary neurulation

Secondary Induction of Tailbuds

Epithelial–mesenchymal interactions, which are secondary inductions, initiate or regulate differentiation, growth, and/or morphogenesis of most organs in vertebrate embryos. A further indication of the secondary nature of caudal development is the requirement for an epithelial–mesenchymal interaction to initiate tail development.

Limb outgrowth is controlled through epithelial–mesenchymal signaling involving an apical ectodermal ridge or AER. The tailbud is surmounted by a ventral ectodermal ridge or VER. Is the VER the tailbud equivalent of the AER? Removal of tail ectoderm almost completely eliminates tail growth in embryonic chicks, just as AER removal eliminates limb growth. Induction of limbs and tail in avian embryos share common mechanisms, as shown by experiments in which limb-bud mesenchyme grafted beneath tail ectoderm induces tail ectoderm to form an ectodermal ridge that regulates limb outgrowth, chondrogenesis, and skeletal formation.[18]

Mutations that decrease or eliminate the AER slow or stop limb development. The VER is missing in *vestigial tail* (*vt*) and *Brachyury* (*T*) mutant mice, and tails fail to develop in these mutants. In *repeated epilation* (*Er*) mutant embryos, the VER and the AER are both abnormal and so are tail and limb development. Reduced proliferation in the ventral part of the tail accounts for rostral defects in *curly tail* mutant mice. Tail development, like limb development, therefore occurs by secondary induction and not by primary development from germ layers, a difference indicative of the fundamental distinction between cranial and caudal development. Consequently, there is a fundamental distinction between cranial and caudal neural crest. The neural crest in murine and avian embryos has a dual origin: cranial (primary) neural crest arises from neuroectoderm; caudal (secondary) neural crest arises from the tailbud. If the tailbud is of mixed origin, then neural crest may not be entirely ectodermal and therefore may not arise from a single germ layer. Indeed, the neural crest is a germ layer in its own right.[19]

The Neural Crest as a Fourth Germ Layer

After amassing the evidence for a recent review on germ layers and germ-layer theory, I concluded that the neural crest qualifies as a germ layer. This fourth germ layer is confined to craniates, which are therefore tetrablastic not triploblastic (Fig. 1.1); see Hall (1998a, b) for detailed discussions of the evidence. Indeed, possession of a neural crest is a craniate synapomorphy. Just as the evolution of mesoderm allowed triploblastic organisms to form new (and often unique) body parts, so the evolution of the neural crest allowed craniates to diversify body parts even further. Clinicians and medical geneticists acknowledge the neural crest as a germ layer when they recognize neu-

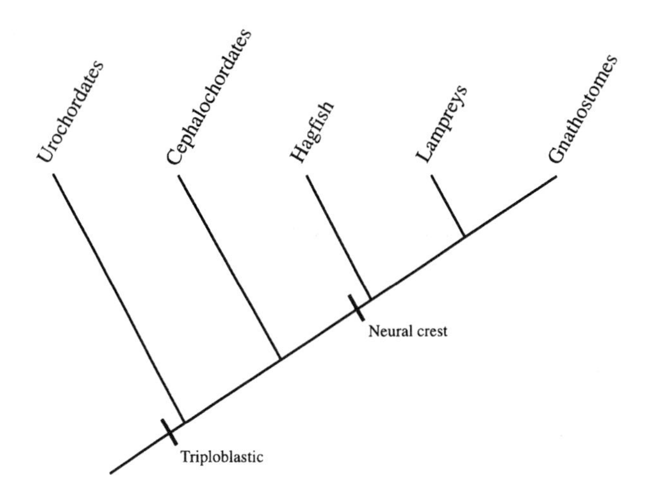

FIGURE 1.1. Urochordates and cephalochordates are triploblastic, having ectoderm, endoderm, and mesoderm as germ layers. The neural crest forms a fourth germ layer in craniates (hagfish) and vertebrates (lampreys and gnathostomes are the extant members).

rocristopathies, in which the common link between affected tissues and organs is their origin from the neural crest; see Chapter 11.

Ectoderm and endoderm are primary germ layers; they were the first to appear in animal evolution and are the earliest to form embryonically, being present in the unfertilized egg. Mesoderm is a secondary germ layer; it is not preformed in the vertebrate egg but arises following inductive interactions between ectoderm and endoderm.[20] Like mesoderm, the neural crest arises very early in development and gives rise to very divergent cell and tissue types. Like mesoderm (and as discussed in Chapter 2) the neural crest arises by secondary induction from a primary germ layer. The neural crest is therefore a secondary germ layer. The three germ layers recognized for almost the past 180 years can be replaced by four germ layers, two of which are primary (ectoderm and endoderm) and two of which are secondary (mesoderm and neural crest). This is a far cry from the days when the suggestion that mesenchyme arose from ectoderm was biological heresy and professional suicide, as Julia Platt found.

When and how the neural crest arises during embryonic development is discussed in the next chapter. When this fourth germ layer makes its first appearance during evolution is discussed in Chapter 3. The remainder of the book is a summary of the evolutionary potential available to craniates by virtue of their possession of neural crest cells.

2

Embryological Origins

The number of cell types known to arise from the neural crest is astonishing, as is the number of tissues and organs to which neural crest contributes either directly, by providing cells, or indirectly, by providing the necessary (often inductive) environment in which other cells develop (Table 2.1). The enormous range of cell types arising from this single embryonic region is an important reason for regarding the neural crest as a fourth germ layer. When and how the neural crest arises during embryonic development is the topic of this chapter.

In one sense the embryological origin of the neural crest is self-evident. Neural crest appears at the junction of neural and epidermal ectoderm in the neural folds of neurula-stage embryos (Fig. 2.1). The very name—neural crest—is indicative of this location. As emphasized by Brun in 1985, neural crest and neural and epidermal ectoderm are all found within the neural folds. Placodal ectoderm also arises from the lateral neural folds or from ectoderm immediately lateral to the neural folds. Indeed it is difficult if not impossible to label neural crest in the neural folds without labeling placodal ectoderm. Placodal derivatives are discussed later in the chapter.

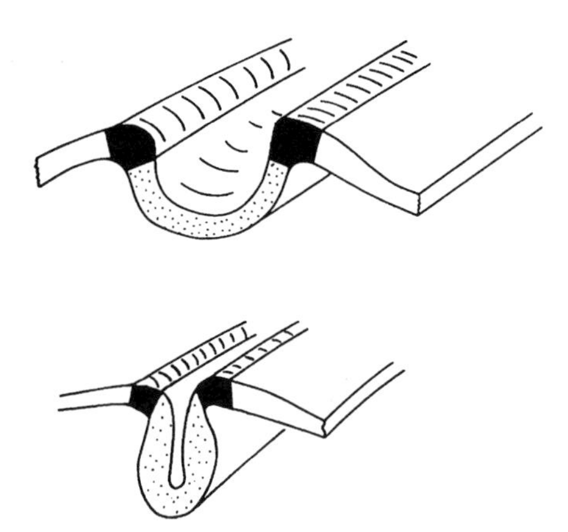

FIGURE 2.1. Localization of the neural crest (black) at open neural plate (above) and closing neural fold (below) stages as seen in an avian embryo. Neural crest is located at the boundary between neural ectoderm (stippled) and epidermal ectoderm.

TABLE 2.1. A list of the cell types derived from the neural crest and of the tissues and organs that are entirely neural crest or that contain cells derived from the neural crest.

Cell types	
Sensory neurons	Cholinergic neurons
Adrenergic neurons	Rohon-Béard cells
Satellite cells	Schwann cells
Glial cells	Chromaffin cells
Parafollicular cells	Calcitonin-producing (C) cells
Melanocytes	Chondroblasts, chondrocytes
Osteoblasts, osteocytes	Odontoblasts
Fibroblasts	Cardiac mesenchyme
Striated myoblasts	Smooth myoblasts
Mesenchymal cells	Adipocytes
Angioblasts	
Tissues or organs	
Spinal ganglia	Parasympathetic nervous system
Sympathetic nervous system	Peripheral nervous system
Thyroid gland	Ultimobranchial body
Adrenal gland	Craniofacial skeleton
Teeth	Dentine
Connective tissue	Adipose tissue
Smooth muscles	Striated muscles
Cardiac septa	Dermis
Eye	Cornea
Endothelia	Blood vessels
Heart	Dorsal fin
Brain	Connective tissue of glands (thyroid, parathyroid, thymus, pituitary, lacrymal)

Because the neural folds contain more cell types than neural crest it can be difficult to isolate and/or graft neural crest from the neural folds. The distinction between grafting neural folds and grafting neural crest is important. Unless neural folds are carefully cleared of neural and epidermal ectoderm, a graft of a neural fold contains neural and epidermal ectoderm as well as neural crest. This becomes important when drawing conclusions about intrinsic patterning of neural crest cells or about neural crest or placodal origins. If ectoderm is included in the grafts, patterning that appears intrinsic to neural crest may, in fact, be imposed by the ectoderm. Furthermore, in situations in which ectoderm is required to induce neural crest cells, grafting neural crest alone will not reveal the differentiative potential of grafted neural crest cells

Nor is extirpation of neural crest a totally satisfactory experimental method. Some neural crest cells may be left behind, while others may have already begun to migrate before the extirpation is performed. Adjacent non-neural-crest cells—neural ectoderm, neural crest rostral or caudal to the region extirpated, or from the contralateral side when neural crest is removed from only one side—may replace the extirpated neural crest through regulation, a topic discussed in Chapter 12. The neural crest cells removed may normally have been involved in the induction of non-neural-crest cells. Absence of a tissue or cell type after neural crest extirpation, therefore, is not unequivocal proof of neural crest origin. All these cautions make extirpation a less desirable technique than those that allow neural crest cells to be removed and replaced with uniquely labeled cells, or that allow neural crest cells to be labeled in situ.[1]

The intimate association between the four presumptive areas—neural crest and neural, epidermal, and placodal ectoderm—although most evident in neurula-stage embryos, does not arise at neurulation, although without special methods, neural crest and the other ectodermal cell types cannot be identified before neurulation. In early amphibian blastulae, vital staining, extirpation, and/or labeling show that future neural crest lies at the boundary between presumptive epidermal and neural ectoderm (Fig. 2.2). Transplantation of ^{3}H-thymidine-labeled regions of chick epiblast into unlabeled early embryos similarly reveals presumptive neural crest at the epidermal-neural ectoderm boundary (Fig. 2.3), although

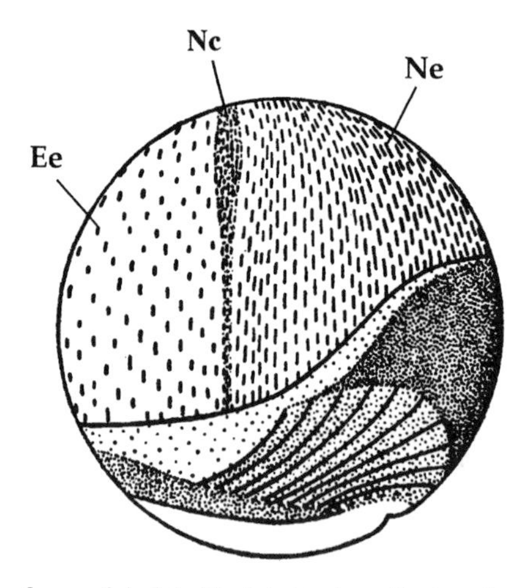

FIGURE 2.2. Fate map of a urodele late blastula to show the location of future neural crest (Nc) at the boundary between epidermal (Ee) and neural (Ne) ectoderm. Modified from Hörstadius (1950).

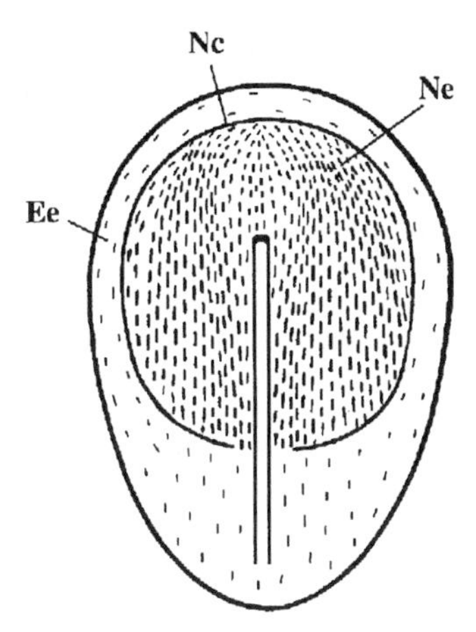

FIGURE 2.3. Fate map of the epiblast of a chick embryo to show the location of future neural crest (Nc) at the boundary between epidermal (Ee) and neural (Ne) ectoderm. Based on data from Rosenquist (1981) and Garcia-Martinez, Alvarez, and Schoenwolf (1993).

epidermal versus neural fate is not determined until after the onset of neurulation. Do we know which events determine that this embryonic region will form neural crest? Are neural crest cells induced, or do they self-differentiate? If induced, is their induction separate from, part of, or secondary to neural induction? The answers to these questions align neural crest with mesoderm as a secondary germ layer arising from a primary layer by induction.[2]

Although they arise at the boundary between neural and epidermal ectoderm, neural crest cells are regarded as derivatives of neural rather than epidermal ectoderm because they produce neurons and ganglia. Neural crest derivatives such as pigment cells and mesenchyme can arise in the absence of neural derivatives, but experimental induction of neural tissue is almost always accompanied by neural crest derivatives. Neural crest cells do not appear when epidermal ectodermal derivatives arise in the absence of neural derivatives. The molecular markers outlined below provide further evidence for this neural connection. Although this list is small, it is growing.

Neural Crest Markers

Cellular or molecular markers are essential if we are to address the origin of the neural crest and neural crest cells.

HNK-1, an antibody against a cell surface sulfoglucuronyl glycolipid, labels avian premigratory and some postmigratory neural crest cells, but not ones that are more fully differentiated (Fig. 2.4). HNK-1 also labels odd-numbered rhombomeres in the hindbrain of embryonic chicks and identifies neural crest cells in embryonic lampreys, fish, birds, and mammals, but not in amphibians.[3]

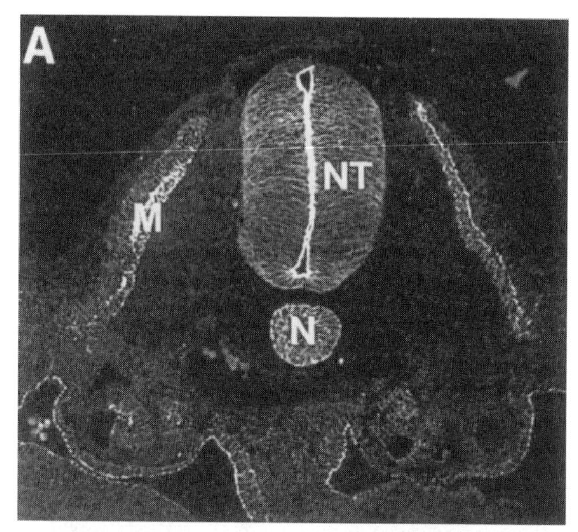

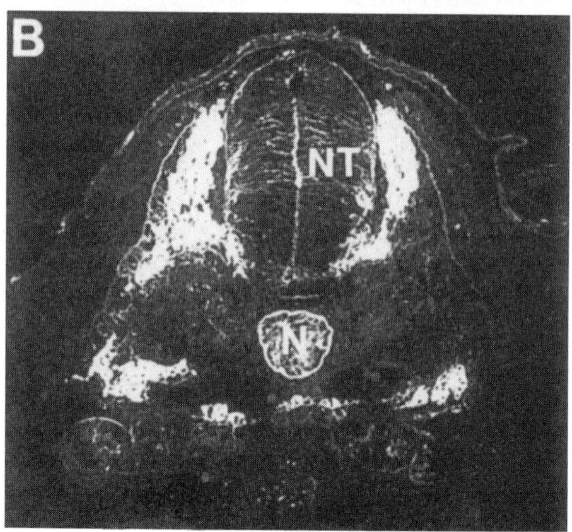

FIGURE 2.4. These two fluorescent micrographs of adjacent thin sections through the trunk of an H.H. stage 18 chick embryo show the comparative distribution of antibodies against a cell adhesion molecule N-cadherin (A) and HNK-1 (B). HNK-1 is expressed strongly in migrating neural crest cells, which appear white in (B). N-cadherin is expressed in the lumen of the neural tube (NT), notochord (N), and myotome (M), but not in neural crest cells. Reproduced from Akitaya and Bronner-Fraser, Expression of cell adhesion molecules during initiation and cessation of neural crest cell migration, *Dev Dyn* 194:12–20, Copyright © (1992), from a figure kindly supplied by Marianne Bronner-Fraser. Reprinted by permission of Wiley-Liss Inc., a subsidiary of John Wiley & Sons, Inc.

Neural crest, neural plate, neural tube, and all differentiated neuronal cells express the cell adhesion molecule N-CAM. Epidermal ectoderm does not express N-CAM, providing a further indication of the links between neural crest and neural ectoderm.[4]

Ncx, a *Hox-11*-related gene, is a marker of neurons and hormonal cells derived from the neural crest and located in dorsal root ganglia, cranial nerve and enteric ganglia, and adrenal medullary cells (Hatano et al., 1997).

The zinc-finger gene *Slug* is another marker for pre- and postmigratory neural crest cells. Expression of *Xslug* has been used to identify neural crest in induction studies in *Xenopus laevis*; see below. However, although *Xslug* is expressed in neural crest in neurula-stage embryos, it is also expressed in mesodermally derived mesenchyme. Consequently, expression at these stages must be interpreted with caution.[5]

All ten of the *Wnt* genes are found in mouse embryos of 8 to 9.5 days gestation; three have sharp boundaries of expression in the forebrain. *Wnt-1* is involved in determination of the midbrain/hindbrain boundary and patterning of the midbrain. As such, *Wnt-1* can serve as a marker for populations of neural crest cells from this brain region. *Wnt* signaling is also required for proliferation of neural crest cells; double mouse mutants (*Wnt-1*$^{-}$/*Wnt-3a*$^{-}$) display defective neural crest and deficient dorsal neural tube. The stapes and hyoid bones (derivatives of hindbrain neural crest) are missing, while thyroid cartilages are abnormal.[6]

Molecular markers often have to be used in conjunction with cellular markers to avoid ambiguity or false positive results. Mesenchyme is a good example of difficulties that can arise when using a cellular marker alone.

Mesenchyme arises from both neural crest and mesoderm, and so is not always a reliable marker of neural crest cells. For example, amphibian gastrula ectoderm exposed to basic fibroblast growth factor (bFGF) forms neural tissue, sometimes in association with mesenchyme. The presumption that this mesenchyme is derived from neural crest seems reasonable, provided that no mesoderm was included in the explant. Neural crest cells certainly can respond to fibroblast growth factor (FGF); it enhances survival, proliferation, and differentiation of neural precursors, while avian neural crest cells exposed to antisense probes against bFGF transdifferentiate from Schwann cell precursors into melanocytes. Other studies with bFGF used the appearance of pigment cells as the marker for differentiation of a neural crest phenotype.[7]

Cartilage (plus neural tissue) was evoked from early gastrula ectoderm of *Rana temporaria* using concanavalin A as the evoking agent. As the starting tissue was embryonic ectoderm, the cartilage was presumed to be neural crest in origin, but induction of mesoderm from the ectoderm cannot be ruled out. Ann Graveson, working in my laboratory, visualized induction of neural crest in Japanese quail (*Coturnix coturnix japonica*) embryos using chick future notochord (Hensen's node) as the inducer. Chondrocytes formed and could be positively identified as neural crest in origin because of the presence of the quail nuclear marker. A similar approach was used by Bronner-Fraser and her colleagues to identify HNK-1-positive or *Slug*-expressing cells that arise in association with the grafted neural plate.[8]

Do the associations between neural crest and neural tissues mean that neural crest, like neural ectoderm, arises as an ectodermal response to neural induction, or is the neural crest set aside as a determined "tissue" earlier in development? To answer such questions, we need to take a brief look at neural induction.

Neural Induction

Neural induction could be the topic of a book in its own right, but I provide only an outline as background for a discussion of the induction of neural crest.

Neural induction occurs by interaction between axial mesoderm or chordamesoderm and overlying ectoderm after interactions at the gastrula stage that establish the dorsal center of the embryo. Ectoderm is dorsalized by growth factors and caudalized by *Hox* genes.

Induction of the nervous system involves cascades of signals that suppress the growth factor bone morphogenetic protein-4 (BMP-4), a growth factor that is also involved in induction of ventral mesoderm. BMP-4 ventralizes neural ectoderm and the neural tube, but is initially distributed throughout the neural ectoderm. BMP-4 and BMP-7 must be inhibited for ectoderm to become neural, indicating a role via negative control for BMPs in neural induction. Chordin (a protein involved in determination of the dorso-ventral body axis), noggin (a secreted polypeptide), and follistatin (a protein that binds to the growth factor activin and inhibits BMP-7) each bind to BMP-4, prevent BMP-4-receptor interactions, and specify the most rostral[9] (anterior) neural ectoderm associated with fore- and hindbrain. BMP-2 also ventralizes the embryonic dorso-ventral axis, mesoderm, and nervous system and is involved in later organogenesis; in *Xenopus* neurulae, zygotic transcripts of BMP-2 are found in neural crest, olfactory placodes, pineal organ, and heart primordia.[10]

BMP-7 is distributed initially in paraxial and ventral mesoderm adjacent to the future hindbrain in mouse embryos. Overexpression of BMP-7 in ventro-lateral mesoderm dorsalizes the neural tube and promotes the growth of neural ectoderm in *Xenopus* and in the chick. As discussed in the following section, BMPs are also active players in induction of the neural crest.

FGF appears to play a role in neural induction in avian embryos; overexpression of FGF in chick embryos—which Rodríguez-Gallardo et al. (1997) achieved by placing FGF-soaked beads within the primitive streak—induces ectopic neural cells from epidermal ectoderm, providing evidence for the competence of the ectoderm for neural differentiation in avian as well as amphibian embryos. Whether neural crest also arose in these ectopic neural cells was not reported. In *Xenopus*, neural differentiation declines and melanophore differentiation increases when gastrula ectoderm from increasingly older embryos is exposed to bFGF, a finding that is consistent with altered competence of the ectoderm with age and with a progressive shift from neural to neural crest induction (Kengaku and Okamoto 1993).

Induction of Neural Crest

The Roles of Mesoderm and Epidermal and Neural Ectoderm

The classic interpretation of the associations between notochord, neural ectoderm, and neural crest, first demonstrated by Raven and Kloos in 1945, was that neural crest was induced by mesoderm. The argument went as follows: The median roof of the archenteron, which houses the presumptive notochord, contains more inducer than the lateral archenteron roof (future lateral mesoderm). The median roof therefore induces neural structures and neural crest, while the lateral roof induces neural crest alone (Fig. 2.5). This interpretation rests on a lower threshold for induction of neural crest than for neural tissue and on a graded distribution of neuralizing inducer. The outer boundary of the neural plate, however, is determined by a loss of ectodermal competence to respond to neural induction, rather than by a low threshold of induction at the boundary.[11]

Because the outer boundary of the neural plate is determined by loss of competence of the ectoderm to neural induction, rather than by a low threshold of induction at the boundary, it could be argued that one need not seek a specific neural crest inductor independent of neural induction but rather an altered responsiveness of the ectoderm to neural induction. This is also true for induction of epidermal ectoderm, which is promoted by BMP-4. Activin (another growth factor within the TGF-ß superfamily) inhibits neuralization, but does not induce an epidermal cell fate. Receptor mediation is part of the mechanism; injection of a dominant negative BMP-4 receptor into *Xenopus* animal cap ectoderm neuralizes the ectoderm, a neural fate that can be reversed with injection of BMP-4 mRNA. A translation initiation factor downstream of BMP-4 is also involved in the determination of neural versus epidermal fate in *Xenopus*. eIF-4AIII (a member of the eIF-4A gene family previously characterized only in plants) is expressed in gastrula ventral ectoderm and induces an epidermal fate when overexpressed in dissociated cells that would otherwise have become neural.[12]

The studies specifically directed to the origin of the neural crest indicate that neural crest arises between neural and epidermal ectoderm precisely because this is where neuralizing and epidermalizing influences meet, the combined action of these influences generating the neural crest. In perhaps the first of these studies,

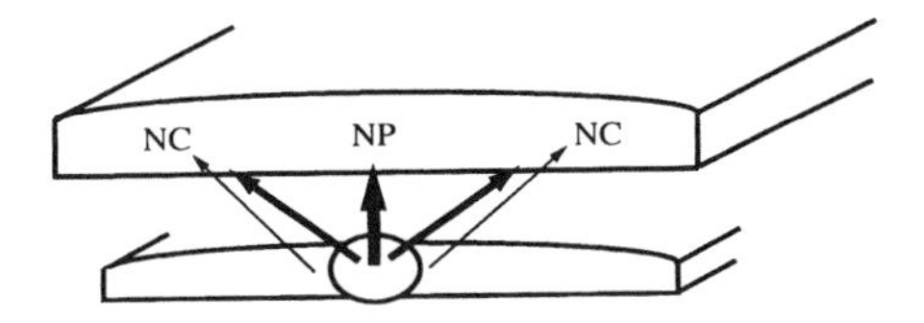

FIGURE 2.5. A model of induction of neural plate (NP) and neural crest (NC), based on differential strength of induction (shown by the thickness of the arrows). Notochord (circle) is a stronger inducer than is lateral mesoderm. See text for details.

Rollhäuser-ter Horst (1980) using *Triturus taeniatus* and *T. alpestris*, grafted gastrula future epidermal ectoderm in place of future neural crest in neurula-stage embryos. The grafted ectoderm formed neural folds that responded to the combined neuralizing induction of the chordamesoderm and epidermalizing induction of the lateral mesoderm by differentiating into neural crest, evidenced by production of classic neural crest derivatives.

Induction of neural crest is now known to be sequential, involving dorsal (notochord) mesodermal induction of neural ectoderm as a first step and epidermal ectodermal induction of neural crest at the epidermal/neural ectodermal border as a second step. This second step may complete the induction of neural crest, or induce additional neural crest cell types. If lateral mesoderm plays a role, it is in association with this second step, follows as a third step (Fig. 2.6), or may be involved in induction of only some neural crest derivatives. For example, chick neural ectoderm associated with mesoderm forms melanocytes but not neurons (Selleck and Bronner-Fraser 1995). In *Xenopus*, however, mesoderm may regulate a gradient of BMP associated with induction of neural crest and inhibition of epidermal differentiation (Marchant et al. 1998).

Although the role of mesoderm was thought not to extend beyond the initial induction of neural ectoderm by the most dorsal mesoderm (notochord; Fig. 2.6), paraxial mesoderm may play a role in neural crest induction. Ventral marginal zone explants of *Xenopus* embryos that have been dorsalized using *Noggin*, produce melanocytes, even though no axial mesoderm is present. In tissue recombination experiments, and on the basis of induction of high levels of expression of *Slug* and differentiation of melanocytes, lateral mesoderm was found to be a more potent inducer of neural crest than dorsal mesoderm. Studies with whole embryos supported this conclusion. Removal of lateral mesoderm reduces markers of neural crest induction/differentiation; removal of axial mesoderm has no effect.[13]

Juxtaposing neural and non-neural ectoderm of avian embryos of H.H. stages 4–10 evokes neural crest, while juxtaposing the same tissues from embryos of H.H. stages 8–10 leads to dorsalization of the ectoderm, evidenced by up-regulation of such dorsal and neural crest markers as *Wnt-1*, *Wnt-3a*, and *Slug*. Recent studies

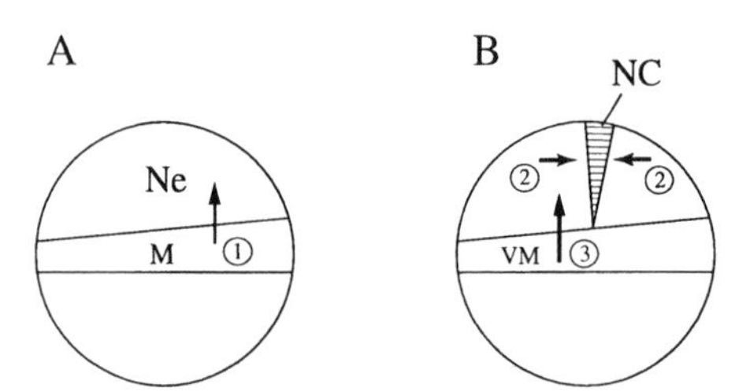

FIGURE 2.6. The sequence of the major steps in neural crest induction as seen in early amphibian embryos. (A) Step 1: the mesoderm (M) induces neural ectoderm (Ne). (B) Step 2: neural and epidermal ectodermal induce neural crest (NC) at the neural-epidermal boundary. Step 3: ventral mesoderm (VM) may play a role in epidermal ectodermal induction of neural crest.

demonstrate that *Xwnt7B* is involved in induction of *Xenopus* neural crest: *Xwnt7B* induces neural crest markers (*Xslug*, *Xtwist*) in ectoderm cotreated with noggin and in neuralized ectoderm in vitro, while exogenous *Xwnt7B* enhances expression of *Xtwist* in vivo. A role for *Wnt-8* in neural crest induction in *Xenopus* has also been demonstrated. La Bonne and Bronner-Fraser (1998) found that although *Slug* is a marker for neural crest, *Slug* alone is not a sufficient signal to induce neural crest in *Xenopus*, and they favor a two-step model involving *Wnt* and FGF. *Slug*, which is not expressed in murine premigratory neural crest cells but is expresed in migrating cells, has been shown through generation of a targeted null mutation not to be required for neural crest (or for mesoderm) formation.[14]

A Role for Bone Morphogenetic Proteins (BMPs)

Other recent studies indicate that BMP-4 and BMP-7 mediate neural crest induction and migration in chick embryos. Acquiring or maintaining a dorsal fate involves interaction between neural ectoderm and signals from epidermal ectoderm. Genes that are preferentially expressed in the dorsal neural tube, from which neural crest cells arise, initially have a more uniform distribution throughout the neural tube but are inhibited ventrally by genes such as *Shh*. Cells of the ventral neural tube can also be switched into becoming neural crest cells if they are grafted into the migration pathway taken by neural crest cells.

BMP-4 is initially expressed in the lateral neural plate and subsequently in the dorsal neural tube and midline ectoderm. It now appears that *Shh* mediates regionalization of the medial portion of the neural plate, from which neurons and neural crest arise, while BMP from the adjacent epidermal ectoderm regionalizes the lateral neural plate, from which placodes arise (Fig. 2.7). BMP-4 and BMP-7 are both expressed in the epidermal ectoderm adjacent to the neural tube (Fig. 2.8); either of the two BMPs can substitute for that ectoderm in promoting migration of neural crest from the neural tube and activating such neural crest markers as *Slug*. The BMP-4 found at the edges of the neural plate and in the dorsal neural tube also signals to paraxial mesoderm; grafting BMP-4-producing cells into paraxial mesoderm

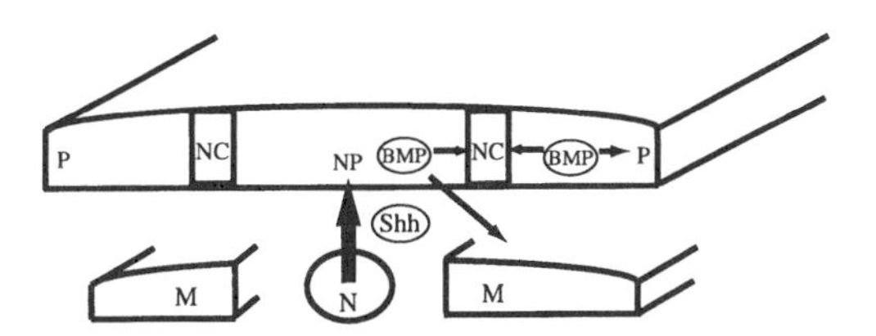

FIGURE 2.7. The role of sonic hedgehog (*Shh*) and BMP-4 and BMP-7 in neural crest induction. *Shh* in the notochord (N) induces neural ectoderm from the neural plate (NP). BMP in the neural plate and epidermal ectoderm induces neural crest (NC) at the neural-epidermal ectoderm boundary. Epidermal ectodermal BMP diffuses laterally to induce placodal ectoderm (P). Neural plate BMP diffuses to the mesoderm (M) to induce somitic mesoderm.

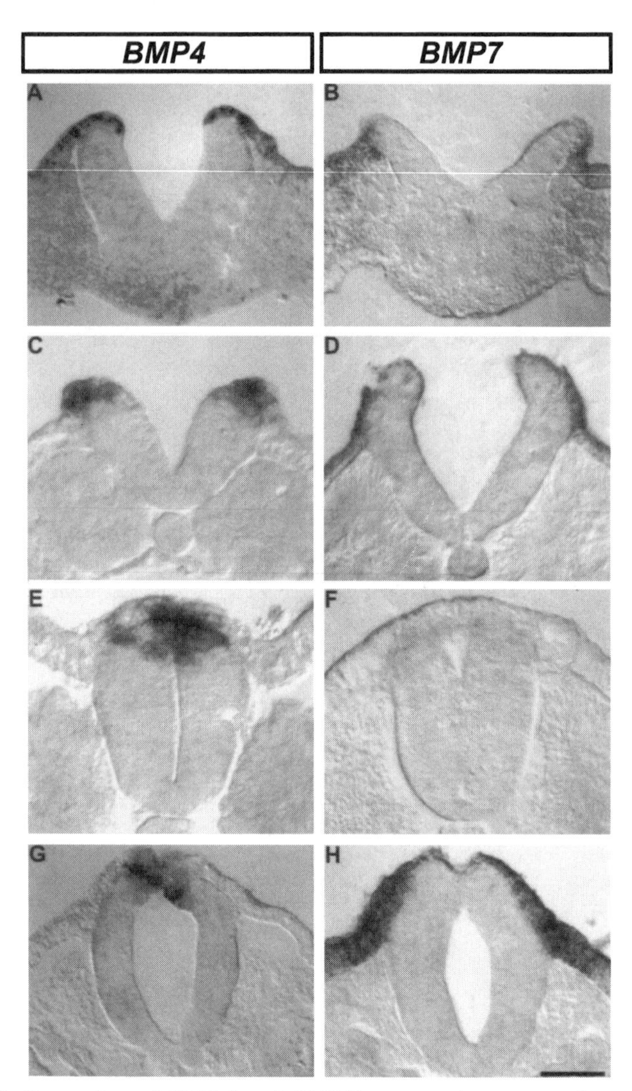

FIGURE 2.8. Expression of BMP-4 and BMP-7 as seen in cross-sections through the developing neural folds, dorsal neural tube, and epidermal ectoderm in a chick embryo of H.H. stage 10. At the level of the open neural folds (A–D), BMP-4 is expressed in neural folds and in the epidermal ectoderm flanking the neural folds (A, C), while BMP-7 is only expressed in epidermal ectoderm (B, D). At the level of the closed neural tube (E–H), BMP-4 is concentrated in the dorsal midline of the neural tube (E, G), while BMP-7 is concentrated in the epidermal ectoderm, especially in the region of the future forebrain (H). Bar = 80 μm (A–D); 100 μm (E–H). Reproduced from Liem et al. (1995) from a figure kindly provided by Karel Liem and with the permission of the publisher. Copyright © Cell Press.

induces expression of *Msx-1* and *Msx-2* and leads to formation of ectopic cartilages. Similarly, suppression of BMP inhibits its ventralizing action in dorsal locations.[15]

Dorsalin-1, which like BMP is a mitogen and member of the TGF-ß superfamily, is expressed in the dorsal neural tube and exerts both positive and negative control over neural crest induction. *Dorsalin* enhances by some 15-fold the numbers of neural crest cells that migrate from isolated neural tubes, perhaps indicative of a general mitogenic effect. It promotes differentiation of melanocytes but inhibits neuronal differentiation, indicating differentiative effects that are specific to subpopulations of neural crest cells. *Dorsalin-1*, like BMP-4 and BMP-7, can substitute for the signaling normally provided by lateral epidermal ectoderm, strongly suggesting a major role for this gene product in neural crest induction.[16]

A code of homeobox-containing (*Hox*) genes patterns both cranial and visceral regions of vertebrate embryos; see the last section of this chapter. Links between BMP and *Hox* genes in the induction of neural crest are being uncovered, the homeobox-containing gene *Msx-1* mediating the role of BMP-4 in epidermal induction and neural ectodermal inhibition in *Xenopus*.

The pair-rule family of homeobox genes in *Drosophila* is responsible for subdivision of the embryonic body into regions. *Zic-3*, a homologue of the pair-rule gene *odd-paired*, is expressed in neural ectoderm and neural crest, appearing first in the neural plate at gastrulation (Fig. 2.9). *Zic-3* is one of the earliest genes so far identified as involved in neural ectoderm and neural crest induction. Expression of *Zic-3* is blocked by BMP-4. A role in neural crest induction/proliferation

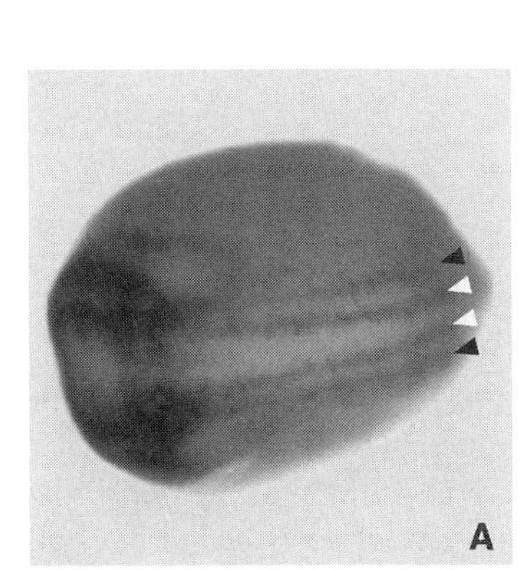

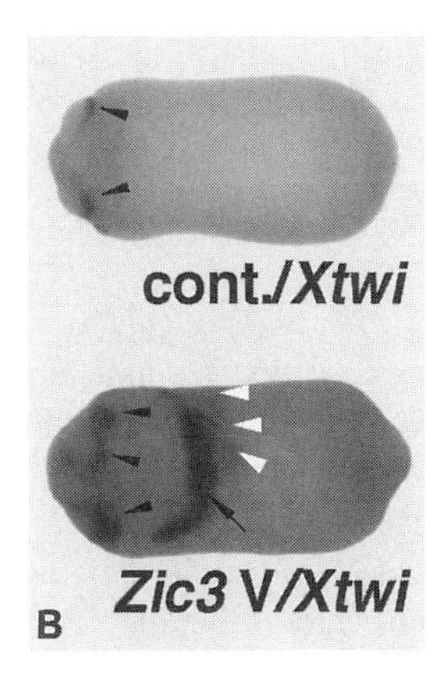

FIGURE 2.9. Expression of *Zic-3* in *Xenopus laevis*. (A) Expression in a stage-16 neurula (anterior to the left) is in the lateral edges of the neural plate (white arrowheads) and in the neural crest (black arrowheads). (B) A control embryo (cont.) and an embryo injected with *Zic-3* mRNA at the 8-cell stage (*Zic3*). *Xtwist* (*Xtwi*) is used as a marker for neural crest cells. In the control embryo, *Xtwist* is confined to cranial neural crest cells (black arrowheads). In the embryo in which *Zic-3* was overexpressed, *Xtwist* visualizes an expanded cephalic neural crest (black arrowheads, arrow) and ectopic clusters of pigment cells (white arrowheads). Reproduced from Nakata et al., *Xenopus Zic3*, a primary regulator both in neural and neural crest development. *Proc Natl Acad Sci USA* 94:11980–11985, from a figure kindly supplied by Jun Aruga. Copyright © (1997) National Academy of Sciences, USA.

is supported by evidence that overexpression of *Zic*-3 leads to expansion of neural crest cells (Fig. 2.9) or induction of neural crest cell markers in animal cap explants. Both events can also be induced by BMP-4 or BMP-7.[17]

Once neural crest is specified, neural crest cells reuse BMPs at different times, in different places, and in different ways. BMP-2 is found in distinct fields of expression in facial epithelia and in mesenchyme of neural crest origin, but not in somatic or prechordal mesoderm. Later in development, BMP-2 and BMP-4 play important roles in the development of the heart, teeth, and skeletal tissues (some of which are discussed in Chapter 10). In concert with *Msx* genes, BMP-2 and BMP-4 regulate apoptosis of neural crest cells, discussed in Chapter 12.[18]

Three *Msx* genes have now been characterized from mouse embryos. (Zebrafish have at least five *Msx* genes [*Msx*A–*Msx*E], but these are not orthologous to *Msx-1* and *Msx-2* found in amphibians, birds and mammals, a finding that is consistent with separate gene duplications in fishes.) *Msx-1* and *Msx-2* have similar patterns of expression in early mouse embryos, initially in the dorsal neural tube and in migrating neural crest cells, subsequently in visceral arches, facial processes, teeth, hair, and limb buds. *Msx-3* is confined to the dorsal neural tube in mouse embryos. In embryos with 5 to 8 pairs of somites, Msx-3 is expressed segmentally in the hindbrain in all rhombomeres except rhombomeres 3 and 5. By the 18-somite stage, expression is no longer segmental but is uniform within dorsal hindbrain and dorsal rostral spinal cord (Figs. 2.10 [color plate] and 2.11 [color plate]). *Msx-3*, like the other two *Msx* genes, can be up-regulated by BMP-4 and the normal dorsal expression extended into the ventral neural tube.[19]

Homoiogenetic Induction

Homoiogenetic (planar) induction is the spreading of an induced state by cells previously exposed to the inducer. Neural induction proceeds homoiogenetically in *Xenopus*, i.e., additional neural tissue is induced from already induced neural tissues through a signal traveling across the ectoderm, rather than from continued induction from the notochord below. Homoiogenetic induction was demonstrated by neuralization of ectoderm transplanted adjacent to the neural tube or placed in culture with neural tube, and by replacement of early neural plate ectoderm by gastrula ectoderm in transplantation between *Ambystoma mexicanum* and *Triturus alpestris*. There is loss of ectodermal competence, homoiogenetic spread of neural induction along the ectoderm, and placode formation in association with weak competence at the boundary. In chick embryos, trunk but not head neural ectoderm is induced homoiogenetically; induction of cranial neural tube requires contact with Hensen's node.[20]

It is not clear if neural crest spreads by homoiogenetic induction. Although evidence for separate inductions of neural tube and neural crest in a microculture assay of *Xenopus* early gastrula cells has been claimed by Mitani and Okamoto (1991), close-range and/or homoiogenetic inductions cannot be ruled out in such an experimental approach. These workers used antibody markers for neurons, melanophores, and epidermal cells, but no neural crest markers. Mayor and colleagues, using the genes *Slug*, *Snail*, and *Noggin* as neural crest markers, claimed that neural crest was induced independently of the neural plate. *Noggin* is an important inducer of anterior neural tissues and associated structures such as the cement gland but does not induce hindbrain or spinal cord; induction of postotic and trunk neural crest is under control of genes other than *Noggin*. Hindbrain from the zebrafish can induce ventral epidermis to become neural crest, a finding that is consistent with homoiogenetic induction (Woo and Fraser 1998).[21]

Placodal Ectoderm

Neural crest appears at the neural/epidermal boundary, whether that boundary is at the normal site in vivo or created when a neural tube is induced ectopically within epidermal ectoderm. Placodes form in association with weak competence at the boundary, although the precise relationship between placodes and neural crest is yet to be determined.

Epidermal (sensory) placodes develop as thickenings of the head ectoderm, either adjacent to the neural crest or from the neural folds themselves (Figs. 2.12 and 2.13).

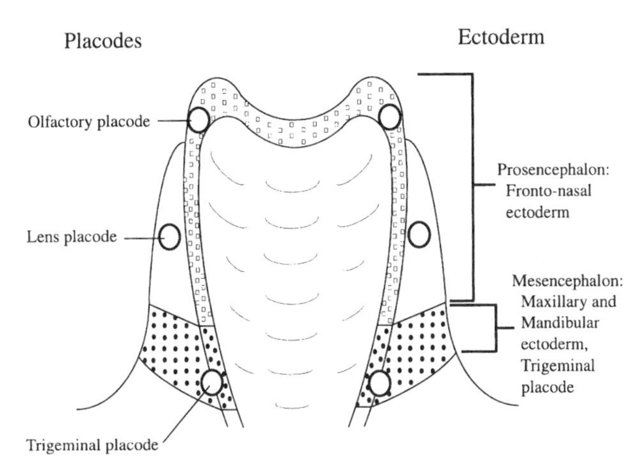

FIGURE 2.12. A diagrammatic representation of the origin of the most rostral placodes and craniofacial ectoderm in the embryonic chick as seen from the dorsal surface. Placodes such as the olfactory may arise from the prosencephalic neural folds (open squares), from ectoderm adjacent to the neural folds (lens) or from both neural folds and adjacent ectoderm (trigeminal). Ectoderm from the prosencephalic neural folds gives rise to the ectoderm of the fronto-nasal processes but not to neural crest or neural ectoderm. Ectoderm from the rostral mesencephalon and adjacent ectoderm (black dots) gives rise to ectoderm of the maxillary and mandibular processes and to the trigeminal placode. The mesencephalic contribution to the trigeminal placode includes neural crest. Based on data from Couly and Le Douarin (1985, 1987, 1990) and Dupin et al. (1993).

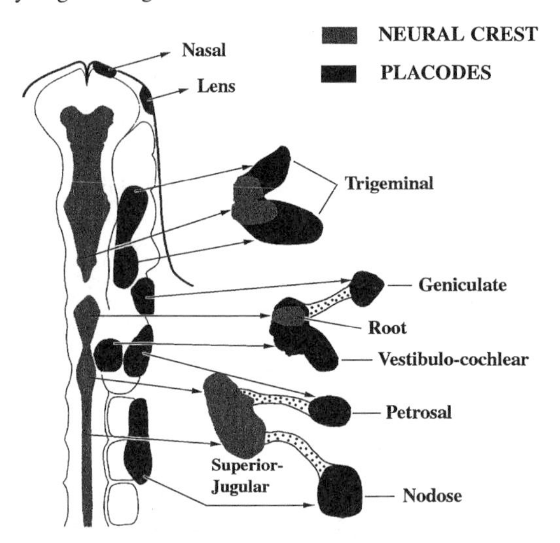

FIGURE 2.13. A diagrammatic representation of the location of neurogenic placodal ectoderm (black) adjacent to the neural tube and of neural crest (gray) that contributes to placodes in a chick embryo as seen from the dorsal side. Only the placodes on the right hand side are shown. The most rostral placodes are the nasal and lens. The cranial sensory ganglia are placodal (Geniculate, Vestibulo-cochlear, Petrosal, Nodose ganglia), neural crest (Root, Superior Jugular ganglia) or of mixed placodal and neural crest origin (Trigeminal ganglion). Note that ganglia can have a proximal component that is neural crest (Superior Jugular) and distal components that are placodal (Petrosal as distal cranial ganglion IX; Nodose as distal cranial ganglion X). Adapted from Webb and Noden (1993).

In axolotls and mice at least, placodes arise from lateral neural fold ectoderm, neural crest arising from median neural folds. Placodes give rise to cranial sense organs such as nose and ear and represent an important source of neural tissue, especially for the central ganglia of the cranial nerves (Fig. 2.13) and for the lateral line and mechanosensory systems of teleosts and amphibians. Olfactory placodes are the most rostral, being succeeded as one moves caudally by the lens, trigeminal, and otic placodes. Six ventro-lateral epibranchial placodes, from which the lateral lines and lateral line nerves arise, are found in fish and many amphibians. The lens placode segregates from head ectoderm in 30-hour embryos, and there is an intimate association between the developing placode, subjacent neural crest cells, and associated extracellular matrix. Meier was convinced that the placode/neural crest association was so intimate that the two must interact developmentally. Non-neurogenic trunk ectoderm can form placodal (nodose) neurons if transplanted into the cranial region from which placodal ectoderm would normally arise.[22]

One difficulty with assessing placodal development is the lack of specific markers.

- Zebrafish placodes express tenascin-C, but so do neural crest, regions of the brain, and mesodermal derivatives.

- The nasal placodes of chick embryos express N-CAM, heparan sulfate proteoglycan (syndecans), and various lectins that bind sugar residues in membrane glycoproteins, but so does the adjacent ectoderm. With further maturation, N-CAM expression is restricted to the olfactory domain, but markers with such temporally regulated modes of expression have to be used with great care.
- Keratan sulfate is expressed in the olfactory, lens, and otic placodes in chick embryos, but also in notochord, pharyngeal endoderm, craniofacial mesenchyme, endocardium, and pronephric tubules. Keratan sulfate and type II collagen co-localize during early nasal development, but it is cumbersome to have to use such a combination of markers to ensure identification of placodes or placodal ectoderm.
- FGF-3 is expressed in the otic placodes of murine embryos but is expressed also in brain, cranial surface, and second arch ectoderm.
- The tyrosine kinase receptor is expressed in the placodal ectoderm of avian embryos, but is not unique to placodal ectoderm.
- The transcriptionally regulated oncoprotein Mybp75 may be more promising. It is expressed in otic and epibranchial placodes of avian embryos and correlates with placodal proliferation and neurogenesis.[23]

Rostro-Caudal Patterning of the Neural Crest

Fate mapping and lineage analysis of Hensen's node in H.H. stage 4 chick embryos indicates that the node consists of presumptive notochord, somitic mesoderm, and endoderm and that individual cells within the node can produce all three layers. Fate is restricted later; additional primitive streaks can be induced in ectopic locations following injection of chick Vg1, a protein localized in the posterior marginal zone of the epiblast, i.e., Vg1 changes the fate of epiblast cells from epidermal to neural. Regression of Hensen's node and accompanying induction of notochord and neural ectoderm during avian development may impart rostro-caudal patterning onto neural crest. The mechanism may involve TGF-ß; once neural crest cells begin to migrate, TGF-ß regulates cell-substrate adhesion.[24]

Immortalized Hensen's node cells secrete a TGF-ß-dependent factor that enhances cranial but suppresses trunk neural crest. Treating trunk neural crest with TGF-ß enhances cranial neural crest markers—exposure to 400 pM TGF-ß decreases the number of melanocytes, while increasing the number of fibronectin-positive (cranial) cells—while blocking TGF-ß down-regulates cranial and up-regulates trunk markers in cranial neural crest. Although both cranial and trunk crest have similar amounts of TGF-ß mRNA, cranial crest is more sensitive to exogenous TGF-ß. Such a mechanism, tied to primary neural induction and notochord formation, supports imposition of rostro-caudal patterning onto neural crest early during primary neurulation. Indeed, co-culture with Hensen's node modifies the fate of neural crest cells, the determining factor being the age of the embryo from which Hensen's node was derived. Nodes from young (H.H. stage 4) embryos respecify trunk neural crest cells as cranial; cranial markers such as fibronectin and actin are up-regulated, while a trunk marker (melanin) is down-regulated. This ability is lost from Hensen's node by H.H. stage 6, in line with timing of neural induction and regionalization by Hensen's node *in ovo*. Nodes from H.H. stages 2 to 4 induce both anterior and posterior nervous system, while nodes from H.H. stages 5 and 6 induce only posterior nervous system. In part this reflects declining competence of the epiblast at H.H. stage 4. Similarly, the ability of specific rostro-caudal regions of the neural ectoderm to induce lens formation, and of specific regions of the epidermal ectoderm to respond to those inductions, is determined during neural induction and primary axis formation.[25]

A possible approach to an analysis of the differing properties of trunk and cranial neural crest cells is differential sensitivity to mutation or to environmental agents such as drugs. *Xenopus* cranial neural crest cells are especially sensitive to the steroidal alkaloid cyclopamine; most craniofacial cartilages are missing after its administration. Other cranial neural crest derivatives are normal, as is the trunk neural crest and its derivatives, indicating a preferential action of cyclopamine on mesenchymal cranial neural crest cells. Similarly, *Xenopus* cranial neural crest is very sensitive to a monoclonal antibody against a lectin that is specific to the neural crest stage of development; expression of endogenous galactoside-binding lectins increases during neural crest cell migration and decreases with cell adhesion, coincident with expression of N-CAM and cadherins. Galactose and sialic-acid-containing cell surface carbohydrates also regulate adhesion of migrating neural crest cells and melanophores. Almost 20% of embryos exposed to the antibody show major deformities of the lower jaw and reduced numbers of chondrocytes in the head.[26]

Cell Lineages Within the Neural Folds

Neural crest cells and cells of the central nervous system are very closely related, indeed so closely related that they share a common lineage, central neurons, and neural crest derivatives arising from the same cloned cells. Nevertheless, not all the cells in the neural folds form neurons or even neural crest.[27]

Intracoelomic grafting of prosencephalic ectoderm from chick embryos of H.H. stages 4 and 5 demonstrates that much of the neural tube arises from medial rather than lateral neural fold ectoderm; the origin of placodes from lateral neural folds has already been noted. Neural crest does not arise from neural folds in the region of the future forebrain. Surprisingly, this most rostral "neural" ectoderm forms facial ectoderm (Fig. 2.12). Quail/chick chimeras were used to demonstrate that the facial ectoderm of chick embryos arises from the neural folds of the forebrain and is patterned into regions or *ectomeres*, the epidermal ectodermal equivalent of the neural ectodermal neuromeres discussed in the following section. While the fate of prosencephalic neural crest is committed early (certainly by H.H. stages 10 to 14), the fate of the more caudal mesencephalic/metencephalic neural crest is not committed until later.[28]

Hox Gene Codes

The branchial region of vertebrate embryos forms by coordinated interactions between neural crest cells, visceral arches, and the developing brain under the direction of a *Hox* code based on overlapping expression boundaries of *Hox* genes. Alternate (odd-numbered) rhombomeres (neuromeres) in mice have characteristic boundaries of expression of *Hox* genes and express *Krox-20*. Other gene products are also expressed segmentally; odd-numbered rhombomeres of the hindbrain of chick embryos bind to HNK-1, and, as discussed in Chapter 12, express *Msx-2* and BMP. Using knowledge of the expression boundaries in the rhombomeres of the chick hindbrain as a basis, Paul Hunt and colleagues demonstrated that rostral limits of expression of *Hoxb-1* to *Hoxb-4* from the Antennapaedia complex coincide with particular rhombomere boundaries. These expression patterns, like cranial nerve patterning, are intrinsic to each rhombomere; they are maintained if rhombomeres are transplanted to another site along the neural axis, being driven by the *Hox* code they carry with them (Figs. 2.14 and 2.15 [color plate]).[29] For example:

- *Hoxb-1* (formerly *Hox-2.9*) is expressed only in rhombomere 4 (r4) , even if r4 is allowed to form more rostrally within the neural tube. (Fig. 2.14)

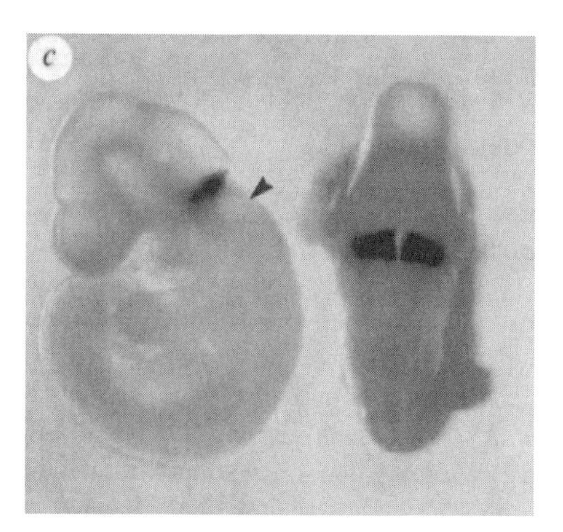

FIGURE 2.14. Expression of *Hoxb-1* (visualized with ß-galactosidase) is restricted to rhombomere 4 in mouse embryos of 9.5 days of gestation. The arrow marks the position of the otic vesicle. Reproduced in black and white from the colored original in Guthrie et al. (1992) from a figure kindly supplied by Andrew Lumsden. Reprinted with permission from *Nature* 356:157–159. Copyright © (1992), Macmillan Magazines Limited.

- *Hox-a3* has its rostral boundary of expression at the border between rhombomeres 4 and 5. This expression boundary is reflected in both neural tube and neural crest, is autonomous, and is retained if rhombomeres 4 and 5 are transplanted.[30]

Hox group 3 paralogous genes (*Hoxa-3*, *Hoxb-3*, and *Hoxd-3*) act in a combinatorial fashion to pattern neurectoderm and mesenchyme from both neural crest and mesoderm. Indeed, it appears from the studies by Manley and Capecchi (1997) that the identity of specific *Hox* genes may be less critical than the number of genes functioning in a region or developmental field. Vielle-Grosjean et al. (1997) have described a branchial *Hox* code with a high degree of conservatism of *Hox-1* to *Hox-4* in the hindbrain and visceral arches of human embryos. Interestingly, individual *Hox* genes are differentially down-regulated in different human tissues later in development.

Neural crest cells migrating from the hindbrain express a combination of *Hox* genes appropriate to their rhombomere of origin and may carry this combination to the visceral arches. Common patterns of *Hox* gene expression in hindbrain, neural crest, and visceral arches into which neural crest cells migrate represent a fundamental unity of mechanism patterning this region of the head. A similar pattern is seen in the zebrafish, in which *Krox-20* is expressed in r3 and r5 and in migrating neural crest cells. Interestingly, *Krox-20* expression in the branchial neural crest in chick embryos does not equate with segmentation of the hindbrain, highlighting the existence of different postmigration patterning mechanisms in different species. Ectoderm of the visceral arches expresses the same combination of *Hox* genes as the neural-crest-derived mesenchyme of that arch. Indeed, similar expression boundaries are detected in early mouse embryos in surface ectoderm, cranial ganglia, migrating neural crest cells, and the mesenchyme of the visceral arches.[31]

Hox genes pattern neural crest cells in the visceral arches and so pattern the arches themselves. The midbrain/hindbrain boundary is regulated by members of at least five gene families—*Otx-2*; *Wnt-1*; *FGF-8*; *En-2*, *-5*, and *–8*; and *Pax-2*, *-5*, and *-8*. *En-2* is up-regulated during neural induction in a region-specific manner. Thus we find expression of the *Engrailed* gene *En-2* at the boundary between mid- and hindbrain, and in the mandibular arches, optic tectum, and anterior pituitary of *Xenopus laevis* embryos. *En-2* is expressed in the mandibular arches before mandibular processes form, although the studies did not resolve whether expression was in the mesenchyme or in the associated ectoderm. Although *Engrailed* is also restricted to the boundary between mid- and hindbrain in avian embryos, it can be induced elsewhere in association with repatterning neural ectoderm. Mutations in at least some of these patterning genes delete both mid- and hindbrain. Similarly, in zebrafish, injection of antibodies against *Pax-2* leads to malformations of the midbrain/hindbrain boundary, down-regulation of *Pax-2* transcripts in the caudal midbrain, and alterations of both *Wnt-1* and *En-2*, two genes that are regulated by *Pax* genes. Of evolutionary interest is the expression of *Engrailed* at the midbrain/hindbrain boundary in embryos of the lamprey *Lampetra japonica*. In the lamprey, however, *Engrailed* is expressed not in the neural crest, but in one muscle of the mandibular arch, the velothyroideus; see Chapter 4.[32]

Another component of the genetic cascade governing visceral arch development and specification lies in the differential expression of *Distal-less* (*Dlx*) genes in mouse visceral arch ectoderm. *Dlx* is a marker for forebrain and rostral craniofacial development. Specific arch deficiencies occur in embryos carrying *Dlx* mutations. Expression boundaries of *Hox* and *Dlx* pattern the visceral arches in orthogonal A-P and P-D directions that correspond to the embryonic A-P axis and the P-D pathway of neural crest cell migration.[33]

A Role for Mesoderm?

Initially thought to reflect a transfer of the *Hox* code from neural ectoderm to neural crest to visceral arches to ectoderm, the similarity of the expression boundaries of *Hox* genes in these tissues may not reflect simple transfer of a *Hox* code; e.g., separate enhancer elements are present in *Hox* gene clusters in the neural tube and neural crest. A study by Frohman, Boyle, and Martin (1990) based on pattern of expression of the murine homeobox-containing gene *Hoxb-1* indicates that the primary pattern lies with head mesoderm and not within rhombomeres. According to this scenario, the *Hox* code arises in mesoderm, is transferred to rhombomeres of the hindbrain, then to the visceral arch mesenchyme (via migrating neural crest cells), and finally to pharyngeal endoderm and superficial ectoderm. This study raises an important point. Does the primary rostrocaudal regionalization of neural ectoderm and neural crest derive from neural induction, or is it secondarily imposed onto the neural tube from mesoderm?

The cranial nerves of avian embryos are patterned by proximo-distal signals that are in part cranial mesodermal and in part rhombomeric in origin. A role for paraxial mesoderm is supported further by studies in which rhombomeres were transplanted either more anteriorly or more posteriorly along the neural axis than their normal locations, and the resulting patterns of *Hox* genes expressed analyzed. The boundary of *Hox* gene expressed is controlled, in part, by paraxial mesoderm, in part by signals from the neural epithelium itself, within the constraint that posterior properties and posterior *Hox* genes overrule anterior properties and anterior genes. Thus transplantation of rhombomeres from caudal to rostral does not alter the pattern of *Hox* genes expressed or the fate of the cells, while transplantation from rostral to caudal modifies both *Hox* code and cell fate to that appropriate to the new location.[34]

3

Evolutionary Origins

Invertebrates are organized around a ventral nerve cord, chordates around a dorsal nerve cord and notochord. Despite this fundamental difference, the molecules that establish the dorso-ventral axes and pattern the dorsal and ventral nervous systems are conserved between *Drosophila* (and by inference and extension, in all arthropods) and *Xenopus* (and by inference and extension, in all vertebrates). The short gastrulation gene *sog*, which specifies ventral in *Drosophila*, is equivalent in sequence similarity to the gene *chordin*, which specifies dorsal in *Xenopus*. Amazingly, *Drosophila sog* mRNA injected into *Xenopus* eggs specifies dorsal, while *Xenopus chordin* mRNA injected into *Drosophila* eggs specifies ventral. Furthermore, by antagonizing *chordin*, BMP establishes dorso-ventral polarity of *Xenopus* neural ectoderm. Decapentaplegic protein, a homologue of BMP, plays the equivalent role in *Drosophila*, providing further evidence for conservation of genetic control of invertebrate and vertebrate nervous systems. The most parsimonious explanation for such conservation is that these signaling molecules were present in the common invertebrate/chordate ancestor.[1]

Given that neural crest arises as a consequence of neural induction, the existence of common genetic elements for specification of invertebrate and vertebrate nervous systems increases the likelihood that a search for neural crest precursors in craniate ancestors might be fruitful. The neural crest as an essential component of the dorsal nerve cord that arises as a result of neural induction is a quintessential craniate characteristic, or, according to some, *the* quintessential craniate characteristic (Fig. 3.1). Maisey, for instance, groups many craniate features as consequences of the "presence of neural crest, and all that that entails" (1986, p. 241). No comparable cells, not even a protoneural crest (whatever a protoneural crest might be), occur in any invertebrate. What of those invertebrate chordates most closely allied to the craniates, the cephalochordates and urochordates (Fig. 3.1 and see Fig. 1.2)? Do they show any hint of this future fourth germ layer? Would we expect such hints to be cellular, genetic, or both?[2]

I approach the question of whether the cellular or molecular basis for the neural crest was present in chordate ancestors through an examination of the closest living relatives to the chordates, the cephalochordates and urochordates, and by a brief examination of fossil cephalochordates. The balance of the chapter deals with two issues: the evolutionary origin of the features associated with the craniate/vertebrate head, many of which are either derived from the neural crest or depend on the neural crest for their initiation; and how skeletogenic tissues could

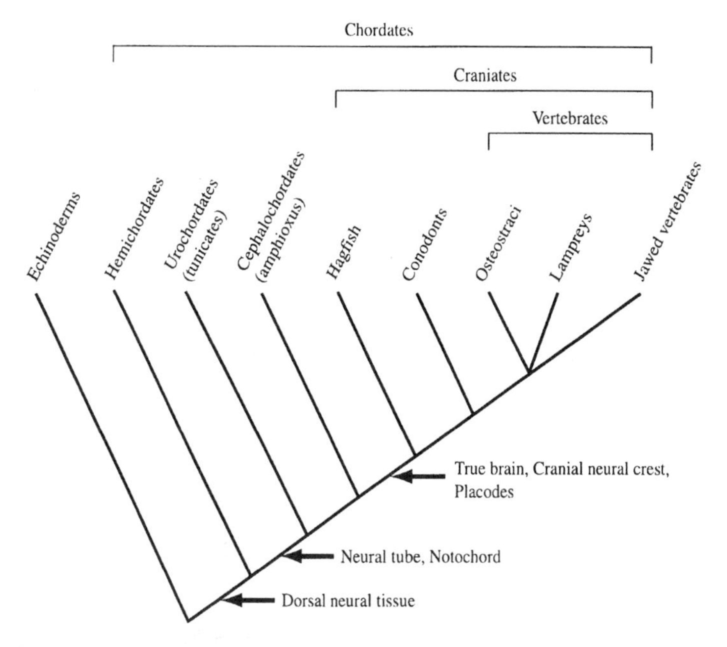

FIGURE 3.1. Features associated with the evolution of the chordate/craniate dorsal neural tube are shown on this phylogenetic tree. Neural crest and placodes are craniate features that followed the evolution of a neural tube and notochord. Modified from P. W. H. Holland and Graham (1995).

have arisen from neural crest cells whose developmental connections are to neural ectoderm and neurons, rather than to mesoderm, which is the source of much of the vertebrate skeleton.

Cephalochordates

Amphioxus (*Branchiostoma*),[3] in the subphylum Cephalochordata, is generally acknowledged to be the least derived chordate and therefore the closest chordate relative of the vertebrates (Fig. 3.1). In the past, it was argued that *Branchiostoma* does not possess anything resembling a neural crest. Cephalochordates lack even such basic neural crest cell types as pigment cells, although they do possess Rohon-Béard cells, which are thought to be of neural crest origin in vertebrates (see Chapter 5).

Recently, however, Nicholas Holland and his colleagues (1996) argued that two cellular features of amphioxus dorso-lateral ectoderm are suggestive of neural crest cells. One is topographical—ectodermal cells lie at the lateral border of the neural plate. The second is that these epidermal cells migrate over the neural tube and toward the midline (Fig. 3.2b). Although the direction of migration is unlike that taken by neural crest cells, both the location and ability to migrate are features expected of protoneural crest cells, although neither is an exclusive property of neural crest cells; many invertebrate nerve cells are migratory and like neural crest cells they transform from epithelial to mesenchymal to initiate migration.

What of the genetic machinery in amphioxus?

Members of the *Snail* family of zinc-finger transcription factors are markers for neural crest (Chapter 2). The isolation and expression of an amphioxus (and an ascidian) *Snail* gene at the lateral borders of the neural plate and in the dorsal neural tube is precisely where we would expect putative neural crest markers to lie, and is suggestive evidence for a protoneural crest, or at least for the precursor cells from which neural crest could have arisen. The fate of the amphioxus *Snail*-expressing cells has not yet been determined. In the ascidian *Ciona* they form the ependyma, the cells lining the spinal cord. In ascidians, therefore, these cells have already diverged from a strictly neuronal lineage.[4]

Expression patterns of amphioxus *Hox* (*AmphiHox*) genes provide a further source of data for the problem of the origin of the neural crest. Gene duplication allows one copy of the gene to retain the original function and the second to diverge for a new function. The chordate ancestor is postulated to have had a single cluster of *Hox* genes. The absence of multiple *Hox* clusters in amphioxus, but their presence in vertebrates (including lampreys) suggests that the acquisition of new or expanded gene functions at the outset of vertebrate evolution may have been instrumental in permitting the origin and diversification of neural crest cells. (Other genes that exist as separate genes in vertebrates occur as single genes in amphioxus; e.g., AmphiBMP2/4, the amphioxus equivalent of BMP-2 and BMP-4 (Panopoulou et al., 1998). A similar situation is seen in ascidians; see the following section.)[5]

Amphioxus has at least 12 *Hox* genes (*AmphiHox-1* to *-12*), with rostro-caudal boundaries of expression reminiscent of those seen in vertebrates and described in Chapter 2. In vertebrates, *Hoxa-1*, *Hoxb-1*, and *Hoxd-1* have their most rostral boundaries of expression between rhombomeres 3 and 4, while *Hoxa-3*, *Hoxb-3*, and *Hoxd-3* have their most rostral boundaries of expression between rhombomeres 4 and 5. *AmphiHox-1* and *AmphiHox-3* are expressed in amphioxus, but only in a very circumscribed region of the nerve cord. *Hox-3* genes are expressed in all but the most rostral region of the vertebrate neural tube. *AmphiHox-3* has a similar pattern of expression in the nerve cord. These shared expression boundaries have been used to support the homology of much of the amphioxus nerve cord with the vertebrate hindbrain. It has particularly been argued, in part from expression of *AmphiHox-3* in posterior mesoderm and the rostral dorsal nerve tube, that amphioxus has homologues of the pre- and postotic hindbrain found in vertebrates. This is an important boundary in vertebrates; postotic neural crest defines the boundary between head and trunk.[6]

AmphiDll, the amphioxus homologue of the vertebrate gene *Distal-less*, is expressed in the rostral neural plate, in two dorsal clusters beside the neural tube near the cerebral vesicle, and in premigratory and migratory epidermal cells (Fig. 3.2). *Dlx* is expressed in a similar position in vertebrates but in neural crest cells, suggesting that *AmphiDll* may presage the location of vertebrate neural crest cells.

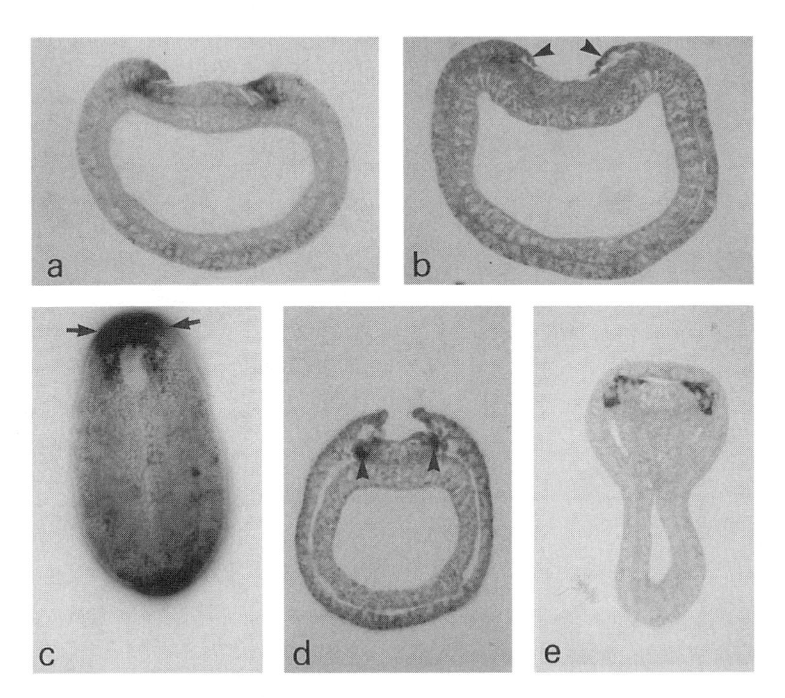

FIGURE 3.2. Expression of *AmphiDll* in neurula-stage embryos and in larvae of *Branchiostoma*. (a) Cross section through the rostral region of an early neurula, showing expression (black) in epidermal ectoderm adjacent to the neural plate, which also contains *AmphiDll*-positive cells. (b) Cross section through the caudal region of the same embryo as in (a), showing *AmphiDll*-positive epidermal ectoderm growing over the neural plate (arrow heads). (c) Expression in the rostral epidermal ectoderm (arrows) covering the neural plate as seen in a dorsal view of a whole mount at the end of the early-neurula stage; anterior to the top. (d) Cross section through the rostral region of the embryo shown in (c), showing strong expression in lateral cells of the neural plate (arrow heads). (e) Cross section through a hatching neurula-stage embryo, showing the strongly *AmphiDll*-positive cells on either side of the central neural plate. Modified from N. D. Holland et al. (1996), *Development* 122:2911–2920, with the permission of The Company of Biologists Ltd., from figures kindly provided by Nicholas Holland.

Orthodenticle (*Otx*) genes are essential for the development of the rostral mammalian head. *Otx-1* and *Otx-2* have nested expression domains in the murine rostral hindbrain. *Otx-2*—which is expressed initially throughout the epiblast and subsequently in first arch mesenchyme—mediates fore- and midbrain regionalization through expression in the visceral endoderm. Mice that carry homozygous mutations for *Otx-2* fail to form any structures rostral to rhombomere 3. Heterozygotes display otocephaly and lack eyes and mandibles. *AmphiOtx* has a pattern of expression surprisingly similar to *Otx*; initially in the rostral neurectoderm and mesendoderm, and subsequently in the rostral tip of the cerebral vesicle and the frontal eye (Fig. 3.3).[7]

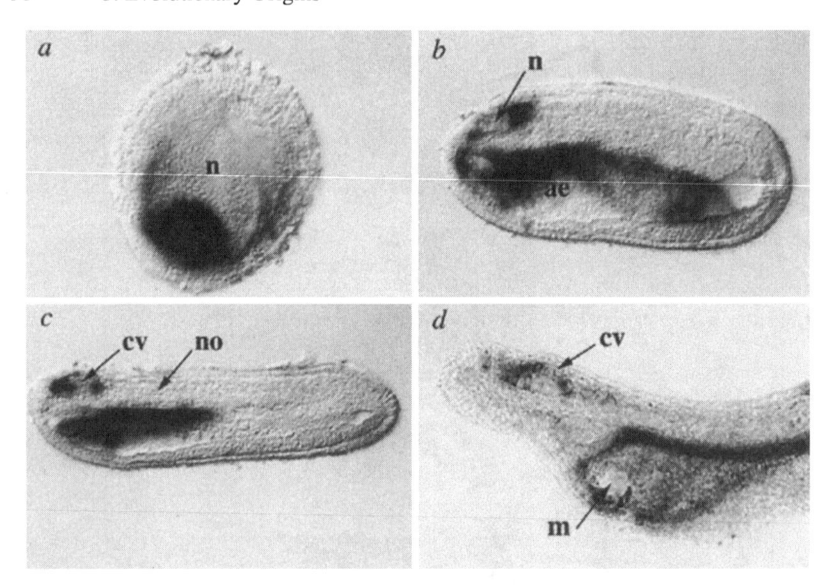

FIGURE 3.3. Expression of *AmphiOtx* in whole-mount embryos of *Branchiostoma.* (a) Dorsal view of an early neurula showing strong expression in the anterior neural plate. (b, c) Mid- and late-neurula stages, seen in lateral view (anterior to the left), show localized expression in the neural plate (n), cerebral vesicle (cv), and anterior endoderm (ae). (d) An early larval stage with neural expression confined to clusters of cells in the anterior cerebral vesicle (cv) and a few epidermal ectodermal cells rostral to the vesicles. Expression continues to be strong in the endoderm of the anterior pharynx. m, mouth region. Reprinted with permission from Williams and Holland, *Nature* (383:490), Copyright © (1992), Macmillan Magazines Limited, from a figure kindly provided by Nick Williams.

These expression patterns of *AmphiHox*, *AmphiDll*, and *AmphiOtx* certainly indicate the availability of genes in amphioxus known to be important for neural tube and neural crest development in craniates. It is controversial to claim that such expression patterns indicate that amphioxus has a differentiated forebrain, or that the frontal eye is an unpaired homologue of the vertebrate eye, as indeed, is the whole issue of determination of structural homology from expression patterns of single genes. Clearly, the question of whether gene expression patterns provide evidence for structural homology is an important issue with respect to our understanding of the origins of the craniate body plan.[8]

Application of excess retinoic acid to *Branchiostoma floridae* elicits effects comparable to those elicited by excess retinoic acid in vertebrates (see Chapter 12), but in vertebrates the effects involve neural crest cells. Expression of *AmphiHox-1* is extended more rostrally than normal and is associated with failure of development of the mouth and gill slits in amphioxus embryos exposed to retinoic acid. Similar defects in vertebrates are mediated by defective neural crest or by defective neural crest induction. In amphioxus they are mediated via the pharyngeal endoderm. Given that the neural crest cartilages of all extant vertebrates examined develop in response to inductive interactions from pharyngeal endoderm or other ectodermal or endodermal epithelia (Chapter 10), it seems reasonable to conclude that these interactions are ancient, having co-evolved with the neural crest. Pharyngeal endodermal induction and induction/retinoid interactions were therefore features of the craniate ancestor. Linda and Nicholas Holland, who published the amphioxus study in 1996, argue that class-1 paired genes such as *AmphiPax-1* originated in pharyngeal endoderm and were co-opted by neural ectoderm and neural crest in craniates.[9]

Urochordates

Ascidians (tunicates, subphylum Urochordata) possess a band of pigmented cells along the dorsal aspect of the neural tube that has been regarded as presaging the neural crest origin of vertebrate pigment cells. Ascidians also possess a population of neuronal and neuroendocrine cells associated with the dorsal strand of the nervous system and which may arise from (and therefore migrate away from) the dorsal strand. The third hint of a protoneural crest is that cells in the floor of the ascidian pharynx are reactive for calcitonin, the vertebrate calcium-regulating hormone known to be of neural crest origin. While not evidence for an ascidian neural crest, these cells are certainly evidence of ascidian cellular activity comparable to activity that is a function of neural crest cells in craniates.[10]

Bi- or multipotentiality and conditional specification of cell fate are basic, although not exclusively, properties of neural crest cells (Chapter 10). Ascidians display determinate development and autonomous specification, meaning that the fate of the majority of their cells is fixed through the inheritance of specific cytoplasmic constituents, rather than by cell–cell interactions. One would not expect such determinate cells to be capable of changing their fate. A few ascidian muscle cells, however, have their fates set following inductive interactions with adjacent cells, rather than autonomously by lineage and cytoplasmic inheritance. J. R. Whittaker has shown that ascidian presumptive pigment cells can be induced to express a different cell fate in response to interaction with endoderm. Does such bipotentiality and ability to respond to inductive interaction mean that ascidian pigment cells share features expected of protoneural crest cells? In considering the answer to such a question, recall that as discussed in Chapter 2, the neural crest arises in association with, but secondary to, neural induction.[11]

BMP-7 plays an important role in induction of the neural crest (Chapter 2). A homologue of BMP-7 is expressed in neuroectoderm at the boundary of neural and epidermal ectoderm in the ascidian *Halocynthia roretzi*. Should ascidians be shown to use this homologue of BMP-7 to establish neural ectoderm, then this patterning mechanism would predate the ascidian/chordate split. Homologues of BMP-2 and BMP-4 are also present in ascidians, where they may inhibit specification of dorsal neural plate as they do in vertebrates. (In amphioxus, a single gene, AmphiBMP2/4, which is equivalent to BMP-2 and BMP-4 in vertebrates, is expressed in many early tissues consistent with a role in establishing the dorso-ventral polarity of ectoderm but not of mesoderm.)[12]

Other important molecular regulators of neural tube patterning are shared by ascidians and vertebrates. The ascidian *Ciona* has a homologue of the vertebrate gene *HNF-3β*, which, like the vertebrate gene, is expressed in ventral midline cells of the developing neural tube. *Snail*, the ascidian homologue of the chick gene *Slug* that is a marker for dorsal neural cells and neural crest, is expressed in *Ciona* at the border of neural and dorsolateral ectoderm, an analogous position to *Slug* in vertebrates.

In vertebrates, the homeobox-containing gene *Pax-7* is expressed in mesencephalic neural crest and in rhombomeres 1, 3, and 5 of the hindbrain (Fig. 3.4 [color plate]). The importance of *Pax-7* is illustrated by the fact that its disruption is associated with malformations of facial structures of neural crest origin. Furthermore, *Pax-7* and *Pax-3* are redundant functionally. Two ascidian genera, *Ciona* and *Halocynthia*, express the genes *Pax-3/7* and *Pax-2/5/8* in neural tube cells as do vertebrates; *Pax-3/7* is a single ascidian gene equivalent to *Pax-3* and *Pax-7* of vertebrates. *HrPax-3/7* (*Hr* for the species *Halocynthia roretzi*) is expressed in future dorsal neural tube, dorsal epidermis, sensory neurons, and in 15 segmental spots within the neural tube. Wada, Holland, and Satoh (1996) saw a similarity in this distribution to expression of *Pax-3* in the vertebrate neural crest, and thought that the 15 spots might be remnants of a segmental ancestral pattern. Wada et al. (1998) used expression domains of *HrPax-2/5/8* (which corresponds to *Pax-2*, *Pax-5*, and *Pax-8* of vertebrates) to divide the brain of *Halocynthia roretzi* into regions that they consider to be homologues of vertebrate fore-, mid-, and hindbrain and spinal cord, but, as already noted, establishing structural homology on the basis of expression boundaries of single genes, even important single genes, is not universally accepted.[13]

Fossil Cephalochordates

Does the fossil record provide any useful information about neural crest origins? Although at least three putative cephalochordates have been identified in Cambrian fauna, none display any obvious neural crest derivatives. This does not, of course, prove that such elements were not present.

The earliest known fossil chordate, *Pikaia gracilens* from the Burgess Shale inshore marine fauna of British Columbia, is thought to be a cephalochordate because of the presence of a notochord and segmental muscle blocks and its amphioxus-like appearance. The chordate affinities of *Pikaia*, however, although highly suggestive, have yet to be subjected to a rigorous analysis.

Yunnanozoon lividum, from the Early Cambrian Chengjiang fauna of Yunnan Province, China, has also been interpreted as a cephalochordate with notochord, filter-feeding pharynx, endostyle, segmented muscles, branchial arches, and gonads. Other workers interpret *Yunnanozoon* as a hemichordate, the branchial arches being branchial tubes and the notochord the upper portion of the pharynx.[14]

A single 22-mm-long specimen of another cephalochordate from the Chengjiang formation was described in 1996 by Shu, Conway Morris, and Zhang. *Cathaymyrus diadexus* has a pharynx, gill slits, segmented "myomeres," and a putative notochord. One can only speculate whether presence of a pharynx and gill slits means presence of a pharyngeal arch (neural crest) skeleton. It need not. As discussed below, a collagenous skeleton could have functioned equally well in the early chordates. Of course, the collagenous skeleton could have had a neural crest origin.

The First Craniates and the Evolution of Cartilage

The evolutionary history of the neural crest and the nature of the first craniates are tied closely to the evolution of the muscular pharynx, brain, skull, and sensory apparatus of the craniate head. An innovative synthesis that prompted much investigation was elaborated in the 1980s by Carl Gans and Glen Northcutt, both then of the University of Michigan, who built on foundations established by E. S. Goodrich, N. J. Berrill, A. S. Romer, B. Schaeffer, and others.[15]

The Gans and Northcutt scenario drew from and integrated three areas of biology:

- Comparative anatomy, especially the homology of neural crest neuronal derivatives with the epidermal nerve net of hemichordates. This suggestion was made before the close connection of neural crest and neural induction was appreciated, although a dorsal nervous system could have arisen from the coalescence of an epidermal nerve net around the dorsal midline.
- Developmental biology, especially detailed knowledge of the developmental origins and fates of neural crest cells, and findings on the age-old problem of segmentation of the vertebrate head, some of which were discussed in the previous chapter.[16]
- The phylogenetic systematic principle that organisms should be grouped, not on the basis of primitive or archaic characters, but on the basis of shared/derived characters known as synapomorphies since the pioneering study by Hennig, first available in English in 1966.

The scenario runs as follows. The precraniate was likely soft-bodied, lacked a mineralized skeleton, and indeed may have lacked any skeleton at all. The fossils discussed in the previous section bear this out. It probably lived in a near-shore or estuarine marine environment, obtaining its food by filter feeding using a ciliated branchial basket. One can imagine an animal like *Pikaia* fitting this description. The evolution and elaboration of a notochord, central nerve tube, and axial muscular system allowed these animals to become more mobile, use muscles rather than cilia for locomotion, use the notochord as a stiff rod to avoid antero-posterior telescoping during bending of the body, and to use the dorsal nerve cord to integrate sensory input and coordinate movement.

With increasing muscularization of the body came the development of muscles in the wall of the pharynx, facilitating the pumping of water and allowing the pharynx to function in gas exchange as well as in filter feeding. The neural tube or peripheral nerve net would at this stage have produced neuronal and perhaps pigmented cells. The advantage of storing energy by deforming the muscularized pharynx, allowing pumping by an elastic recoil mechanism, is hypothesized to have led to the differentiation of cartilage from cells that were previously only neural. A number of similarities of cartilage and nerve cells support such a switch:

- Chondromucoid, a basic extracellular matrix component of primitive cartilage, is very similar to molecules found in mechano- and electroreceptors.
- Type II collagen and aggregan (two of the major protein and proteoglycan components of cartilage), and the link proteins that link proteoglycans to other matrix components, are found in embryonic central nervous systems.
- The S-100 acidic protein, a possible regulator of Ca^{++}-mediated cell functions, initially thought to be unique to nerve cells, is present in both cartilage and pigment cells.
- "Cartilage-specific" chondroitin sulfate proteoglycan (CSPG) is also found within cerebellar astrocytes of the rat (unless they differentiate into oligodendrocytes when they stop synthesizing CSPG), providing a nice example of a component shared by chondroblasts and neurons, and of the modulation of a matrix product with altered differentiation.[17]

What we now know as the neural crest could thus have existed initially as an epidermal nerve plexus or net controlling ciliary function during movement and filter feeding. With increasing muscle-based locomotion, the dorsal nerve cord took over innervative control of locomotion, freeing the epidermal nerve cells (or their precursors) for other functions. Those new functions allowed modifications of existing cell products involved in mechanoreception to deposit an extracellular cartilaginous matrix in the pharynx.

Whether this scenario can accommodate origin of the neural crest in direct association with the dorsal neural tube depends on the origin of the neural tube and its relationship to the epidermal nerve net. One of the two major theories for the origin of the chordates—from direct-developing acorn worms (hemichordates)—posits that the chordate neural tube did arise by the rolling up and coalescence of the hemichordate epidermal nerve net. The alternate major theory sees the neural tube arising from the ciliary bands of a dipleurula larva. We cannot yet choose between these (or other) competing hypotheses.[18]

Other Skeletal Tissues

Cartilage is a primary skeletal tissue. It is found surprisingly commonly in a variety of invertebrates—coelenterates (medusas and polyps), annelids (sabellid or feather-duster worms), molluscs (snails, cephalopods), and arthropods (*Limulus polyphemus* [the horseshoe crab] and imagos of locusts)—and is the only skeletal tissue in lampreys and hagfishes.

An issue that has been contentious for decades is whether cartilage preceded bone in vertebrate evolution. The two tissues appear equally early in the fossil record, but that may give us a false impression; any unmineralized cartilage preceding bone would have fossilized only under very unusual and rare circumstances. Gans and Northcutt saw collagen as the primitive structural molecule of connective tissue, cartilage as the first skeletal tissue, and bone—whether neural crest or mesodermal—as a later evolutionary development(s). Cartilage enhances pharyngeal gas exchange while a bony exoskeleton limits diffusion. Therefore, Gans and Northcutt argued that cartilage must have existed before early animals could have afforded to ossify the skin; the early vertebrates could not give up gaseous diffusion until they possessed an alternative gas exchange mechanism utilizing the muscular, cartilage-supported pharynx.

The early association of neural crest skeletal tissues with the pharynx established what was thought to be a dichotomy between a skeletogenic cranial and a nonskeletogenic trunk neural crest. Chapter 5 discusses whether this is a fundamental feature of the neural crest, or whether trunk neural crest originally possessed skeletogenic potential that was secondarily lost.[19]

What of the origins of other neural crest derivatives such as dermal bone and dentine? The conclusion emerging from the data now available is that cartilage arose as the primitive skeletal tissue in the endoskeleton. Deposition of bone around and then within cartilage (as perichondral and endochondral bone) were later evolutionary events. Cartilage is therefore the primary tissue of the endoskeleton. Dermal bone and dentine (in association with enamel) arose as the primary skeletal tissues in the exoskeleton as a dermal armour. Cartilage in the exoskeleton is secondary.[20]

Northcutt and Gans outline the following sequence for the development of dentine and bone following the appearance of cartilage:

- elaboration of dentine and its availability for use in electroreception because of the presence of hydroxyapatite rather than calcite in the dentinal matrix;
- appearance of layers of bone and its utilization both as armor for protection and as a Ca^{++} storage reservoir; and
- elaboration of a more generalized bony skeleton.

The earliest fossilized chordate skeletal remains are the scales of jawless ostracoderm "fishes" and conodont elements from the Upper Cambrian and Lower Ordovician. Assigning conodonts to the chordates raises some interesting issues for the origin of skeletal tissues.

Conodonts have a geological history from the Lower Cambrian to the Upper Triassic. They are represented as fossils by minute, toothlike, phosphatic structures known as conodont elements, which were part of an elaborate "pharyngeal" feeding apparatus. Increasingly, workers are coming to recognize conodonts as chordates on the basis of identification of bone, dentine, and enamel in conodont elements and from the discovery that the animal containing conodont elements

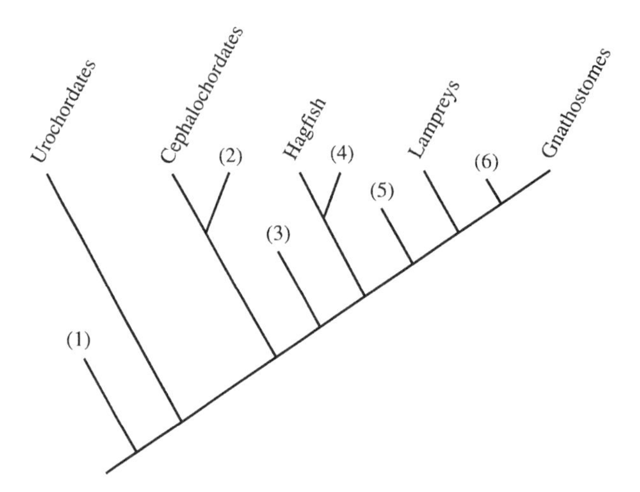

FIGURE 3.5. The relationship of conodonts to urochordates, cephalochordates, craniates, and vertebrates is unresolved. Six possibilities (1–6) are shown on this phylogeny. See text for details. Adapted from Aldridge and Purnell (1996).

had a notochord, segmental muscle blocks, tail fin, and paired eyes.[21]

Where conodonts fit within the chordates remains unresolved, but has implications for our understanding of the evolution of the neural crest and of endo- and exoskeletal tissues. Six possibilities are shown in Figure 3.5. Locating conodonts basal to the vertebrates (positions 1–4 in Figure 3.5) requires that elements of both muscular and skeletal systems evolved independently in conodonts and vertebrates, or that they were lost independently in hagfishes (position 4). Locating conodonts basal to position 5 would require secondary loss of skeletal tissues in lampreys.[22]

Neural Crest and Mesodermal Skeletal Tissues

Mesodermal bone is entirely endoskeletal. Bone that arises from neural crest cells is endoskeletal where it replaces primary cartilage and exoskeletal (dermal bone) where it forms by direct ossification without going through a cartilaginous phase. The visceral skeleton in all vertebrates examined is neural crest not mesodermal in origin (see chapters 4 to 8). Given that cartilage preceded bone in the endoskeleton, this cartilaginous pharyngeal skeleton derived from neural crest would be the original (primitive) craniate endoskeleton. Mesoderm- and neural-crest-derived bone supporting the skull (both mesoderm and neural crest contribute to cranial bones in modern-day vertebrates; see Chapter 7), and mesodermally derived trunk appendicular and axial cartilage and bone, were later evolutionary events. Northcutt and Gans comment that "the ability of true mesodermal tissues to form calcified tissues requires either an independent evolution of biosynthetic activity by mesodermal cells or the transfer of the mechanism

Box 3.1
Neural Crest Versus Mesodermal Bone

Differences between bone derived from neural crest and bone derived from mesoderm emerge in clinical situations requiring bone grafts, in which skeletal grafts of neural crest origin fare better than skeletal grafts of mesodermal origin when used to replace defective or damaged craniofacial skeletal tissues. Bone that originates in neural crest is better for healing defects in the human mandibular skeleton; it is incorporated more effectively and suffers less resorption than mesodermal bone. Chin (mandibular symphysis) grafts are superior to rib or iliac crest grafts for secondary bone grafting of defects of the bone associated with the teeth, such as alveolar cleft defects. Such differences could be attributed to embryological origin (neural crest versus mesoderm), differences in the three-dimensional architecture of different bones, or differing histogenesis; most mesodermal bones develop endochondrally, while most bones with a neural crest origin develop intramembranously. Lamellar and woven bone each have distinctive patterns of noncollagenous matrix proteins, further reinforcing differences that have a histogenic basis.[23]

The phenotypes of human bone cells maintained in vitro can be correlated with their site of origin. When Kasperk et al. (1995) analyzed osteogenic cells from the mandible (which arise from neural crest) or from the iliac crest (which are mesodermal in origin) of four patients they found that mRNA for bFGF and IGF-II increased in mandibular cells while mRNA for alkaline phosphatase decreased. mRNA for TGF-ß increased in iliac crest cells, which also divided more slowly than did mandibular cells.

---

from ectodermal to mesodermal cell lineages," and clearly favor the latter possibility: "this ectodermal cellular program was later transferred to mesodermal tissues" (1983, pp. 19, 21). In a superb overview of chordate phylogeny, Maisey (1986) presented the considerable evidence supporting neural crest cells rather than mesoderm as the primordial vertebrate skeletal tissue. Indeed, mesoderm- and neural-crest-derived bone are not the same; see Box 3.1 and the section on germ-layer specificity in Chapter 5.

In summary, despite the apparent difficulties of knowing what really happened half a billion years ago, we have made surprising and exciting advances in our understanding of the origin of the neural crest, of neural crest skeletal derivatives, and of the craniates and vertebrates themselves. These advances demonstrate the central role played by the neural crest and the power of evolutionary developmental biology, the renewed synthesis of comparative anatomy, developmental biology, systematics, and evolution (see Hall 1998a). For the insights that neural crest origins give to craniate/vertebrate origins we eagerly await future developments. In the meantime, we can examine the neural crest and its derivatives in agnathan craniates (hagfishes) and jawless vertebrates (lampreys), the topic of the next chapter, before going on to consider the neural crest in jawed vertebrates (gnathostomes).

4

Agnathans

Living jawless vertebrates were formerly included in the cyclostomes, a group comprised of lampreys (petromyzontids), hagfishes (myxinoids), and various groups of extinct jawless fishes. Cyclostomes, however, are no longer considered a natural group. Researchers have also grappled with whether lampreys and hagfishes are representatives of a monophyletic group of vertebrates, having shared a common ancestor, or whether they represent two separate lines of jawless craniates. A further issue is the relationship between extant agnathans on the one hand and jawed (gnathostome) vertebrates on the other: are lampreys more closely related to jawed vertebrates than are hagfish? Although debate continues, the weight of evidence supports the view that lampreys and hagfishes have quite different and separate evolutionary histories, and that only lampreys share a close evolutionary relationship with the jawed vertebrates (Fig. 4.1, also see Fig. 1.1). Indeed, while lampreys and hagfishes are both craniates, hagfish lack an axial vertebral skeleton and so are not vertebrates (see Fig. 3.1).[1]

Once thought to be primitive craniates little changed since the appearance of the first jawless vertebrates in the Ordovician some 500 million years ago, lampreys and hagfishes are as diverse, highly specialized, and well adapted to their modes of life as we are to ours. Both are speciose groups. Hagfishes, which appear to be the more "primitive" of the two, are structurally diverse, not structurally conservative. For example, the Pacific hagfish, *Eptatretus stoutii*, has a well-developed lateral line system, but myxinids such as the Atlantic hagfish, *Myxine glutinosa*, lack any component of a lateral line system.[2]

Given the powerful role that knowledge of developmental processes has played in elucidating relationships both within and between other groups, that

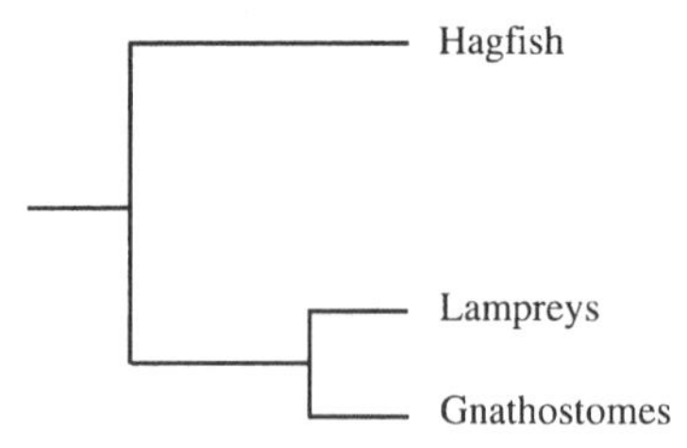

FIGURE 4.1. Lampreys are more closely related to gnathostomes (jawed vertebrates) than either are to hagfishes, which are craniates but not vertebrates.

lampreys and hagfishes are not closely related to one another, and the role that the neural crest played in craniate and vertebrate evolution, it is important to understand as much as we can about the neural crest in embryos of these two agnathans. The available information is summarized in this chapter.[3]

Hagfish

Hagfishes can be dealt with very quickly for we know virtually nothing of their neural crest. Indeed, we know very little about any aspect of hagfish embryonic development. The major study—based on perhaps the only collection of live hagfish embryos ever made, in the summer and fall of 1896 in Montery Bay off Pacific Grove, California (where Julia Platt was elected mayor in 1931)—is that by Bashford Dean on the Californian hagfish, *Eptatretus stouti* (formerly *Bdellostoma stouti*), published a century ago (Dean 1899). However, Dean made no comment on the neural crest.[4]

There is some information on hagfish neural crest in a 1942 paper by Conel, based on embryos collected by Dean. In a study largely devoted to the neural crest in sharks, and with only one figure and two paragraphs of text, Conel reported that the neural crest in *Eptatretus stouti* evaginates from the dorsal neural folds as a hollow sac connected to epidermal ectoderm above and to the dorsal surface of the neural tube below, with a lumen that is continuous with the cavity of the neural tube. This pouch or ganglionic vesicle, as he termed it, was continuous along the developing spinal cord but segmental, with thin intervening intersegmental connections. These are clearly pouches of neural crest cells, which Conel argued give rise to the segmentally arranged dorsal root cranial of both cranial and spinal nerves.

In the absence of a supply of embryos for experimentation, Ronald Strahan (1958) applied the technique of coordinate transformation devised by D'Arcy Thompson to deduce developmental sequences for various living and fossil agnathans. In this technique, a grid placed over the outline of one species is compared with a second species by positioning the intercepts of the grid on the same topographical landmarks and examining the transformation of the grid. Strahan described hagfish development by transformations from a generalized agnathan embryo, concluding that the pattern of development of the hagfishes is so divergent from that of the other Agnatha—living lampreys, fossil cephalaspids and, anaspids—that hagfishes most likely represent an independent evolutionary line. In coming to this conclusion, Strahan was following the reasoning that if the patterns of development of two apparently closely related groups are very different from one another, they must either have diverged from one another early in their evolution, or represent separate evolutionary lines. A recent examination of hagfish development also argues for fundamental differences between lampreys and hagfishes. In lampreys and all other vertebrates, for example, the pituitary develops by inpouching of the ectoderm. The pituitary develops by delamination from endoderm in the Californian hagfish.[5]

We also know a little about the hagfish skeleton. The Atlantic hagfish, *Myxine glutinosa*, has two types of cartilage within the lingual skeleton. One is quite unlike any vertebrate cartilages (remember that hagfishes are not vertebrates), but is similar to some invertebrate cartilages in consisting of hypertrophic chondrocytes filled with cytoplasmic filaments; Glenda Wright and her colleagues regard it as the most primitive of all the agnathan cartilages. The second type, although histologically similar to vertebrate cartilage, possesses an extracellular matrix protein, myxine, that differs from collagen, elastin, lamprin (found only in lamprey cartilage; see below), and from the major matrix protein of the other form of hagfish cartilage. It is not known whether either of these cartilages are of neural crest origin or whether they are homologous with lamprey or gnathostome visceral arch cartilages.[6]

Lampreys

Rather more is known about the neural crest of representative species of lamprey. Koltzoff, for instance, described a neural crest from histologically sectioned embryos of *Petromyzon planeri* as long ago as 1901, while in the 1930s, Bytinski-Salz concluded that the pigment cells that differentiated in grafts of lamprey cranial neural folds must have originated from neural crest cells included in the grafts. Extirpation of neural crest also results in pigment deficiencies; see below.[7]

The bulk of our knowledge, however, is of the lamprey skeleton, which I relate below. The chapter concludes with a short description of molecular analyses of lamprey neural crest and of the homology of lamprey, hagfish, and gnathostome visceral arches.

The Skeleton

As indicated in the previous section, the cartilages of hagfishes and lampreys are structurally and biochemically distinct. Although presence of type II collagen is a *sine qua non* of cartilage, in the sea lamprey, *Petromyzon marinus*, type II is only found in the perichondrium and most peripheral extracellular matrix of cartilages. The bulk of the cartilaginous extracellular matrix consists of branched fibrils and fibers, 150–400 Å in diameter, composed of a structural protein, lamprin, unique to lampreys. The amino acid composition of lamprin is:

- reminiscent of elastin, in containing only small amounts of acidic amino acids;
- unlike elastin, in containing significant amounts of tyrosine and histidine; and
- unlike both elastin and collagen, in containing only traces of hydroxyproline.

The amino acid composition of representative lamprey cartilages is compared with type II collagen and elastin in Table 4.1.

Lamprin constitutes some 50% of the dry weight of lamprey annular and neurocranial cartilages. Glycosaminoglycans, which constitute the bulk of the extracellular matrix of vertebrate cartilages, make up less than 5% of the dry weight of lamprey cartilages. Clearly, these are unusual vertebrate cartilages. Indeed, the skeleton of *Petromyzon marinus* has three distinct types of cartilage:

- the annular and neurocranial cartilages, which are based on lamprin and are neural crest in origin;
- the branchial basket and pericardial cartilages, which are based on a second protein with affinities to elastin and are also derivatives of the neural crest (Fig. 4.2);
- the mucocartilages, which are distinctive from the remainder of the skeleton, both histologically and in not being of neural crest origin.[8]

TABLE 4.1. The amino acid composition (residues/1000) of the major protein of four cartilages of the sea lamprey, *Petromyzon marinus*, in comparison with type II collagen and elastin.

	Cartilages					
Amino Acid	Annular	Neurocranial	Branchial Basket	Pericardial	Type II collagen	Elastin
HYP	1	3	22	17	74	9
ASX	21	23	66	70	47	6
THR	31	32	14	14	27	15
SER	27	29	43	46	38	12
GLX	38	36	60	61	99	19
PRO	96	95	109	102	115	113
GLY	282	268	323	313	330	313
ALA	156	140	169	176	104	244
VAL	79	106	30	33	12	128
MET	<1	<1	<2	<2	<1	<1
ILU	20	21	15	16	11	18
LEU	119	115	107	33	29	54
TYR	53	50	47	50	2	19
PHE	14	13	16	12	12	33
LYS	7	7	13	13	16	5
HIS	38	38	14	13	4	1
ARG	17	21	29	30	51	8

Based on data summarized from Robson et al. (1997).

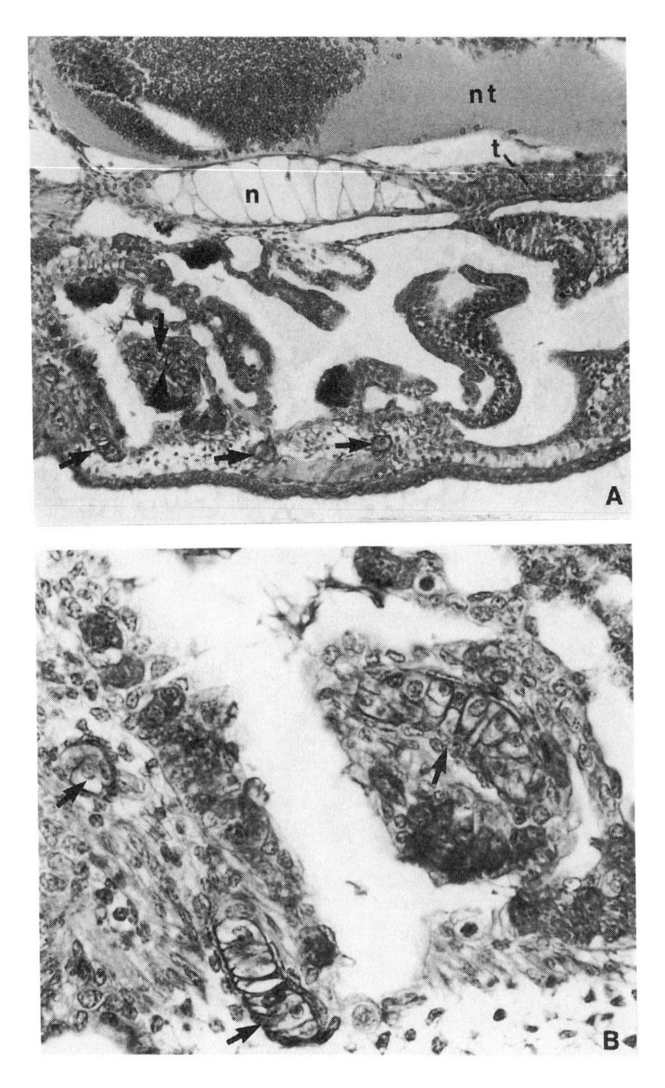

FIGURE 4.2. The distinctive histology of lamprey branchial arch cartilages (arrows) can be seen in these low (A, x 265) and high (B, x 460) magnification sagittal sections through a stage-17 larval *Petromyzon marinus*. The trabecular cartilage (t) adjacent to the notochord (n) and beneath the developing neural tube (nt) is also evident. From a specimen kindly provided by Robert Langille.

The Neural Crest Origin of the Skeleton

de Beer (1947) offered an evolutionary explanation for the neural crest origin of visceral arch cartilages in the axolotl, *A. mexicanum*—the possibility that the vertebrate visceral skeleton had always developed from neural crest cells, i.e., that a visceral arch skeleton of neural crest origin was the evolutionarily primitive condition. de Beer cited various early workers who claimed such a neural crest origin and emphasized the importance of knowing the source of the visceral skeleton in such vertebrates as cyclostomes and sharks, which were then regarded as "lower," "primitive," or "ancestral."

It is difficult to determine from static histological evidence whether lamprey cartilages are derived from cranial mesoderm, neural crest cells, neuroepithelium, or epidermal placodes. Johnels (1948) concluded that the basicranial skeleton, including the parachordals, was mesodermal, but that neural crest cells contributed to the skeleton of the visceral arches. Damas (1944, 1951) believed that the head cartilages were of neural crest origin and supported this interpretation of normal development with an experiment in which he exposed ammocete larvae to light. Light-treated larvae developed abnormalities, including deficiencies in the visceral skeleton. Noting that defective visceral chondrogenesis was preceded by a failure of contact between branchial ectoderm, branchial endoderm, and neural crest mesenchyme, Damas concluded that the cartilages were derived from neural crest cells (as amphibian visceral cartilages had been shown to be; Chapter 5), and that interactions with embryonic epithelia similar to those between neural-crest-derived cells and pharyngeal endoderm in amphibian embryos, occurred in lamprey embryos.

In the 1950s, David Newth performed a series of neural crest extirpations and transplantations on embryos of the brook and river lampreys (*Lampetra planeri* and *L. fluviatilis*) to determine the cellular origin of the head cartilages. Because these studies illustrate the potential pitfalls of moving neural crest cells away from their normal inductive environment, I discuss them in some detail.[9]

Newth (1950) first removed the cranial neural crest from either one or both sides of the neural tubes of *L. planeri* embryos at the 3-somite stage. Ten of the 15 embryos that survived lacked all or some of the cranial ganglia, but none showed any deficiencies in the cartilages of the visceral skeleton or in the trabeculae cranii (the cartilage of the cranial floor). Newth performed similar extirpations in his second study, but grafted the extirpated neural crests into the flanks of donor embryos. As *no* embryos exhibited any skeletal deficiencies and as cartilage did not form in any of the grafts, Newth concluded that the neural crest was *not* the source of the chondrogenic cells. However, the absence of pigment cells, dorsal root ganglia, and some head mesenchyme in the operated embryos led Newth to conclude that these three cell types arose from the neural crest.

Thus, the first neural crest extirpation and grafting studies in lamprey embryos consistently affirmed a *lack* of involvement of neural crest cells in the development of the cartilages of the cranial and visceral skeletons. These results were surprising. The abundant evidence from studies on amphibian embryos discussed in Chapter 5 had demonstrated the fundamental involvement of cranial neural crest cells in the cranial, facial, and visceral skeletons.

Two possibilities could explain David Newth's results:

- Because lampreys are primitive vertebrates, went one argument, perhaps their cranial skeleton has a different cellular origin from that of jawed vertebrates. If this is so, a non-neural-crest origin of these cartilages would represent the original (primitive) vertebrate condition (a position argued against in Chapter 3).
- Alternatively, if lampreys are specialized or derived vertebrates (which we now know them to be), a non-neural-crest origin could be taken as evidence of a secondary loss of the original primitive vertebrate condition of a cranial skeleton formed by neural crest cells and the subsequent acquisition of a mesodermal origin.

Newth's third paper was published in Russian in 1955. In an English abstract he records that he excised and transplanted neural crests in *L. planeri* and obtained evidence that the visceral arch cartilages of the branchial basket *are* of neural crest origin. His fourth paper provided the details substantiating this reversal.[10]

Newth discussed two potential difficulties with his initial studies that could have produced false negative results. One was the possibility of compensation for the extirpated neural crest cells by regulation from adjacent regions of the neural crest. If only a portion of the neural crest is removed—if extirpation is unilateral or from only one cranio-caudal region of the neural axis—neural crest cells from the intact side, or from more cranial or caudal regions, could migrate to fill the wound and replace the extirpated neural crest cells. If these cells were capable of forming the same cell types as the extirpated neural crest cells, development of the operated embryos could be completely normal. Regulation of the neural crest is taken up in greater detail in Chapter 12.[11]

The second potential difficulty arises from grafting neural crest cells into the flanks of host embryos. Newth realized that placing the neural crest cells in an ectopic site might have removed them from necessary inductive influences emanating from cell layers in their normal location. Failure of flank-grafted neural crest cells to chondrify might then not speak to the inability of the neural crest cells to become cartilage cells, but to the absence of the necessary inductive environment at the graft site. Indeed, we now know that axolotl trunk neural crest cells fail to migrate when grafted in place of cranial neural crest and thus fail to come into contact with the inductive signals required to activate chondrogenesis. Consequently, such a grafting technique does not provide an adequate test of the chondrogenic potential of such cells (Graveson, Hall, and Armstrong 1995).

A third potential explanation, suggested by Hall (1987a), is that migration of neural crest cells might already have occurred, or at least begun, before the late neurula or 3-somite stage from which Newth extirpated the neural folds. In this case, even if the skeleton was neural crest in origin, extirpation would neither result in skeletal deficiencies nor speak to neural crest origin.

Newth overcame the first two potential problems by grafting lamprey neural crest cells into the flanks or branchial regions of neurula-stage embryos of the urodeles *Triturus cristatus* or *T. helveticus*. Neural crest cells differentiated into chondroblasts and formed cartilages when grafted into the branchial region, but not when grafted into the flank. Newth concluded:

- that his earlier flank grafts had indeed removed neural crest cells from the necessary inductive environment provided by branchial ectoderm or endoderm; and
- that lamprey visceral skeleton is formed from neural crest.

This study, and the results obtained by Damas cited above, aligned both the cellular origin of lamprey cartilages and the requirement for activation via inductive interaction with embryonic epithelia, with results previously obtained in embryonic amphibia (discussed in Chapter 5) and established a neural crest origin for the visceral arch skeleton as the basic vertebrate condition.

No further extirpations of lamprey neural crest were reported until a study on the anadromous North American sea lamprey, *Petromyzon marinus*, in the mid-1980s by Rob Langille, who removed one of seven regions of cranial neural crest (each 250 μm in cranio-caudal extent) from early neurula-stage embryos (shown as I to VII in Fig. 4.3) and allowed the embryos to develop into larvae when cranial and visceral skeletons would normally be present.[12]

Pigmentation of the heads of these embryos was reduced substantially, indicating an involvement of the neural crest in the differentiation of pigment cells, as noted by Newth. The trabeculae and branchial arch cartilages were either entirely absent or substantially reduced, depending on the region of neural crest extirpated. These extirpation studies do not allow us to conclude that the trabeculae are entirely of neural crest origin (although they may be), only that they are substantially derived from the neural crest. The trabeculae of urodele amphibians do receive some cells from mesenchyme derived from head mesoderm (Chibon 1966, 1967, and Chapter 5).

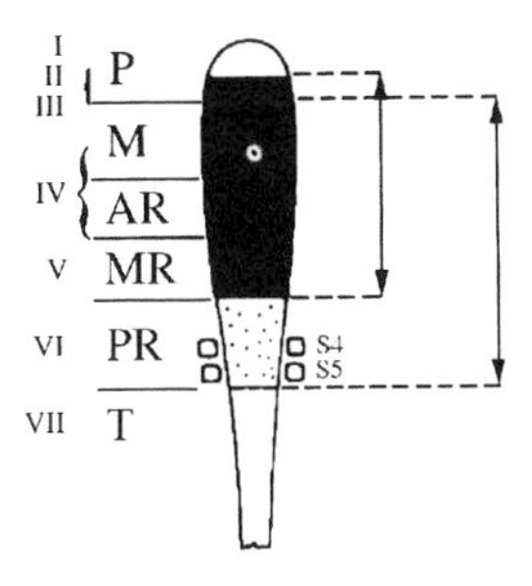

FIGURE 4.3. Regionalization of the cranial neural crest showing the cranio-caudal extent of the neural crest contribution to the chondrocranial (black; double-headed arrows) and pharyngeal skeleton (stippled; double-headed arrows) in the lamprey *Petromyzon marinus*, viewed from the dorsal surface. The skeletogenic cranial neural crest extends from mid-prosencephalon (P) caudal to the level of somites 4/5 (S4, S5). AR, anterior rhombencephalon; M, mesencephalon; MR, mid-rhombencephalon; PR, posterior rhombencephalon; T, trunk neural crest cells. Numbers I to VII represent the boundaries of regions excised to generate the fate map described in the text and in Tables 4.2 and 4.3.

In no embryo examined by Langille were the parachordal cartilages, the cartilage of the otic capsule, or the cranial mucocartilages missing or deleted. This led to the conclusion that unlike the trabeculae and visceral arches, these cartilages do not receive any contribution from the neural crest. A mesodermal origin of the parachordals in lampreys is consistent with the situation in other vertebrates. On the basis of their analyses of normal development, Damas and Johnels concluded that lamprey otic capsular cartilage was mesodermal in origin; otic capsules develop in lamprey embryos from which the neural crest is removed. The otic capsule in amphibians is also mesodermal. However the otic capsule of birds is chimeric, receiving cells from both neural crest and head mesoderm. Therefore, in lampreys, otic cartilages are either entirely mesodermal or partly neural crest in origin, in the latter case with the missing neural crest cells in extirpation experiments compensated for by an additional mesodermal contribution to the otic capsule. Studies with labeled neural crest cell grafts will be required to resolve this issue.[13]

The regional extirpations of the neural crest performed by Langille allowed a chondrogenic portion of the neural crest to be identified and mapped into zones. Based on a correlation of the cranio-caudal region of the neural crest ablated and the skeletal deficiencies he observed, Langille came to the following conclusions concerning the skeletogenic capability of the seven zones shown in Figure 4.3:

- Zone I, the rostral prosencephalon, provides no cells to either trabecular or branchial arches and is nonchondrogenic (Table 4.2). This is equivalent to the transverse neural folds in amphibians discussed in Chapter 5.
- Zone VII, which lies caudal to the level of the sixth pair of somites, provides no cells to either trabecular or branchial arches and is nonchondrogenic.
- Only neural crest cells extending from the caudal prosencephalon to the level of the fifth pair of somites (regions II to VI in Tables 4.2 and 4.3) are chondrogenic.
- Trabeculae arise from rostral cranial neural crest, extending from the caudal prosencephalon caudally to the mid-rhombencephalon (regions II to V, Table 4.2).
- Branchial arch cartilages are derived from more caudal cranial crest, extending from the rostral mesencephalon caudally to the level of the fifth pair of somites (regions III to VI, Table 4.2 and Fig. 4.3).

The patterns of localization and regionalization of the skeletogenic neural crest are consistent with those found in the jawed vertebrates illustrated in Figure 5.4, and for the hyoid arch in *Chelydra serpentina*, a turtle (Toerien 1965a).

Langille also determined the contribution of cells within the caudal cranial neural crest (regions III to VI in Table 4.2) to the branchial arches. The more caudal the branchial arch, the more caudal is the cranial neural crest that produces it (Table 4.3):

- Arches 1 to 5 develop from neural crest cells derived from regions III to VI.
- Arches 1 to 3 arise predominantly from neural crest cells in regions III to V.
- The most caudal arches (6 and 7) develop from neural crest cells from regions V and VI.

TABLE 4.2. Percentages of larvae with missing or deficient trabeculae and/or branchial arches following removal of neural crest from embryos of the lamprey *Petromyzon marinus*.[a]

Region removed[b]		Trabeculae	Branchial arches
I	anterior prosencephalon	0	0
II	posterior prosencephalon	78	0
III	anterior mesencephalon	50	70
IV	posterior mesencephalon/anterior rhombencephalon	25	75
V	mid-rhombencephalon	75	100
VI	somites 4–5	0	70
VII	somite 6	0	0

a. Based on data in Langille and Hall (1988b).

b. See Figure 4.3 for the seven 250 μm regions of neural crest extirpated. N = 4 to 10 specimens removed. The levels of the neural tube and/or adjacent somites corresponding to each region are indicated.

TABLE 4.3. Percentages of larvae showing missing or deficient branchial arches 1–7 following removal of 250 μm regions of neural crest from embryos of the sea lamprey, *Petromyzon marinus*, to illustrate the rostro-caudal level of the cranial neural crest contributing cells to the branchial arch cartilages.[a]

		Branchial arches[c]		
Region removed[b]		1 to 3	4 and 5	6 and 7
I	anterior prosencephalon	0	0	0
II	posterior prosencephalon	0	0	0
III	anterior mesencephalon	71	28	0
IV	posterior mesencephalon/anterior rhombencephalon	33	66	0
V	mid-rhombencephalon	50	50	25
VI	somites 4–5	14	51	71
VII	somite 6	0	0	0

a. Based on data in Langille and Hall (1988b).

b. The levels of the neural tube and adjacent somties corresponding to each region are indicated. See Figure 4.3 for the seven regions. N = 3 to 7 specimens/region removed.

c. Branchial arch 1 is the most rostral, 7 the most caudal. The arches are grouped as rostral (1 to 3), mid (4 and 5), and caudal (6 and 7).

Thus, regionalization of the skeletogenic cranial neural crest in *P. marinus*, which is very similar to that described in Chapter 5 for urodele and anuran amphibians, represents a patterning of the neural crest established early in vertebrate evolution. We would love to know the situation in hagfishes, for without that knowledge we cannot use the neural crest or the visceral arch skeleton to draw any meaningful conclusions regarding the origin and fate of the neural crest in nonvertebrate craniates or lamprey/gnathosome relationships; see Langille and Hall (1989) for further discussion.

Molecular Analyses

Lampreys have attracted the attention of some recent molecular investigations that shed light on some aspects of the neural crest.

HNK-1 labels premigratory and some postmigratory neural crest cells but is down-regulated with differentiation of neural crest derivatives. HNK-1 was used to demonstrate the absence of a ventral pathway of trunk neural crest cell migration in lamprey embryos, an absence that was correlated with the absence of ganglia of the sympathetic chain in lampreys. The same marker was used to follow neural crest mesenchyme into the dorsal fin and to identify neurons within the dorsal neural tube that were taken to be homologues of Rohon-Béard cells and HNK-1-positive cells in the heart. Interestingly, no HNK-1-positive cells were found in the head, presumably because of the age of the embryos and the down-regulation of HNK-1 in differentiated cells.

Whether the HNK-1-positive cells in the heart are evidence of a cardiac neural crest in lampreys as documented in some other vertebrates remains to be determined. HNK-1 was detected in heart primordia of avian embryos by Luider and colleagues (1993), but was not regarded as evidence of neural crest derivatives. The same laboratory (Luider et al. 1992) has reported HNK-1 cells in the avian embryonic gut before it is colonized by neural crest cells and emphasizes that the various antigens of HNK-1 glycoproteins are likely to recognize different antigenic sites. In rat embryos, HNK-1-positive cells do correlate with conductive tissue within the heart.[14]

The neural and neural crest cells of jawed vertebrates are sensitive to excess vitamin A; exposure produces defects in cells of neural crest origin. Treating lamprey embryos with vitamin A results in loss of the pharynx and truncation of the rostral neural tube that is both dose- and stage-dependent, indicating that sensitivity to vitamin A predated the separation of agnathan and jawed vertebrate (see Chapter 12).

Lampreys possess a rostral velum that is derived from the mandibular segment but not from a gill arch and that functions as a pharyngeal pump. The rostral arches of hagfishes are also associated with a velum rather than with the gills, but some question the homology of hagfish and lamprey (and therefore hagfish and jawed vertebrate) visceral arches. The *Engrailed* gene in the lamprey, *Lampetra japonica*, is expressed in a set of mandibular arch muscles, the velothyroideus,

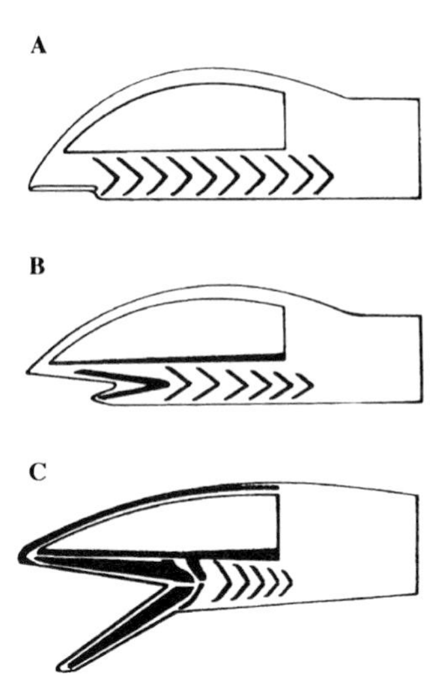

FIGURE 4.4. (A) An agnathan, showing the serially arranged gill arches. (B) Transformations of the first of the serial gill arches to produce the jaws in gnathostomes. (C) The jaws were further strengthened, elongated and supported by the skeleton of the second pair of gill arches. Modified from du Brul (1964).

which activate the velum. This muscle has been homologized with the levator arcus palatini and dilator operculi muscles of vertebrate mandibular arches and the shared gene expression patterns used to argue for homology of lamprey velar and gnathostome mandibular arches. Furthermore, some argue that the velum rather than a gill arch is the homologue of jaws, i.e., that rather than jaws arising from gill arches, jaws led to the development of new gill arches.[15]

The traditional view of the origin of jaws in vertebrates is that they evolved from the most anterior gill arches of a jawless craniate/vertebrate (Fig. 4.4). If hagfish arches are homologous with the visceral arches of lampreys and jawed vertebrates (which are of neural crest origin), we would expect hagfish visceral arch skeleton to be neural crest in origin. Hagfishes are not vertebrates, however, and a neural crest origin for branchial cartilages may postdate the hagfish/lamprey split, although evidence concerning the origin of the chondrogenic neural crest presented in Chapter 3 makes this unlikely.

The next four chapters outline the evidence for neural crest derivatives in the major groups of jawed vertebrates. Because the arrangement best fits how the data on the neural crest were gathered, I have organized some of the information into such chapters as Amphibians and Fish even though these are no longer regarded as monophyletic groups. The arrangement is to facilitate discussion of the state of knowledge concerning the neural crest in commonly studied vertebrates.

Part II

Derivatives and Diversity

5

Amphibians

Because so much of the early work on the neural crest was undertaken using amphibian embryos, I begin the series of chapters devoted to gnathostomes with a discussion of studies on the amphibian neural crest, providing evidence for the neural crest origin of pigment cells, various nerve cells, craniofacial mesenchyme, and craniofacial skeletal tissues.

Pigment Cells

Pigment cells provide excellent markers for neural crest cell migration and are still used with profit to detect subpopulations of amphibian neural crest cells, especially when wild-type (pigmented) tissues are grafted to albino embryos. It is important to remember, however, that although pigment cells are prominent neural crest derivatives, not all pigment cells are of neural crest origin. Eye pigments, for example, have a dual origin; pigment of the choroid coat and outer iris is neural crest in origin, but pigmented retinal epithelium arises from the neuroepithelium of the optic cup.[1]

In the first quarter of this century, Wiedenreich, Harrison, Holtfreter, and Mangold all posited a neural crest origin for pigment cells from their in vivo analyses and the appearance of pigment cells in cultured frog spinal cord. In the 1930s and 1940s, DuShane, Raven, and Twitty convincingly demonstrated the neural crest origin of these cells using extirpation and transplantation of the amphibian neural crest. DuShane's 1935 study is representative of evidence based on transplantation between individuals of a single species. Pigment cells appeared at the graft site after transplantation of neural folds to such ectopic positions as the ventral body wall, while transplanted limb buds acquired pigmentation through migration of neural crest cells into the transplant. Bilateral extirpations of trunk neural crest by these workers resulted in a local absence of pigment cells, spinal ganglia, dorsal fin, and Rohon-Béard cells in regions corresponding to the ablated neural crest.[2]

Studies by Twitty, Raven, and Bytinsky-Salz (1937, 1938) and Baltzer (1941) are typical of those in which neural crest was transplanted between embryos of different species, usually between urodeles, or from anuran to urodele embryos; the neural crest origin of pigment cells was determined because of the ability to distinguish donor from host tissues. Niu (1947) showed that trunk neural crest is a more potent source of pigment cells than cranial crest.

Labeling the neural crest of larvae of the salamander *Hynobius lichenatus* with vital dyes or india ink, coupled with transmission and scanning electron microscopy, allowed Hirano and Shirai (1984) to track migrating neural crest cells. In a subsequent study in which somites were unilaterally extirpated, Hirano (1986) demonstrated that neural crest cells follow one of two paths. Pigment cells migrated between epidermis and neural tube. Prospective ganglia migrated lateral to the neural tube. These results imply that separate cell populations give rise to these two neural crest cell types. As we will see in Chapter 10, they do.

When they metamorphose from tadpoles to adults, many amphibians display characteristic color changes as previously pigment-free cells undergo pigmentation. The cells that provide the pigmentation seen in adults are also of neural crest origin, the hormonal changes associated with metamorphosis triggering onset of pigmentation. This was demonstrated by Twitty and Bodenstein (1939), who transplanted neural crest from *Ambystoma punctatum* (in which only adults are pigmented) into the ventral regions of embryos of *Triturus torosus*, which has pigmented larvae. Pigmentation did not appear in the grafts during larval development, but pigment spots characteristic of the donor, *A. punctatum*, appeared with metamorphosis. More recently, Parichy (1998) has shown that patterns of pigmentation in adults of *A. tigrinum* are largely independent of both larval patterns and of the lateral line cues that help generate the larval patterns.[3]

Neuronal Cells

Neuronal derivatives of the neural crest include cranial and spinal ganglia; adrenergic and cholinergic sensory neurons of the sympathetic, parasympathetic, and peripheral nervous systems; and Schwann, glial, and Rohon-Béard cells (see Table 2.1). In short, the peripheral nervous system. Our understanding of these neuronal derivatives owes much to pioneering studies using amphibian embryos. Indeed, many of the basic elements of modern-day neurobiology were established in studies in which amphibian neural crest cells were labeled, excised, or transplanted. More recent studies on the neuronal derivatives of the neural crest have concentrated on avian rather than amphibian embryos (Chapter 7).[4]

Spinal and Cranial Ganglia

Identification of spinal ganglia as neural crest derivatives goes back to Wilhelm His's discovery of the crest. Indeed, so readily was the claim accepted that this embryonic region was known as the ganglionic crest.

His made his observations on static specimens; it was not until the 1930s that techniques were developed to visualize cells in living embryos. Detwiler stained local regions of neural crest with pieces of agar impregnated with vital dyes to show that spinal ganglia arise from trunk neural crest cells that migrate between somites and spinal cord/notochord. Experimental proof was provided by Harrison and DuShane, who both observed that spinal ganglia and sensory nerves fail to form in tadpoles that develop from embryos from which the neural crest has been removed at the time of neural fold closure. Extirpation of the ventral portion of the neural tube results in larvae lacking motor nerves, but has no effect on spinal ganglia or sensory nerves; i.e., the neural crest is a dorsal neural derivative. Further confirmation was provided by Raven, who transplanted neural crest between embryos of different species of amphibians, using such species-specific characters as differential cell or nuclear size to distinguish host from donor cells.[5]

Spinal ganglia do not self-differentiate but require influences from other cells and tissues, primarily from the somites. This conclusion goes back at least to the studies by Detwiler (1934), in which removal of somites led to local absence of spinal ganglia and spinal nerves, while addition of somites led to supernumerary ganglia and nerves. The signals that direct differentiation of spinal ganglia and other neural crest cells are discussed in Chapter 10.

Acceptance of the neural crest origin of cranial ganglia did not come quite as readily as did acceptance of spinal ganglia as neural crest derivatives. Some argued for a neural crest origin, others for a placodal origin. Both were correct. As established by Stone (1922) and Yntema (1943), many cranial ganglia receive contributions from both neural crest and placode (see Figs. 2.12, 2.13, and Chapter 7).

The Sympathetic Nervous System

As with the origin of cranial ganglia (but for different reasons), equally authoritative figures differed on whether the neural crest contributed to the sympathetic nervous system. In part the differences were technical; extirpation and transplantation studies gave less reproducible results than vital staining. The clearest interpretation of the early studies was that both neural crest and ventral neural tube contribute cells to the sympathetic nervous system; see Hörstadius (1950). Resolution would have to await studies on other vertebrates, notably chick embryos, as discussed in Chapter 7.

The suspected dual origin of the sympathetic nervous system also applied to the origin of chromaffin cells of the adrenal gland and to ganglia associated with the aorta. (Interestingly, lampreys lack both chromaffin cells and peripheral sympathetic neurons, perhaps indicating a long evolutionary linking of these two derivatives within a single neural crest population.) Enteric and visceral ganglia were more convincingly shown to arise from neural crest.[6]

Schwann Cells

Peripheral motor nerves arise from neural ectoderm. Indications that the Schwann cells that sheath peripheral motor nerves arise from neural crest go back to the extirpation and tissue culture studies of Harrison, and Müller and Ingvar in the early years of this century. Indeed, Harrison's were the first tissue culture experiments performed with any tissue. Subsequent vital dye staining of *Ambystoma* embryos by Detwiler and Kehoe confirmed the neural crest origin of Schwann cells.[7]

Rohon-Béard Cells

Giant, transient, ganglionic cells form a network associated with the dorsal neural tube in some amphibian embryos and tadpoles and in the embryos of all cartilaginous and bony fishes, e.g., as extramedullary neurons in *Raja batis*, a skate, and in the cichlid *Oreochromis mossambicus*. The cells are called Rohon-Béard cells after Rohon (1884) and Béard (1896), who described them most fully. Balfour had discovered these cells in elasmobranchs in 1878, while Béard first described cell death as a normal part of neuronal development in skate Rohon-Béard cells. Functionally, Rohon-Béard cells are replaced by spinal nerve ganglia. Why and how they disappear is poorly understood.[8]

Béard hypothesized that Rohon-Béard cells arise from the neural crest in amphibian embryos. The cells fail to form if the trunk neural crest is removed, but because Rohon-Béard cells and neural crest originate from separate lineages that can be identified as early as the 512-cell stage in *Xenopus*, Marcus Jacobson (1991) questions whether Rohon-Béard neurons are of neural crest origin. Separate lineages, combined with origination at the gastrula stage before neural crest has been induced, is certainly inconsistent with what we know of neural crest lineages. Given that the two lineages lie adjacent to one another in the blastula, that placodal and neural crest ectoderm lie in very close proximity, and that some pladocal ectoderm arises from the neural folds (see Fig. 2.12), a placodal origin is not ruled out; the extirpation studies could have removed ectoderm other than neural crest from the neural folds. Further study of these interesting cells is clearly merited.[9]

Craniofacial Mesenchyme and Cartilage

The skeletogenic cranial neural crest is the source of much cranial mesenchyme, skeletal, and dental tissues, and of the controversy over the veracity of the germ-layer theory outlined in Chapter 1. Hörstadius (1950) devoted the longest chapter of his book—the 37-page Chapter 3, which has 26 figures—and his monographic 1946 paper with Sven Sellman, to the contribution of neural crest to the axolotl cranial skeleton.

Evidence from Histology

On the basis of a study of shark embryos, Kastschenko (1888) made the original suggestion that cranial mesenchyme was derived from the neural crest. Some five years later, Goronowitsch came to a similar conclusion from studies of teleost fishes and birds, but the first suggestion of a neural crest origin of cartilage was in amphibians. Based on her studies on the mud puppy, *Necturus*, Julia Platt (Box 5.1) proposed in 1893 that ectoderm rather than mesoderm contributes the mesenchyme and cartilage of the visceral arches and the dentine of the teeth. Taking advantage of the fact that yolk droplets in ectodermal cells can be distinguished in histological sections from those in mesodermal cells, Platt mapped the cellular origins of the mesenchyme and followed the fate of migrating neural crest cells into early embryonic development. While regarding cranial mesenchyme as ectodermal, she believed that the mesenchyme arose from placodal lateral cranial ectoderm as well as from neural crest. Platt coined the terms *mesectoderm* for mesenchyme derived from ectoderm and *mesendoderm* for mesenchyme derived from mesoderm. Nowadays, we use *mesenchyme* for ectodermally and mesodermally derived mesenchyme. When an ectodermal origin needs to be distinguished, the term *ectomesenchyme* is sometimes used.[10]

Twenty-four years elapsed before Landacre (1921) demonstrated that neural crest, not placodal ectoderm, is the source of the visceral cartilages of *Amblystoma (Ambystoma) jeffersonianum*. Landacre took advantage of the differences in size, staining properties, pigment, and yolk contents of neural crest cells to follow their migration along and into the visceral arches. In addition to confirming a neural crest origin, he mapped those parts of the skeleton that arise from neural crest—the anterior trabeculae cranii and visceral arch skeleton, except for the second basibranchial, which is mesodermal. As discussed in Chapter 1, the germ-layer theory was so entrenched that a neural crest origin of skeletal tissues was

---

Box 5.1
Julia Platt (1857–1935)

American-born Julia Platt was a remarkable woman. Educated at Harvard, Radcliffe College, Chicago, Freiburg, and Munich, she worked with many of the leaders of the day—C. O. Whitman, E. B. Wilson, Robert Wiedersheim, Carl von Kupffer, Richard Hertwig, and Charles B. Davenport. One of the first women to receive a Ph.D. degree from a German university—from Freiburg in 1889, where August Weismann was the principal referee for her thesis—Platt was one of the first women neuroscientists, and one of the first two women to join the American Society of Morphology (later the American Society of Zoologists, now the Society for Integrative and Comparative Biology [SICB]). The Julia Platt Club, a forum for the presentation of results of research in evolutionary morphology, had its inaugural meeting at the SICB meeting in Boston in January 1998. Platt's outstanding training was followed by the publication of 12 papers between 1889 and 1899—one on axial segmentation in the chick, three on the development of head cavities, one on amphioxus, six on the differentiation of *Necturus* (including the ectodermal origin of head cartilages and development of the lateral lines), and one on the specific gravity of protozoa and tadpoles.[11]

---

not readily accepted. Edwin Goodrich, thought by many to be the leading comparative anatomist of his day, believed that "this doctrine of the formation of special 'mesectoderm' in the head is, however, almost certainly founded on misinterpretations and erroneous observations on unsuitable material" (de Beer 1947, p. 308). Contention continued until Stone, Raven, and Holtfreter confirmed that the neural crest is a major source of mesenchyme, connective tissue, and cartilage (see below). de Beer provided further evidence for a neural crest origin of the visceral cartilages and odontoblasts in *Ambystoma* and also thought it probable that neural crest cells differentiated into osteoblasts of dermal bones, although his evidence was less convincing than for the neural crest origin of cartilage and teeth.[12]

Evidence from Extirpation and Transplantation

Leon Stone (Box 5.2) was the first to extirpate selected regions of the neural crest in a search for derivatives. Extirpation of neural crest from *Ambystoma punctatum*[13] and *Rana palustris*, and analysis of subsequent deficiencies, confirmed Landacre's descriptive study of the neural crest origin of the skeletal cartilages and enabled Stone to delineate a fate map of the cranial neural crest (Fig. 5.1).

Raven went beyond Stone's approaches when he grafted trunk neural crest from *A. mexicanum* in place of cranial neural crest of *Triturus* embryos. Transplanted trunk neural crest failed to produce the cartilages that would have arisen from cranial neural crest. In his monographic treatment of the development of the vertebrate skull, and with reference to Raven's studies, Gavin de Beer asked: "Can heteroplastic grafting experiments demonstrate the derivation of visceral arch cartilage from neural crest cells in other groups besides Amphibia?" (de Beer 1937, p. 514). It would be 14 years before David Newth answered that ques-

---

Box 5.2
Leon Stansfield Stone (1893–1980)

Leon Stone was educated at Lafayette College and at Yale University, where he spent a 42-year career in the anatomy department, retiring as Bronson Professor of Comparative Anatomy Emeritus in 1961. Stone pioneered the use of cinematography and time-lapse techniques in experiments on regeneration of the amphibian iris, lens, and retina and on eye development in fetal (pouch) possums.[14] E. S. Crelin of Yale University, who was a Ph.D. student of Stone's (Stone himself having been a student of Ross Harrison), told me that after the publication of the 1922 paper mapping the neural crest, Stone visited Germany, where the heretical notion of cartilage arising from an ectodermal derivative led to him being dubbed "Neural Crest Stone."

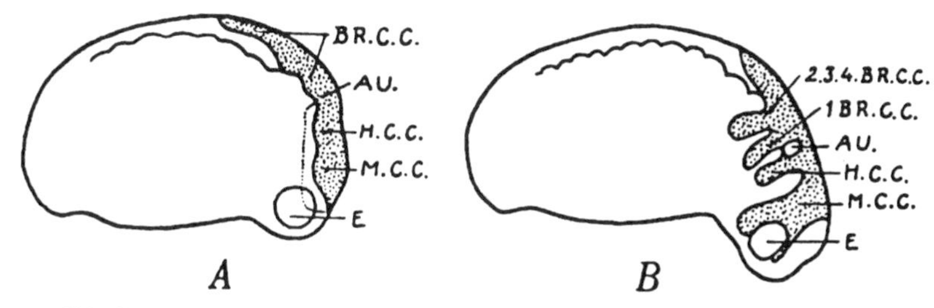

FIGURE 5.1. Stone's (1926) fate map showing the origin of the craniofacial mesenchyme from premigratory (A) and late migratory (B) cranial neural crest of *Ambystoma punctatum* (now *A. maculatum*). Distinctive mandibular (M.C.C.), pre- and postotic, and branchial arch populations (BR.C.C.) are evident in B, the mandibular population in two streams rostral and caudal to the eye. AU, auditory vesicle; 1-4 BR.C.C., mesenchyme of branchial arches 1-4; E, optic vesicle; H.C.C., hyoid arch mesenchyme; M.C.C., mandibular mesenchyme. Reproduced from Hörstadius (1950) with the permission of Dagmar Hörstadius Ågren.

tion in the affirmative for lampreys (Chapter 4) and 34 years before Catherine Le Lièvre undertook the same studies with avian embryos (Chapter 7).[15]

Holtfreter's contribution—one of many he made to amphibian development—related to tooth development following transplantation of tissues from one region of an embryo to another. Teeth form from ectoderm (neural crest), but only if endoderm is transplanted along with the ectoderm. Similarly, removal of endoderm is sufficient to prevent tooth formation, even though the teeth arise from ectoderm, and even though ectoderm is left intact. Clearly, tissue interactions are required to evoke tooth-forming potential from neural crest cells, a topic discussed in Chapter 10.[16]

Ross Harrison of Yale University followed up the pioneering studies of Stone and Raven. In 1929, Harrison published his now classic experiments on intrinsic determination of the size of organ rudiments. In grafting eye primordia between different amphibian species, he showed that growth of the eye follows the pattern set by the donor, not the host embryo. Twitty followed with experiments confirming and extending Harrison's findings.[17]

Harrison subsequently exchanged hindbrain neural folds—including premigratory neural crest—between *Ambystoma punctatum* and *A. tigrinum* or *A. mexicanum.* Two days later, when the neural crest cells had migrated down onto the wall of the developing pharynx, Harrison transplanted the pharyngeal wall with grafted neural crest cells to an equivalent position in another specimen (Fig. 5.2). The skeletal elements of the visceral arches of these species are quite distinctive in size. By using the species-specific sizes of the visceral arches and their skeletal derivatives as markers, Harrison confirmed the origin of the visceral skeleton from neural crest cells of the donor embryo. When the donor was the larger species, the visceral arches that developed from the grafted tissues were larger than the arches on the unoperated side. When the donor was the smaller species, the reverse was true. In introducing the session on the evolution and morphogenesis

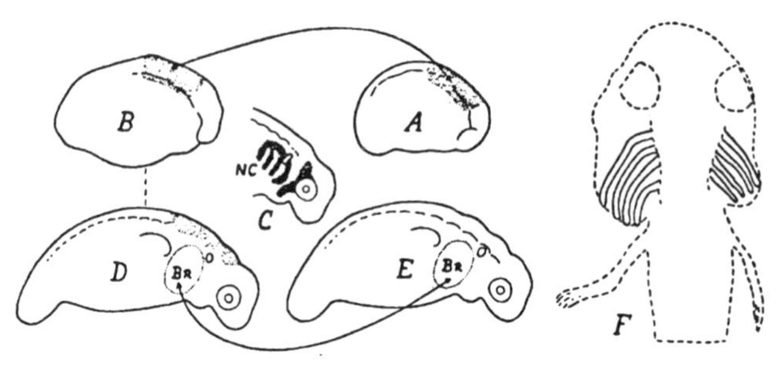

FIGURE 5.2. Transplantation of branchial arch neural crest and adjacent ectoderm between embryos of *Ambystoma punctatum*, *A. mexicanum*, and *A. tigrinum* (A to B) results in migration of the grafted neural crest into the visceral arches of the donor embryo (NC in C). The branchial region (BR) was transplanted to the normal post-migration site two days later (D to E). Harrison used size differences between graft and donor species (F) to demonstrate the neural crest origin of the visceral arch cartilages and that the size of these elements was an intrinsic property of the neural crest cells. Reproduced from Hörstadius (1950) with the permission of Dagmar Hörstadius Ågren.

of the head at the September 1987 International Symposium on Craniofacial Development, J. Z. Young recounted the story of an interaction between himself, Harrison, and Goodrich at the latter's house in Oxford. Being cognizant of Harrison's contributions to the neural crest and knowing that Goodrich vigorously opposed the neural crest origin of skeletal tissues, Young asked of Goodrich, "I suppose you believe that branchial cartilage is of neural crest origin?" Goodrich's stony silence spoke volumes.[18]

As reported in their now classic papers from the 1940s, Hörstadius and Sellman performed an extensive series of experimental studies on cranial development in *A. mexicanum*, the details of which, illustrated with 16 figures, form the bulk of Chapter 3 in Hörstadius (1950). This pioneering study included:

- staining selected regions with vital dyes and mapping the levels of the neural crest that contribute to the cartilaginous trabeculae and the mandibular, hyoid, and branchial arch skeletons (Fig. 5.3);
- an extensive series of extirpations of neural crest and analysis of subsequent deficiencies; and
- an analysis of factors governing the migration of neural crest cells.

Their extirpations of regions of the neural folds added to the knowledge accumulated by Stone, Raven, and Harrison, but more importantly, demonstrated regional specificity in the neural crest. Through these studies, Hörstadius and Sellman mapped the rostro-caudal levels of the neural crest that produce the trabeculae and the mandibular, hyoid, and visceral arch skeletons (Fig. 5.3). They divided the urodele neural crest into three regions:

- neural crest rostral to the mid-prosencephalon, a region that does not produce skeletal tissues;

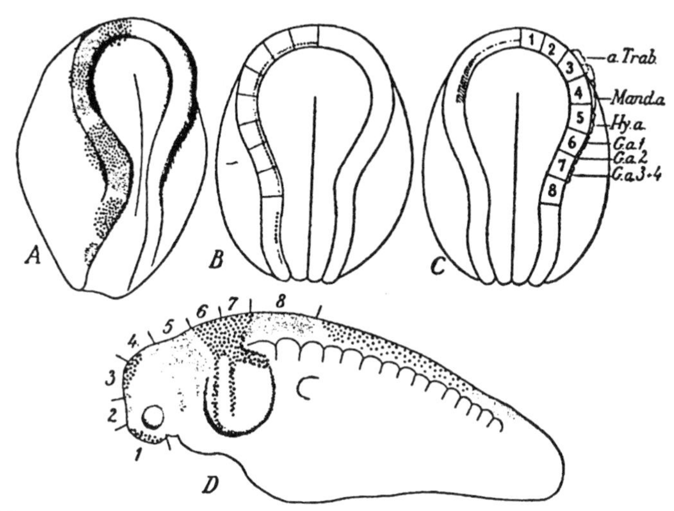

FIGURE 5.3. By applying two different vital dyes to the neural crest before cell migration — neutral red and Nile blue, shown in coarse and fine stippling in A — Hörstadius and Sellman mapped the rostral-caudal extent of the skeletogenic cranial neural crest in *Ambystoma mexicanum* into eight regions, shown in B and C. Regions 1, 2 and 8 (rostral to the mid-prosencephalon and the trunk neural crest respectively) are nonskeletogenic. Regions 3–7 show the rostro-caudal patterning of the skeletogenic neural crest, which gives rise to the anterior trabeculae (a. Trab.), mandibular, hyoid and gill arch skeletons (Mand.a, Hy.a, G.a.1–4). D shows the pattern of migration of populations of neural crest cells into the same eight regions. Reproduced from Hörstadius (1950) with the permission of Dagmar Hörstadius Ågren.

- cranial neural crest caudal to the mid-prosencephalon, a region that gives rise to the trabeculae and the visceral arch skeletons; and
- trunk neural crest, a region that does not produce skeletal tissues.

Hörstadius and Sellman further subdivided the cranial skeletogenic crest (Fig. 5.3). Crest destined to form trabeculae did not form visceral cartilages when transplanted into the more caudal level of the neural folds from which the neural crest that forms visceral arch cartilages normally arose. Conversely, crest destined to form visceral arch cartilages did not form trabeculae when transplanted into the more rostral level of the neural folds from which neural-crest-forming trabeculae normally arose. The skeletogenic neural crest was patterned along the rostro-caudal axis and such patterning was intrinsic. The same regionalization holds true for the chick, the Japanese medaka (a teleost), and the lamprey (a jawless vertebrate) and is a fundamental feature of the organization of the neural crest in vertebrate embryos (Fig. 5.4).

Inherent morphogenetic specificity within the skeletogenic neural crest was further confirmed in Wagner's classic 1949 study, in which he grafted neural crest from *Bombina variegata* into the equivalent position in the neural folds of

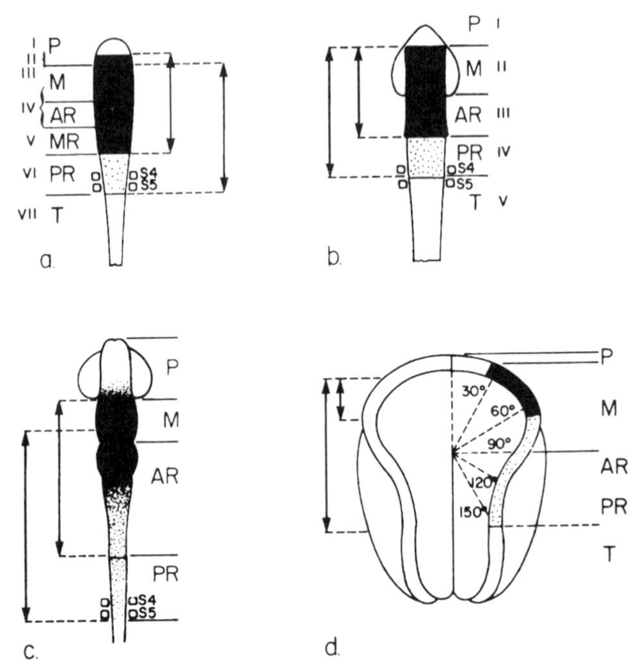

FIGURE 5.4. Regionalization of the cranial neural crest showing the cranio-caudal extent of the neural crest contribution to the chondrocranial skeleton (black; double-headed arrows) and to the visceral skeleton (stippled; double-headed arrows) in (a) a lamprey *Petromyzon marinus*, (b) a teleost, the Japanese medaka, *Oryzias latipes*, (c) a bird, the common fowl, *Gallus domesticus*, and (d) representative urodele amphibians, *A. mexicanum* and *P. waltl*, as seen in dorsal views of the developing neural tubes. The skeletogenic cranial neural crest extends from mid-prosencephalon caudal to the level of somites 4 or 5 (S4, S5) in all species. P, prosencephalon; M, mesencephalon; AR, anterior rhombencephalon; MR, mid-rhombencephalon; PR, posterior rhombencephalon; T, trunk neural crest cells. Numbers I to VII in (a) and I to V in (b) represent boundaries of regions excised to generate the fate maps for lamprey and Japanese medaka skeletogenic neural crests (see Chapters 4 and 6 for details). The angles from the midline in (d) represent sectors of the neural crest extirpated from *P. waltl* embryos by Chibon (1966, 1967) which I have projected onto the fate map generated for *A. mexicanum* by Hörstadius and Sellman (1946). [See the following for the rostro-caudal patterning of the neural crest: for urodele amphibians, Hörstadius and Sellman (1946), Chibon (1966, 1967), Moury and Hanken (1995), and Olsson and Hanken (1996); for the embryonic chick, Le Lièvre and Le Douarin (1975), Le Lièvre (1978), Dupin et al. (1993), and Le Douarin, Ziller, and Couly (1993); and for the Japanese medaka, *Oryzias latipes*, Langille and Hall (1988a). See Langille and Hall (1986, 1988a-b, 1989) for regionalization of lamprey neural crest.]

embryos of the newt *Triturus alpestris*. By grafting neural folds from only one side, Wagner was able to create partially chimeric embryos in which patterning of the skull and visceral skeletal on the grafted side could be compared with that on the intact side (an extension of the approach used by Harrison). The grafted portions of the skull and jaws were frog in type and size, the host structures were newt in size. Clearly, transplanted neural crest cells express a species-specific patterning that is an intrinsic property of the skeletogenic cells.[19]

Holtfreter had found presumptive evidence for tissue interactions in the development of teeth from neural crest. Wagner observed that migrating neural crest cells made multiple contacts with the ectoderm and that the ectoderm responded to these contacts by dividing and thickening. This mitogenic role of the neural crest merits closer study, especially with respect to the developmental and evolutionary links of neural crest and placodes (Chapter 3) and the epithelial–mesenchymal interactions required to initiate neural crest cell differentiation (Chapter 10). The 1946 paper by Hörstadius and Sellman demonstrated the need for pharyngeal endoderm to interact with migrating neural crest cells if they are to differentiate into cartilage cells. This is an example of a class of epithelial–mesenchymal interactions since shown to be necessary to elicit differentiation of cartilage and bone from neural crest cells in all vertebrates examined; Chapter 10 and see Hall (1983a, b, 1987a, 1994a).

Germ-Layer Specificity

In a very interesting study that merits further analysis, Chiakulas (1957) demonstrated lack of recognition of neural crest and mesodermally derived cartilages when he transplanted pairs of larval *Ambystoma maculatum* cartilages into the tail fins of host larvae. Cartilages from the same germ layer fused, femur against femur, Meckel's against Meckel's. Cartilages of different origins placed in contact (femur or humerus against Meckel's) did not fuse. Similar results were obtained by Fyfe and Hall (1979) using embryonic chick mesodermal and neural-crest-derived cartilages maintained in vitro.

Specificity of the amphibian cartilages breaks down during limb regeneration, when chondrocytes dedifferentiate to provide cells for the regeneration blastema. Chondrocytes from Meckel's and the regenerating humerus form a common extracellular matrix if Meckel's cartilage is implanted into fully formed limbs of *A. maculatum* and the limbs are amputated and allowed to regenerate. Transplanted Meckelian chondrocytes also provide cells to the regeneration blastema in *Ambystoma mexicanum*. Dedifferentiation seems to wipe away prior differentiative history by wiping the germ-layer slate clean.[20]

Labeling Neural Crest

The next advance over transplantation came in the 1960s from grafting tritiated thymidine labeled neural folds into unlabeled host embryos. The fate of the migrating cells, now radioactively labeled, could be followed with autoradiography. One of the most detailed analyses of the distribution of the skeletogenic cranial neural crest in amphibians using this approach is the series of studies by Chibon in which ^{3}H-thymidine-labeled neural crest cells were grafted into unlabeled embryos of the European urodele *Pleurodeles waltl*. The fate map for amphibian skeletogenic cranial neural crest produced in this study forms the basis of all subsequent studies dependent on a knowledge of regionalization of urodelean or anuran neural crest. This study, along with those by Weston and Johnson in the embryonic chick (see Chapter 7), pioneered the use of isotopically labeled grafts to map the neural crest.

Chibon divided the cranial neural folds into sectors or wedges, marked out in 30° arcs from the midline of neurula-stage embryos (Fig. 5.4d). Individual sectors were then grafted and their skeletogenic potential determined:

- The most rostral margins of the neural folds lie transverse to the embryonic axis and represent the future forebrain. This 30° sector of transverse neural fold provided no cells to the developing cranial skeleton and was believed not to contain neural crest cells, supporting the findings of Hörstadius and Sellman. The absence of neural crest cells and inability of the transverse neural folds of embryos of *Xenopus laevis* to form cartilage (even when challenged with inductively active epithelia), was confirmed by Seufert and Hall (1990).
- The next sector (30°–60°) provided the cells for the rostral trabeculae, which are the most rostral skeletal elements in the head.

More caudal skeletal elements arise from more caudal sectors of cranial neural crest. The more caudal the element, the more caudal is the crest that produces it:

- palatoquadrate, posterior trabeculae and the basal plate of the skull from the next most caudal sector (60°–90°);
- Meckel's and hyoid arch cartilages from the 70°–100° sector;
- hyobranchial and rostral branchial arch cartilages from the sector at 100°–120°; and finally
- caudal visceral arch cartilages from sector 120°–150°, the most caudal skeletogenic neural crest.[21]

Chibon therefore demonstrated in *P. waltl* that the chondrogenic neural crest extends in an arc along the cranial neural folds from 30°–150° (Fig. 5.4d). This precise fate-mapping confirmed Raven's conclusions of a separation of chondrogenic cranial from nonchondrogenic trunk neural crest in *A. mexicanum* and *Triturus* and the study by Hörstadius and Sellman demonstrating a regionalized skeletogenic neural crest in the axolotl.

Sadaghiani and Thiébaud (1987) utilized interspecific transplantation of neural crest between *Xenopus laevis* and *X. borealis* to map migrating neural crest cells, the cranial (skeletogenic) and vagal neural crest, and the origin of dorsal fin mesenchyme from trunk neural crest. Confirming the pattern described by Stone for the axolotl (Fig. 5.1), Sadaghiani and Thiébaud described three waves of migration of cells that form skeletal tissues:

- a mandibular crest consisting of cells that arise from the mesencephalon and form Meckel's cartilage and the quadrate, ethmoid, and trabeculae;

- a hyoid crest consisting of cells that arise in the hindbrain and form the ceratohyal; and
- a branchial crest, from which the visceral arch (gill) cartilages arise.

This anuran fate map compares closely with the urodele map prepared by Chibon. The chondrogenic fate of *Bombina orientalis* cranial neural crest has also been mapped and shown to be essentially similar (Fig. 5.5).

Migration of cranial neural crest cells begins before neural tube closure in a number of amphibian species, including the direct-developing Puerto Rican frog *Eleutherodactylus coqui*. In such species, which have eliminated the tadpole stage from the life cycle, most of the cranial cartilages found in the tadpoles of other species fail to form. The skeleton that forms is either typically adult or typical of the mid-metamorphic stages of species with tadpoles and metamorphosis; direct-developers, of course, do not metamorphose. Despite these fundamental changes in skeletogenesis, cranial crest migration in *Eleutherodactylus* is surprisingly conserved and similar to that in *Bombina*, *Xenopus*, and *Rana*, three species with tadpoles in their life cycles (Fig. 5.6). In coqui, migration is in the rostral (mandibular), rostral otic (hyoid), and caudal otic (branchial) streams typical of other amphibians, indeed typical of other vertebrates. The only difference in coqui

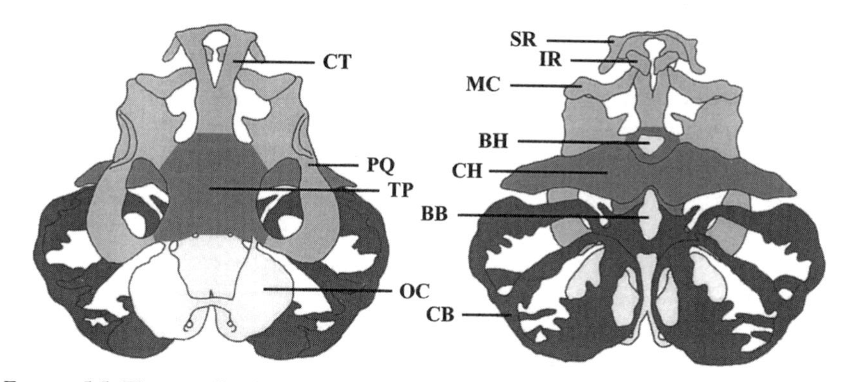

FIGURE 5.5. The contribution of populations of migrating cranial neural crest cells to the cartilaginous larval skull of *Bombina orientalis* as determined by vital dye labelling and as seen in dorsal (left) and ventral (right) views. The mandibular stream (pale gray) gives rise to the trabecular horn (CT), palatoquadrate (PQ), suprarostral (SR), infrarostral (IR), and Meckel's cartilage (MC). The hyoid stream (medium gray) gives rise to the trabecular plate (TP) and ceratohyal (CH). The branchial streams (darkest shading) give rise to the ceratobranchials (CB). Cranial mesoderm (lightest shading) gives rise to the otic capsule (OC), basibranchial (BB), and basihyal (BH). Reproduced from Olsson and Hanken, Cranial neural-crest migration and chondrogenic fate in the Oriental Fire-Bellied toad, *Bombina orientalis*: Defining the ancestral pattern of head development in anuran amphibians. *J. Morph.* 229:105-120. Copyright © (1996). From a figure kindly supplied by Lennart Olsson. Reprinted by permission of Wiley-Liss Inc., a subsidiary of John Wiley & Sons, Inc.

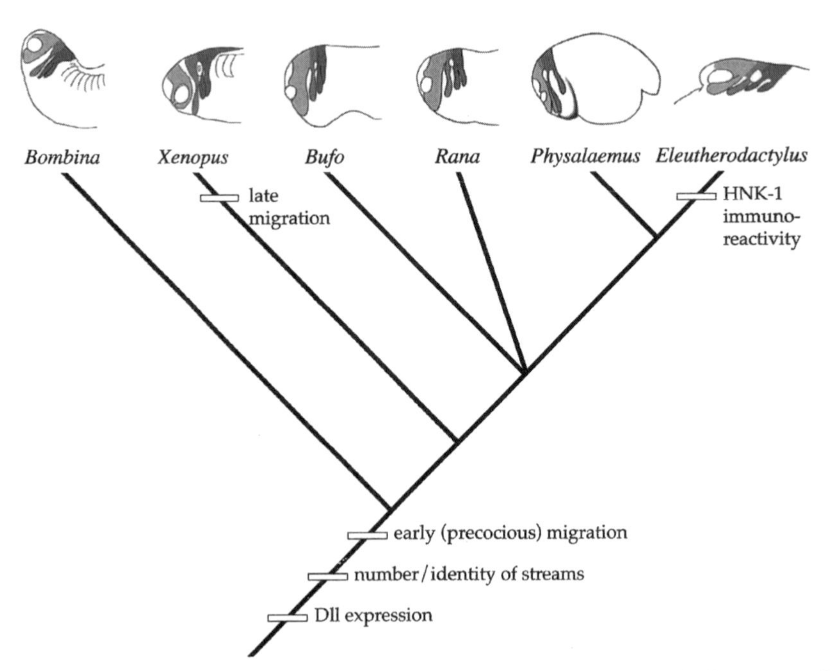

FIGURE 5.6. The three major streams of migrating neural crest cells (mandibular, pre-, and postotic; lightest to darkest shading respectively) are highly conserved evolutionarily in amphibians with tadpoles and in the direct-developing species *Eleutherodactylus*. Early migration is the primitive condition. Late migration as seen in *Xenopus* is derived. *Dll* is expressed in all species examined so far, while reactivity with the antibody HNK-1 is a derived feature in *Eleutherodactylus coqui*; *Physalaemus pustulosus*, which like coqui is a leptodactylidid frog, has a small-yolked egg, produces a tadpole, and has neural crest cells that do not react with HNK-1. I thank Lennart Olsson for providing this figure and the data on which it is based.

is some enhancement of the mandibular stream, the patterns of neural crest migration having hardly been modified with the loss of the tadpole in this species.[22]

Migration is conservative, but there are subtleties of cell populations and cell lineages within the neural crest that we do not yet fully understand. Rocek and Vesely (1989) studied the development of the ethmoid in adult *Pipa pipa*. The ethmoid and trabecular arise from the same region of the neural crest; i.e., elements arising from the same region of the crest need not be equivalent. Atchley and Hall (1991) developed a model and an experimental approach to such cell lineages, using mammalian neural-crest-derived skeletal elements as the paradigm.

Distal-less (*Dlx*) patterns the vertebrate forebrain and more caudal portions of the developing brain and is expressed in the mandibular stream of *Xenopus* neural crest cells. Indeed, *Dlx* is expressed in the neural crest cells of all amphibians with tadpoles examined so far (Fig. 5.6). Is *Dlx* expressed in coqui? The four *Dlx* genes cloned from coqui are expressed in a pattern similar to expression in the indirect developing species. From the data available so far it does not appear that major alterations in cellular or genetic patterning of neural crest migration are responsible for the cranial changes in such direct developing amphibians as coqui.[23]

Trunk and Cranial Neural Crest

Chibon's findings were extended to other vertebrates as the now current dogma that the cranial neural crest is skeletogenic but the trunk neural crest not. It has now been demonstrated that the most rostral trunk neural crest of the axolotl can form teeth (but not cartilage) if challenged with inductively active epithelia. So too can murine rostral trunk neural crest; see Chapter 10. Scale formation, including the development of dentine on the scales, is widespread along the trunk in some modern and many fossil fishes. Some retention of odontogenic ability in the trunk neural crest should not be totally surprising. Nevertheless, it is.[24]

Trunk neural crest cells induce the median fins, as DuShane (1935) demonstrated following extirpation or transplantation (Fig. 5.7). Fin mesenchyme is also of neural crest origin. Collazo, Bronner-Fraser, and Fraser (1993) injected vital dyes into groups or single trunk neural crest cells in *Xenopus laevis* and demonstrated that the mesenchyme of the medial, unpaired fins is of neural crest origin. Fin mesenchyme, pigment cells, spinal ganglia, adrenal medullary cells, and cells of the pronephric kidney were shown to arise as progeny of single trunk neural crest cells. Most clones produced fin mesenchyme, indicating that the majority of trunk cells are multipotential for mesenchyme and the other neural crest derivatives.

Cranial and trunk neural crest differ in other ways. Administration to embryonic *X. laevis* of the steroidal alkaloid cyclopamine destroys cranial neural crest cells and prevents neural-crest-derived cartilages from differentiating, but has no effect on trunk neural crest cells (Dunn, Mercola, and Moore 1995).

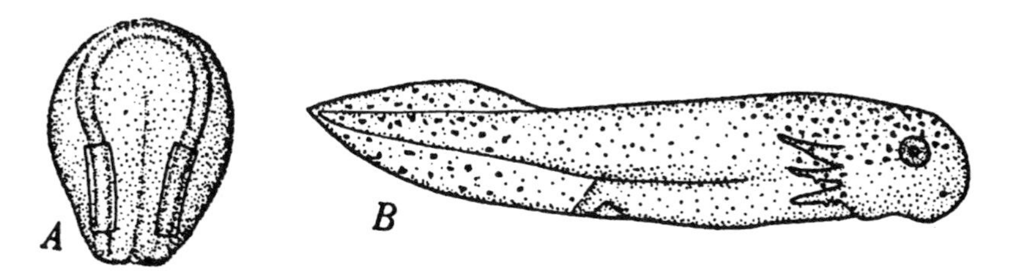

FIGURE 5.7. DuShane (1935) demonstrated involvement of the trunk neural crest in amphibian dorsal fin formation. Bilateral extirpation of trunk neural crest (shown by the boxed areas in the dorsal view of a neurula in (A) resulted in larvae lacking the dorsal fin and pigment cells in the areas that would have been populated by the extirpated neural crest cells, shown in B. Reproduced from Hörstadius (1950) with the permission of Dagmar Hörstadius Ågren.

Cement Glands

Cement glands are found on the ventral surface of the heads of anuran tadpoles and are used for attachment during feeding. In *Xenopus*, cement glands arise following a series of interactions initiated during induction of neural and rostral ectoderm, disruption of any of which can block cement gland formation. Induction is evidenced by differential expression of epidermal and nonepidermal keratins and an antibody against tyrosine hydroxylase associated with the glands.[25]

A gene involved in cement gland induction, *XOtx-2*, the *Xenopus* equivalent of the *Drosophila* gap gene *orthodenticle*, is expressed in anterior neurectoderm during gastrulation. Ectopic expression of *XOtx-2* is a sufficient signal to induce an extra cement gland. *Dlx* is also expressed in the cement glands of *Xenopus*. Coqui lack cement glands, but *Dlx* is expressed in the region of coqui ectoderm equivalent to that from which cement glands arise in *Xenopus*. Fang and Elinson (1996) used cross-species transplantation and tissue recombinations to investigate the potential developmental mechanisms responsible for loss of the cement glands in coqui. Coqui cranial tissues can induce cement glands from *Xenopus* ectoderm, but coqui ectoderm does not respond to inductive signals from *Xenopus*, demonstrating that the competence of coqui ectoderm to respond to induction is modified without modification of the inductive signal. Loss of competence, not loss of induction, leads to loss of cement glands in coqui. Loss of ectodermal competence is also responsible for loss of balancers in some amphibians, loss of limbs in avian mutants such as *limbless*, and loss of teeth in birds.[26]

An important series of messages lies in these examples of how cell and tissue interactions are modified when structures are lost during evolution:

- An organ may be lost without loss of the entire developmental system that produces that organ.
- Loss of organs is often mediated through modification (not loss) of inductive interactions.
- Modification of competence is the usual means by which inductive interactions are altered.
- Inductive signaling persists even when competence to respond is lost.
- The potential for the organ to reappear exists, provided that competence can be restored.

6

Bony and Cartilaginous Fishes

In this chapter, I summarize the little we know about the neural crest in bony and cartilaginous fishes. This does not imply that fish form a natural group of vertebrates or that the various groups of fishes are more closely related to one another than to other vertebrates. The studies available are primarily on teleost fishes and elasmobranchs (sharks, dogfish, and rays).

The Neural Keel

Neurulation is usually thought of as involving the invagination of a flat neural plate that rolls up into tube. In contrast, the neural tube can arise by cavitation of a solid neural keel (Fig. 6.1). Neurulation by cavitation of a neural keel has been described in the bowfin, *Amia calva*, the sea bass, *Serranus atrarius*, the zebrafish, *Danio rerio*, and the lamprey *Petromyzon planeri*. In sharks, neural crest cell migration is also from this solid neural keel except in the midbrain, where cells migrate from the dorsal midline, appearing to accumulate above the neural tube before beginning to migrate, as indeed Balfour depicted almost 120 years ago (Fig. 6.2A).[1]

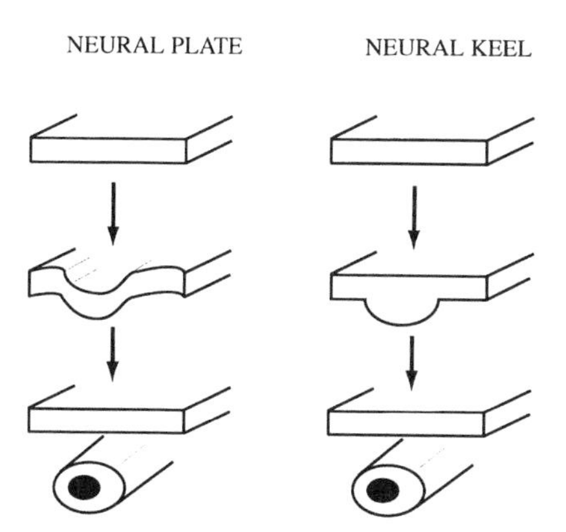

FIGURE 6.1. Neurulation by invagination of a neural plate (left) is compared with neurulation by cavitation of a neural keel at three successive stages. The black circle represents the neural cavity.

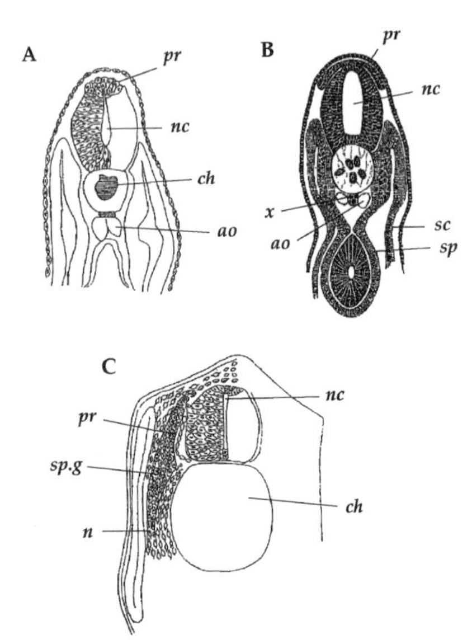

FIGURE 6.2. The neural crest (A) and successive stages in the development of the roots and ganglia of the spinal nerves (B, C) in a shark embryo. A. Origin of the neural crest (pr) by proliferation from the dorsal surface of the neural tube. B. Initial formation of the root of the spinal nerve (pr). C. An older embryo to show the position of the posterior root (pr) and spinal ganglion (sp.g) of the spinal nerve (n). Other abbreviations are: ch, notochord; ao, aorta; nc, neural canal; sc, somatic mesoderm; sp, splanchnic mesoderm; x, subnotochordal rod. Modified from Balfour (1881).

Although textbooks describe a neural keel as typical of neurulation in teleosts, ganoid fishes, and cyclostomes, neurulation by cavitation may only appear typical because so few species have been studied. There is, for example, no solid neural keel in the cichlid *Cichlasoma nigrofasciatum*; involution occurs in tightly opposed neural folds with a very narrow neural groove. Neurulation by involution rather than cavitation in this species was only revealed through high-resolution microscopical analysis, a technique that should be applied to other species.[2]

Distinct populations of migrating neural crest cells can be visualized in embryonic teleost fishes (Fig. 6.3). Sadaghiani and Hirata and their colleagues studied neural crest cell migration in the platyfish *Xiphophorus maculatus*, the swordtail, *X. helleri*, and the Japanese medaka, *Oryzias latipes*, using a variety of techniques including scanning electron microscopy (SEM), reactivity with HNK-1, and immunohistochemistry. HNK-1 labels migrating cells of all three species. Migration of trunk neural crest cells is across the somites rather than through the rostral half of each somite as occurs in other vertebrate groups. Similarly, in the cichlid *Oreochromis mossambicus*, the neural crest cells that form the dorsal root ganglia do not migrate through somitic mesoderm as they do in amniotes, but

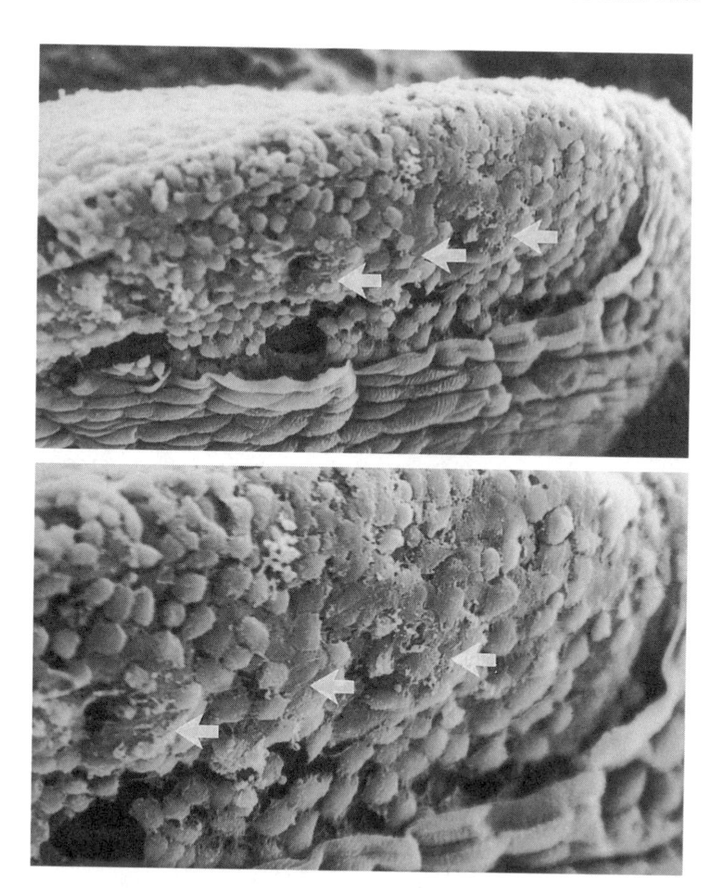

FIGURE 6.3. Scanning electron micrographs at lower (top) and higher (bottom) magnifications of the hindbrain region of a 16-somite-stage (17-hour) embryo of the zebrafish, *Danio rerio*, viewed from the dorsal surface. Ectoderm has been partially removed to reveal streams of migrating neural crest cells (arrows). Figure kindly supplied by Janet Vaglia.

rather migrate as individual cells between neural tube and somites (Laudel and Lim 1993). The components of the extracellular matrix along the pathway taken by migrating neural crest cells in *Xiphophorus* include chondroitin sulfate and fibronectin. The role of these molecules in neural crest cell migration is discussed in Chapter 9.[3]

Bemis and Grande (1992) used SEM to describe mandibular and hyoid migratory populations of neural crest cells in the paddlefish, *Polyodon spathula*, and equated them with similar populations in other vertebrates. Figure 6.4 illustrates populations of migrating cranial neural crest cells in an embryo of the little skate, *Leuroraja erinacea*.

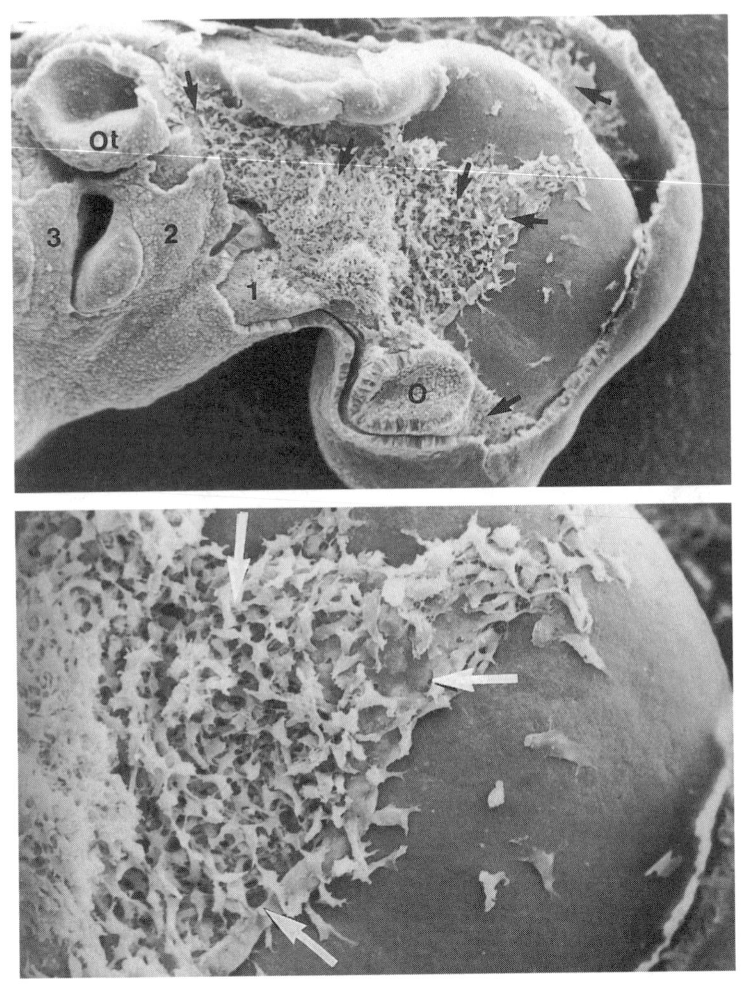

FIGURE 6.4. Scanning electron micrographs at lower (top) and higher (bottom) magnifications of the developing head of an 84–85 somite-stage (60-day-old) embryo of the little skate, *Leucoraja erinacea*, collected off the coast of Nova Scotia. The ectoderm has been removed to reveal populations of migrating neural crest cells (arrows). O, optic vesicle; Ot, otic vesicle; 1–3, the first (mandibular), second (hyoid) and third branchial arches. Figure kindly supplied by Tom Miyake.

Ganglia

The initial observations on the presence of a neural crest and descriptions of neural crest cell derivatives were made in elasmobranchs in the 1870s and 1880, the first only ten years after His discovered the neural crest. Balfour (1876) identified a neural ridge within the neural tube from which spinal ganglia arose in elasmobranchs, and thought that ganglia of the sympathetic nervous system arose as branches of the spinal nerve and therefore as products of the ectoderm. Balfour was adamant that His was wrong in his interpretation of the presence of an intermediate cord, but by 1881 Balfour had changed his view and was illustrating ganglia as having a neural crest origin (Fig. 6.2). Also in the 1880s, Sagemehl and Kastschenko described the dorsal roots of the spinal ganglia arising from the spinal cord in sharks, but made no distinction between cells of the neural tube or neural crest as the source of the neurogenic cells. In what was perhaps the next detailed study, Conel (1942) described the neural crest origin of the dorsal root ganglia of both cranial and spinal nerves in *Squalus acanthias*, a dogfish, and in the electric ray, *Torpedo ocellata* (Fig. 6.5).[4]

The most complete analysis of development of dorsal root ganglia in any teleost is that of Laudel and Lim (1993) on the cichlid *Oreochromis mossambicus*. Using a combination of HNK-1 staining, SEM, and DiI injection, they demonstrated that dorsal root ganglia contain both sensory cells and motor fibers and arise from neural crest cells that lie between the neural tube and the somites.

A major contribution to the early confusion over whether ganglia arose from the neural crest, the dorsal neural tube, or adjacent ectoderm was the fact that

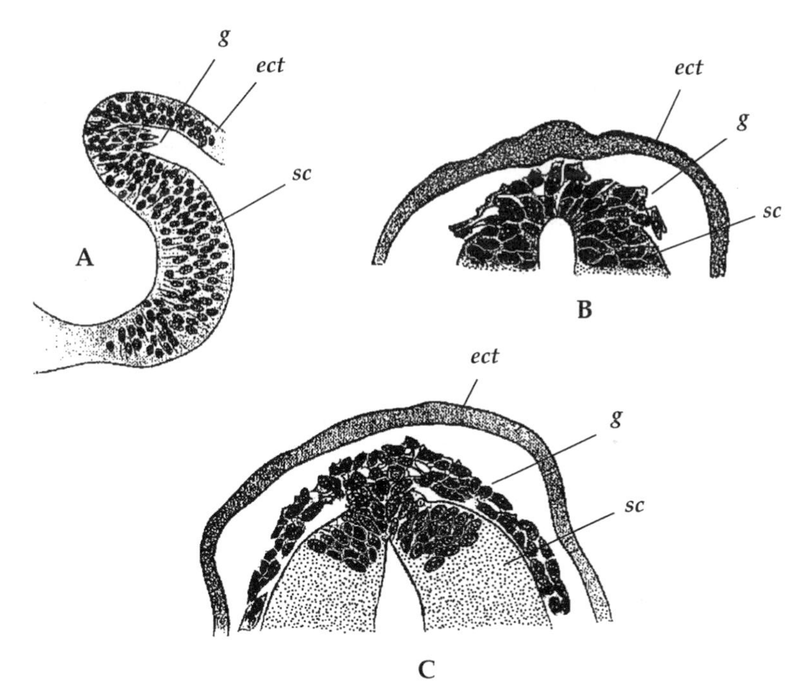

FIGURE 6.5. The neural crest origin of spinal ganglia is shown in these sections of neural folds of the chick (A) and the electric ray, *Torpedo ocellata* (B, C). ect, ectoderm; g, ganglion; sc, spinal cord. Modified from Volume II of *Text-Book of Embryology*, by Kerr (1919).

although neural crest always arises at the lateral borders of neural ectoderm, neural crest is first recognizable at different stages of neurulation in different vertebrates. Neural crest cells do not migrate away from the trunk neural tube in sharks until late in neurulation, i.e., until after the neural tube has invaginated and separated from the epidermal ectoderm (Fig. 6.2A). Neural crest appears at the open neural plate stage in amphibians and rodents but in closed neural folds in birds. There are further complications. The caudal neural crest is associated with secondary neurulation (Chapter 1) and as discussed above, in many teleost fishes neurulation is by cavitation of a neural keel and not by invagination of a flat neural plate.

Pigment Cells

One of the first reports that pigment cells might be derived from the neural crest was a study by Borcea (1909) of the needlefish, *Belone acus*. Having described pigment cells migrating between spinal cord and trunk epidermis, Borcea concluded that they must have arisen from the ectoderm, presumably from trunk neural crest. Lopashov (1944) extirpated and grafted neural tubes from several species of teleosts—the groundling, *Misgurnus fossilis*, the loach, *Nemacheilus barbatulus*, and the perch, *Perca fluviatilis*—to determine whether pigment cells were of neural crest origin. Head pigmentation was reduced or eliminated when regions of the cranial neural tube were removed, and the yolk sacs became pigmented when used as sites for cranial neural tube grafts. Cartilage also differentiated in these grafts, an early indication of the neural crest origin of cranial cartilages in teleosts. Of course, these grafts were not labeled, so that this interpretation of neural crest origin cannot be considered definitive. Removal of neural crest from embryos of the Australian lungfish, *Neoceratodus forsteri*, was claimed by Kemp (1990) not to disturb normal patterns of pigmentation, indicating either regulation of crest cells or that ablation was performed after migration had been initiated.[5]

In the studies by Sadaghiani and Vielkind (1990a, b) and Hirata, Ito, and Tsuneki (1997), neuronal and pigment cells both differentiated from neural crest cells that migrated from neural tubes of *Xiphophorus* and *Oryzias* maintained in vitro.

Labeling the Neural Crest in Bony Fishes

There have been some attempts to label fish neural crest. Lamers, Rombout, and Timmermans (1981) grafted ^{3}H-thymidine-labeled trunk neural tubes from the level of somites 5 and 6 of embryos of the rosy barb, *Barbus conchonius*, either into the hindbrain or into the trunk of unlabeled embryos. ^{3}H-thymidine-labeled cells were seen migrating away from the grafts, both as a superficial population between the developing somites and the trunk epithelium, and as a deeper population medial to the somites and between somites and neural tube/notochord. Similar pathways of migration of trunk neural crest cells have been described in zebrafish and in other vertebrates (Chapter 5) and are an evolutionarily conserved feature of neural crest cell populations. Labeled cells were subsequently seen as differentiated pigment cells, adrenal medullary cells, and neurons of spinal, sympathetic, and enteric ganglia. These three cell types can therefore be assigned to the list of neural crest derivatives in teleosts.

Trunk neural crest cells in the zebrafish are larger and fewer in number than are avian neural crest cells and they migrate along two ventral paths. The technique of labeling premigratory trunk neural crest cells by intracellular injection with lysinated rhodamine dextran has been used to demonstrate the presence of cell populations whose fate is restricted. Indeed, most trunk neural crest cells in the zebrafish are lineage restricted. Time spent in the neural tube is one factor affecting lineage restriction; early migrating cells contribute to dorsal root ganglia and glia and pigment cells, while sensory and sympathetic cells arise only from early migrating cells. Late migrating cells are non-neuronal.

Considerable mesenchyme arises from the trunk neural crest. In zebrafish, it comes only from the medial portion of the neural keel. The lateral portion may form placodes, as it does in amphibians. Although migration of cranial neural crest occurs in waves, each crest segment is restricted (determined) to form individual cell lineages of a single branchial arch, a determination that is set before migration is initiated. These cells give rise to the visceral arch skeleton.[6]

The Visceral Arch Skeleton

Kastschenko (1888) and Goronowitsch (1892, 1893a, b) claimed that neural crest contributes to the cranial mesenchyme in sharks and teleosts, and that dentine of the teeth in sharks and in the carp, *Cyprinus carpo*, originates from neural crest. No fate map of the neural crest of any teleost was available, however, until Langille's study on the Japanese medaka, *Oryzias latipes*, in the mid-1980s. Langille provided a detailed description of the development of the cranial and visceral skeleton and an experimental investigation of the effects on skeletal development of removing 100- to 150-μm-long regions from the cranial neural crest (see Fig. 5.4b). He also determined the regional extent of the skeletogenic neural crest, which extends from the mesencephalon caudally to the rhombencephalon. Neither the most rostral (prosencephalic) neural crest, nor neural crest caudal to the fifth pair of somites, is chondrogenic, a pattern similar to that found in other vertebrates (see Fig. 5.4). His analysis of deletions or deficiencies in operated embryos showed that:

- All the visceral skeleton of the hyoid and branchial arches (basihyal, ceratohyal, symplectic, hyomandibula, hypobranchials I–IV, ceratobranchials I–V,

epi- and pharyngobranchial IV) and the bulk of the rostral neurocranium receive cellular contributions from the neural crest.
- Neural crest cells from the pre- and postotic regions of the rhombencephalon populate the hyoid and branchial arches.
- Neural crest cells from the mesencephalon form the rostral neurocranium.
- Within the developing chondro- and osteocrania, Meckel's cartilage, the quadrate, pterygoid process, trabeculae, epiphyseal cartilages, orbital cartilage, and nasal capsules all develop from neural crest cells.

Table 6.1 contains a full list of cartilages with a neural crest origin in the medaka and the level of the cranial neural crest from which they arise. The visceral skeleton in the swordtail, *Xiphophorus,* has also been shown to be of neural crest origin.[7]

Although the situation in elasmobranchs has not been studied experimentally, Holmgren (1940, 1943) provided substantial evidence from descriptive studies for a neural crest origin of much of the head skeleton in sharks and dogfish (*Squalus acanthias*, *Etmopterus spinax*, *Scyllium canicula*), rays (*Raja clavata*, *Torpedo orcellata*, *Urolophus*), and in a teleost, the salmon *Salmo salar*. Indeed, all vertebrates studied so far possess visceral arch skeletons that develop from neural crest cells. de Beer's 1947 contention "that cartilage can arise from cells derived from two different layers" (p. 380)—visceral cartilage from the neural crest and some cranial and all postcranial cartilages from mesoderm—is now firmly established.

TABLE 6.1. The elements of the cartilaginous skeleton of the Japanese medaka, *Oryzias latipes*, that arise from the cranial neural crest, shown in relation to the region of the neural crest from which they arise.[a]

Region of the neural crest	Cartilaginous element
Prosencephalon	None
Mesencephalon	Orbital cartilage, Meckel's cartilage, quadrate, trabeculae, ethmoid plate, lamina orbitonasalis, pterygoid process, epiphyseal cartilage, basihyal, symplectic
Preotic (anterior to mid-rhombencephalon)	Orbital cartilage, epiphyseal cartilage, basihyal, hyomandibula
Postotic rhombencephalon caudal to the level of the 4th pair of somites	Ceratohyal, hyomandibula, basibranchial, hypobranchials I-IV, ceratobranchials I-V, epi- and pharyngobranchials IV
Caudal to level of 4th pair of somites	None

a. Based on data in Langille and Hall (1987, 1988a). The polar cartilage, hypophyseal, acrochordal, anterior and posterior basicranial commisures, basilar plate, otic capsule, occipital arch and tectum synoticum receive no contribution from neural crest cells and are mesodermal in origin.

Lateral Line System

Bony fishes have an extensive lateral line system. Eight basic patterns of evolutionary change of the lateral line, derived by heterochrony from an ancestral pattern, have been determined in groups such as cichlids whose evolutionary history is well resolved.

As described in the salmon *Salmo* and other genera, preosteogenic cells derived from the neural crest accumulate under the sensory organs (neuromasts) of the lateral line system. Circumstantial evidence suggests that developing neuromasts induce scales or dermal bone in teleosts and cartilage in sharks. Once formed in teleosts, lateral line bone contains a canal for the lateral line nerve and pores for the neuromasts. Canal neuromasts are very distinctive; in the goldfish, *Carassius auratus*, for instance, canal neuromasts are some 50–100 μm in diameter, superficial neuromasts <50 μm in diameter. Surprisingly, some trunk lateral lines of two species of hexagrammid fishes of the genus *Hexagrammis* lack neuromasts. Although amphibians possess neuromasts, they lack the canal neuromasts associated with bone in other groups.[8]

Neuromasts and lateral line nerves arise from ectodermal placodes. Neural crest cells pattern neuromasts but have not been shown definitively to give rise to neuromasts. On the basis of DiI-labeling of zebrafish neural crest, however, it has been suggested that neuromasts have a dual origin; they arise from neural crest and from ectoderm. In the absence of histological evidence and because of the difficulty of labeling neural crest in the neural folds without labeling placodal ectoderm, this claim remains unsubstantiated. As discussed in Chapter 2, placodal ectoderm in at least some species does arise from the lateral neural folds, neural crest arising from medial neural folds. In such species, placodal ectoderm may be induced by neural crest rather than forming from neural crest.[9]

Fins

The proximal fin endoskeleton of teleosts, elasmobranchs, and chondrichthyans is cartilaginous and mesodermal in origin. Bony fin rays, however, are the major component of teleost fins. Mesenchyme of the unpaired dorsal and ventral fins is neural crest in origin; the dorsal fin of swordtail embryos is populated by mesenchymal cells derived from the neural crest, as Hirata, Ito, and Tsuneki (1997) have demonstrated using HNK-1 as a marker. It is presumed, but not yet proven, that the lepidotrichia and actinotrichia that constitute the fin rays of paired fins must also be neural crest in origin, although labeled cells have not been traced sufficiently far into development to demonstrate labeled fin rays.

Two fundamental differences between paired fins and tetrapod limbs are the presence of neural crest distal elements in fins and their absence in limbs, and the presence of mesodermally derived digits in limbs and their absence in fins. Limbs lack any neural crest (exoskeletal, dermal) elements and have expanded the endodermal skeleton distally to form the digits. The transition from fins to limbs is thought to have involved:

- elimination of the dermal fin rays, which are thought to be derived from neural crest;
- distal elaboration of the existing proximal mesodermal endoskeleton;
- elaboration of the new distal endoskeleton as wrist/ankle elements and proximal phalanges; and
- formation of new distal endoskeletal elements as digits.

Developmentally, these changes are essentially the loss of the neural crest component (dermal skeleton) and elaboration of the mesoderm-derived endoskeleton.[10]

Genes and Cell Lines

In an unusually effective approach to the analysis of neural crest derivatives in fish, Vielkind and colleagues (1982, 1983) took advantage of their knowledge of the pathways of neural crest cell migration, DNA technology, and genetic transformation, to inject purified DNA of the *Tu* gene (known to play a role in the differentiation of T-melanophores) into the neural crest of a strain of the swordtail, *Xiphophorus helleri*, that lacks the *Tu* gene and therefore cannot produce melanophores. The appearance and differentiation of T-melanophores in the genetically transformed strain demonstrated incorporation of the *Tu* gene into host cells and the neural crest origin of the T-melanophores. A monoclonal antibody directed against the melanoma gene from *Xiphophorus* cross-reacts with human melanomas but not with other tumors derived from neural crest, demonstrating the utility of fish as model system for human tumors of neural crest origin, a topic revisited in Chapter 11.

A second ingenious experimental approach for delineating neural crest origins was taken by Matsumoto et al. (1983), who established clonal cell lines from melanophore tumors (erythrophoroma and irido-melanophoroma) from the goldfish, *Carassius auratus*, and the Nibe croaker, *Nibea mitsukurii*. Individual clonal cell lines differentiated into melanophores and elements of the dermal skeleton, including dermal bone, scales, fin rays, and teeth. Dermal bone and scales differentiated spontaneously without any supplementation of the culture medium, i.e., without any exogenous factors acting as inducers. Fin rays and teeth were only seen when serum or dimethylsulfoxide (DMSO) was added to the medium. A further study confirmed that goldfish erythrophoroma cells could initiate multiple pathways of differentiation, including melanin-synthesizing cells, platelets, pteridines, neurons, elements of the dermal skeleton (all neural crest cell types), and lens, but that individual cell lines functioned as clonal lineages. On the presumption that these melanophore tumors are homogeneous and do not contain any mesodermally derived elements, several interesting conclusions may be drawn from this study:

- melanophores, bone, scales, and fin rays are neural crest derivatives;
- the cells of melanophore tumors (and of the neural crest itself?) are multipotent; and
- some neural crest derivatives—melanophores, dermal bone, and scales—will differentiate from tumors without induction, while others—fin rays and teeth—require environmental signals to trigger their differentiation.[11]

Zebrafish Mutants

Amphibians, about whose neural crest so much is known, are not ideal animals for genetic screening and the generation of mutations. Zebrafish, however, are ideal animals for saturation mutagenesis, a technique that has opened wide a window onto a vista full of opportunities and uncovered numerous mutants affecting the neural crest or neural crest derivatives. At least 109 mutations affecting the branchial arches and 48 affecting craniofacial development have been described. This cornucopia of mutants is under active investigation. Initial results for several mutants are noted below.[12]

In the zebrafish, as in other vertebrates, *Dlx* exerts specific actions on the craniofacial cartilages of the first and second visceral arches (which arise from hindbrain neural crest) but has little effect on midbrain-derived chondrogenic crest cells. Mutations in murine *Dlx-1* and *Dlx-2* demonstrate that both genes play overlapping roles in patterning skeletal and soft tissues of the first and second visceral arches.

Chinless (*Chn*) disrupts skeletal fate and the interactions between neural crest cells and muscle progenitors; all seven branchial arches lack neural-crest-derived cartilages as a primary defect and lack mesodermal-derived muscles as a secondary defect. Cartilaginous and muscle precursors are present as condensations of cells, indicating that *Chn* affects differentiation and not the origination of these cells. As it affects both neural crest and mesoderm, *Chn* may be acting downstream of neural crest cell initiation and specification, perhaps inhibiting the epithelial–mesenchymal interactions required for neural crest cell differentiation; acting independently but at the same stage of neural crest chondrogenesis and mesodermal myogenesis; or have secondary effects on mesodermal muscle progenitor cells because of defective neural crest/mesoderm interactions. The last seems most likely.[13]

The screening studies that revealed *Chn* used morphological characters to identify mutations. Screening based on molecular markers has also begun. Henion et al. (1996) have produced parthenogenetic diploid embryos and screened them for pleiotropic mutants. Mutations involving both cranial and trunk neural crests have been identified, including mutants in which all neural crest cartilages are missing, in which cartilages are abnormal, or in which only visceral cartilages are missing. The *Spadetail* mutant, in which the somites are partly deleted and tail development is disrupted, was used to investigate migration of trunk neural crest cells. *Snail-2*, a member of the *Snail* gene family expressed at the boundary of neural plate and epidermal ectoderm at gastrulation and in the neural crest, is down-regulated in trunk neural crest of *Spadetail* embryos. Neural crest cells in the medial migration pathway in wild-type embryos are contact-inhibited by somitic cells. Contact inhibition (and therefore somite-based repulsion of migrating neural crest cells) fails to occur in *Spadetail* mutants, resulting in diminished tail development.[14]

7

Reptiles and Birds

Reptiles

We know even less about the reptilian neural crest than we do about that of fishes. Indeed, very few species have even been examined. In the one experimental study in which presumptive neural crest cells have been extirpated from reptilian embryos, Toerien (1965a) used embryos of the snapping turtle, *Chelydra serpentina*, to provide evidence that the mandibular and pharyngeal skeletons arise from neural crest cells and that the skeletogenic neural crest is regionalized as in other vertebrates (see Fig. 5.4). Extirpation of pre-optic cranial neural folds (and with them any resident neural crest cells) at the 4 to 6 somite stage resulted in embryos that lacked Meckel's and quadrate cartilages. Extirpation of postoptic neural folds caudal to the level of the second pair of somites produced embryos lacking hyoid arch cartilages and the columella, and with abnormal tympanic membranes.

Scanning electron microscopy allowed Meier and Packard (1984) to visualize migrating neural crest cells in the snapping turtle in what is now a classic paper. Neural crest cells first emerge from the mesencephalon, then as two streams from the rostral and caudal rhombencephalon. These three streams then fuse to form a continuous mass of cranial mesenchyme.

In the early 1980s, Mark Ferguson documented migrating neural crest cells in embryos of the American alligator, *Alligator mississippiensis*. As in the snapping turtle, migration starts from the mesencephalon as two cell populations. One is more rostral, invading the mandibular processes; the other is more caudal, populating the maxillary processes. A subsequent wave of cells from the prosencephalon provides the mesenchyme of the medial and lateral nasal processes. Ferguson also provided experimental evidence for a neural crest contribution to the mandibular processes. He devised a technique for the culture of whole alligator embryos and administered 5-fluoro-2'-deoxyuridine (FUDR) to coincide with the migration of either the mandibular or maxillary streams of cells. Subsequently these embryos displayed either greatly reduced lower jaws or deficient palates, the teratogen having blocked the migration of neural crest cells destined for these particular embryonic regions. (As discussed in Chapter 12 and by Hall (1987b), FUDR acts at different stages of neural crest cell maturation in reptiles, birds and mammals.[1]

There are perhaps only two studies on the role of epithelial–mesenchymal interactions in the development of reptilian skeletal elements derived from neural crest. Toerien (1965b) showed that the cartilaginous otic capsule (which is probably not of neural crest origin) and the foot plate of the columella (which probably is) fail to develop in *Chelydra serpentina* embryos from which the otic vesicles are removed. Ferguson and Honig (1984) demonstrated reciprocal interaction between alligator mandibular and palatal epithelia and mesenchyme when these tissues were experimentally recombined.

Investigation of the behavior of reptilian neural crest in vitro is similarly limited. Hou and Takeuchi (1992, 1994) cultured neural crest from stage 9 and 10 embryos of the Japanese turtle *Trionyx sinensis japonicas*. Neural crest cells were found to be HNK-1-positive. Trunk neural crest established in vitro underwent limited migration, but did differentiate into neurons and melanophores.

Birds

Some of the very earliest work on the neural crest was undertaken on avian embryos. The first description of neural crest in any vertebrate was of the *Zwischenstrang* lying between epidermal ectoderm and neural tube in the embryonic chick. Among the first claims that cranial mesenchyme was neural crest in origin was that by Goronowitsch (1892, 1893a, b) from studies of avian embryos.

Ganglia and Pigment Cells

During the 1920s and 1930s, a series of experimental studies performed on embryos of the common fowl, *Gallus domesticus*, established the neural crest origin of the sympathetic ganglia. Although not all early workers obtained comparable results, subsequent experimental studies affirmed that these ganglia are indeed of neural crest origin. Enteric and visceral ganglia, chromaffin cells of the adrenal gland, and pigment cells were also shown to be of neural crest origin. Descriptive studies also suggested a neural crest origin of cranial mesenchyme, supporting the claims of Goronowitsch.[2]

The most extensive studies on avian embryos undertaken in the 1930s and 1940s were on the neural crest origin of chromatophores, with feathers as the preferred organ system of study. Dorris found that pigment cells only formed in vitro when neural tube was included in cultures, and that host limb buds became pigmented only when neural crest cells were grafted into them. Such studies pioneered the techniques of grafting between embryos of differently pigmented breeds of domestic fowl, or between embryos of different avian species (neural crest of robin, pheasant, or Japanese quail grafted into white leghorn chick embryos, for example), to follow the fate of grafted neural crest cells. Rawles (1944, 1948) subsequently used this technique to map the location of the presumptive neural crest in chick embryos at the primitive-streak-stage, and to demonstrate the neural crest origin of murine melanophores by intracelomic grafting of neural crest cells from mice into chick embryos.[3]

Subsequent studies greatly expanded our knowledge of the avian neural crest, primarily using the domestic fowl, but with confirmation from studies on the Japanese quail. Consequently, we have more information about the origin, migration, activation, and fate of the neural crest in embryos of the common fowl than for any other species. I highlight our knowledge of the mapping and fate of avian neural crest cells in the sections that follow. Chapters 9 and 10 contain detailed discussions of factors governing migration and differentiation.

^{3}H-thymidine Labeling

The research programs pursued by Jim Weston and Mac Johnston for their doctoral degrees and published in 1963 and 1966, respectively, ushered in an exponentially increasing interest in, and acquisition of knowledge about, the avian neural crest. These papers, along with those on the amphibian neural crest by Chibon (also published in the 1960s and discussed in Chapter 5) are benchmark papers in the study of the neural crest.

Making use of the availability of radioisotopes as markers, Johnston labeled a set of embryonic chicks with ^{3}H-thymidine. Once these embryos had reached the head fold to four pairs of somites stages (24 to 28 hours of incubation), cranial neural folds were removed and grafted into the equivalent position in nonisotopically labeled embryos of the same age. Host embryos were fixed over the ensuing eight days, i.e., until day 9 of incubation. Autoradiographs were prepared to allow the positions occupied by the ^{3}H-thymidine graft cells to be identified. In this way, Johnston demonstrated that the neural folds contained neural crest cells, mapped the migration pathways taken by these cranial neural crest cells as they populated the maxillary, mandibular, and frontonasal processes, and identified ^{3}H-thymidine-labeled cells in much of the developing connective tissue of the head and in some of the developing cartilages in the oldest embryos examined—Meckel's cartilage of the lower jaw, the cartilage of the cranial base, and the hyoid cartilages.

Weston performed similar grafts using trunk neural crest from embryos of 3 to 4 days of incubation and documented two major streams of migrating trunk neural crest cells. One population migrated into the superficial ectoderm, where they differentiated into pigment cells. The other migrated ventrally and medially, moving between the developing spinal cord and somites to become the spinal ganglia and sympathetic neurons.

Migrating neural crest cells express N-CAM and N-cadherin (see Fig. 2.4), cell adhesion molecules that are regulated by the zinc-finger gene *Slug*. The sequence of events that controls epithelial to mesenchymal transformations associated with the initiation of migration of neural crest or mesodermal mesenchyme includes: decreasing cell adhesion, reduction in levels of N-CAM, loss of N-cadherin, and changes in integrins. Exposure of premigratory neural crest to retinoic acid enhances retention of N-CAM so that crest cells, which do emerge from the neural tube in treated embryos, cannot initiate the epithelial to mesenchymal transformation required for migration and so accumulate near the neuroepithelium, leading to the craniofacial anomalies discussed in Chapter 12. Similarly, overexpression of N-cadherin or of cadherin-7 in the neural tubes of chick embryos prevents migration of many neural crest cells, completely prevents melanocytes from migrating along the normal dorso-lateral migration pathway, and leads to an accumulation of melanocytes and melanocyte precursors within the neural tube. The mechanisms that underlie neural crest cell migration are discussed in Chapter 9.[4]

We now also know that migrating neural crest cells receive signals that are required to trigger differentiation, a topic taken up in Chapter 10. As one example among many, BMP-4 and BMP-7 within the dorsal aorta of chick embryos trigger sympathetic neuronal differentiation from migrating neural crest cells. Ectopic expression of these growth factors *in ovo*, or maintenance of neural crest cells in the presence of BMP-4 or BMP-7 in vitro, triggers sympathetic differentiation.[5]

Quail/Chick Chimeras

There are considerable limitations to ^{3}H-thymidine as a marker:

- The only cells labeled are those synthesizing DNA when the ^{3}H-thymidine is applied.
- The label is diluted with each wave of DNA synthesis and can only be followed for comparatively short periods unless proliferation is very slow.
- Labeled DNA from necrotic cells may be "picked up" by unlabeled cells.

Nevertheless, ^{3}H-thymidine remained the best marker until Nicole Le Douarin discovered that nuclei of the Japanese quail can be distinguished from those of the embryonic chick on the basis of the packing of the heterochromatin visualized in Feulgen-stained histological sections. It rapidly became clear that quail cells grafted into chick embryo hosts:

- are incorporated into the host embryo;
- cooperate to produce a normal, albeit chimeric, embryo[6]; and
- can be identified readily and followed late into embryonic and even adult life.

Quail/chick chimeras survive past hatching, although spinal cord chimeras, in which a portion of the trunk neural crest of one species is replaced by that of the other, do show some breakdown in tolerance after hatching. Interestingly, the immune response begins in peripheral nerve ganglia (which are derived from the grafted neural crest), rather than centrally within the spinal cord.[7]

Quail/chick chimeras and other labeling techniques have been used to map the derivatives that arise from cranial, trunk, cardiac, and thymic neural crest. These are the topics of the next four sections.

Mapping the Cranial Neural Crest

Figure 7.1 shows the extent of the cranial neural crest and the boundary between cranial and trunk neural crest.

Le Douarin and her colleagues have made impressive and extensive use of quail–chick grafting to map the entire neural crest. She and Catherine Le Lièvre produced the most detailed and complete map of the cranial neural crest for any species. Another extensive set of data on the craniofacial derivatives of the avian neural crest is the elegant series of very careful studies carried out by Drew Noden of Cornell University. Initially, he grafted ^{3}H-thymidine-labeled neural crest cells into unlabeled host embryonic chicks to analyze the migratory pathways and migratory behavior of cranial neural crest cells. Then he switched to quail/chick chimeras.[8]

Migration of cranial neural crest cells in the chick begins from the mesencephalon at the 5-somite stage. These mesencephalic cells form all the neural-crest-derived mesenchyme of the first visceral arch, and a portion of the ectomesenchyme of the second. The remaining ectomesenchyme of the second arch and of visceral arches 3 and 4 is derived from rhombencephalic neural crest. (The calcitonin-producing cells of the ultimobranchial gland are also derived from the rhombencephalic neural crest.[9]) Noden demonstrated that the pathways of migration are not irreversibly fixed before cell migration commences; patterns of migration are normal when regions of the neural crest are exchanged.

Shigetani, Aizawa, and Kuratani (1995) used DiI-labeling of postotic neural crest cells to identify what they call the circumpharyngeal crest, which is a single

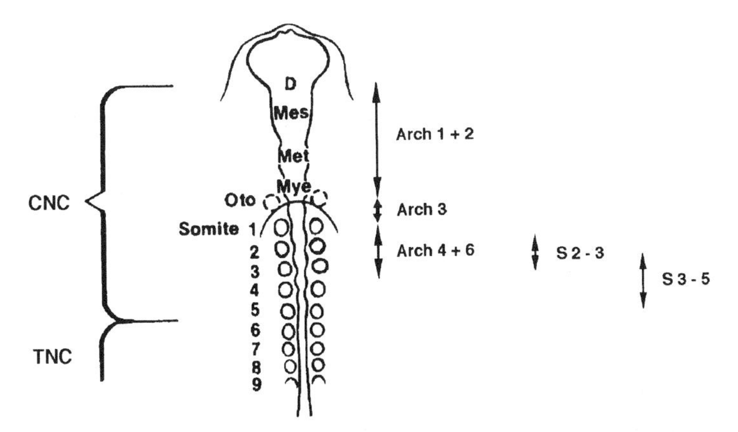

FIGURE 7.1. This dorsal view of an embryonic chick shows the location of the cranial neural crest (CNC) and the boundary between CNC and trunk neural crest (TNC). Contribution of neural crest mesenchyme to each of the six visceral arches and to postotic regions adjacent to somites 2–5 is shown. D, diencephalon; Mes, mesencephalon; Met, metencephalon; Mye, myelencephalon; Oto, otocyst. Modified from Nishibatake, Kirby and Van Mierop (1987).

population of cells that emerges from postotic rhombomeres before subdividing into populations destined for individual pharyngeal arches. A single population of neural crest cells from rhombomere 5 to the level of the boundary of somites 3/4 form, migrates along a dorsolateral pathway to subdivide secondarily into each visceral arch (Figs. 7.2 and 7.3). Circumpharyngeal crest cells from adjacent to three or more somites contribute to each visceral arch (Fig. 7.3), indicating that

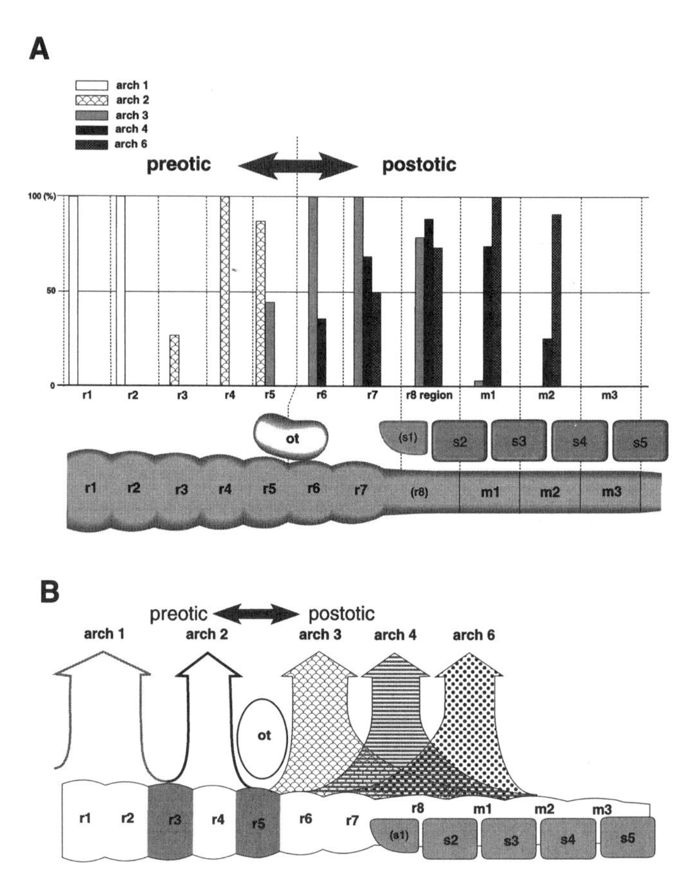

FIGURE 7.2. The origin and distribution of pharyngeal arch crest in the embryonic chick as determined from DiI-labeling. (A) Pre- and postotic crest (ot, otocyst) is shown with reference to rhombomeres 1–8 (r1–r8), myelomeres 1–3 (m1–m3) and somites 1–5 (s1–s5). The vertical axis shows the percentage of embryos showing labeled cells in each pharyngeal arch. (B) A schematic representation of the origin of pharyngeal arch cells along the neural axis to show the registration between rhombomeres and arches in the preotic crest (e.g., r1 and r2 contribute to arch 1; r4 to arch 2), and the absence of registration in the postotic region (e.g., cells from r6 caudal to s3 contribute to arch 3). Reproduced from Shigetani, Aizawa, and Kuratani (1995), Overlapping origins of pharyngeal arch crest cells on the postotic hindbrain. *Devel. Growth Differ.* 37:733–746, with the permission of Blackwell Scientific Pty. Ltd., from a figure kindly provided by Shigeru Kuratani.

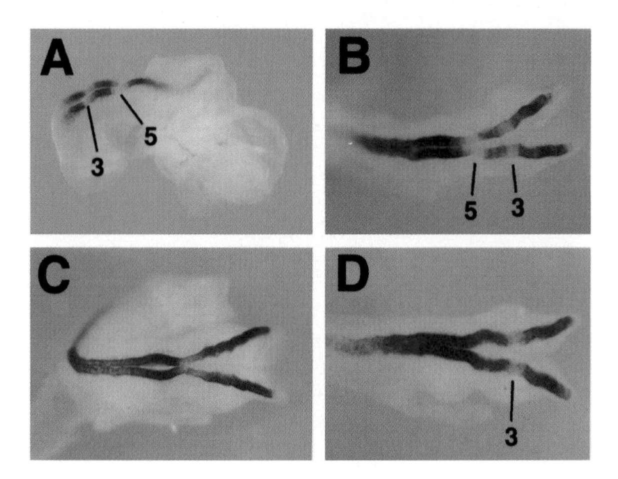

FIGURE 2.10. Expression of *Msx-3* in mouse embryos of 8–9 days of gestation. (A) *Msx-3* is expressed strongly in rhombomeres 1, 2, and 4 and in the spinal cord, and weakly in r3 in this 7-somite embryo seen in lateral view with anterior to the left and rhombomeres 3 and 5 identified. (B) This 10-somite embryo, seen in dorsal view with anterior to the right, shows weak expression of *Msx-3* in rhombomere 3 and lack of expression in r5. (C) There is uniform expression of Msx-3 throughout the hindbrain and spinal cord in this 18-somite embryo seen in dorsal view (anterior to the right) (D) The gap in expression in r5 seen in the normal embryos shown in (B) is not seen in this 10-somite embryo carrying the *Kreisler* (*Krml^{kr}*) mutation. *Kreisler* codes for a transcription factor that regulates rhombomere segment identity through *Hox* genes. Indeed, r5 may not have developed in this embryo. Reprinted from a figure kindly provided by Paul Sharpe from *Mechanisms of Development*, volume 55, Shimeld, McKay, and Sharpe, The murine homeobox gene *Msx-3* shows highly restricted expression in the developing neural tube, pp. 201–210. Copyright © (1996) with permission from Elsevier Science.

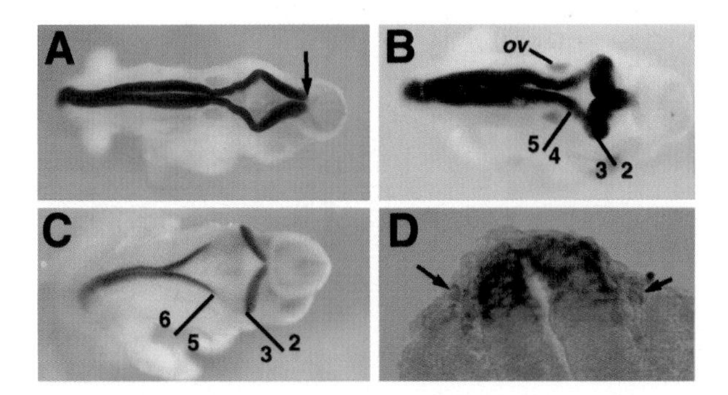

FIGURE 2.11. Expression of *Msx-3* in mouse embryos of 9.5 to 11.5 days of gestation, as seen in dorsal view with anterior to the right (A–C) and in histological cross section (D). (A) Expression is strong in both hindbrain and spinal cord at 9.5 days. The arrow marks the hindbrain/midbrain boundary, expression being negative in the midbrain. (B) Expression is similar at 10.5 days of gestation. 2, 3, 4, and 5, rhombomeres 2, 3, 4, and 5; OV, the otic vesicle, which displays nonspecific trapping of the antibody. (C) At 11.5 days of gestation, expression is restricted dorsally and is absent from rhombomeres 3–5. (D) A transverse section of the neural tube of an embryo of 9.5 days of gestation shows *Msx-3* expression in the dorsal neural tube and in neural crest cells adjacent to the neural tube. Reprinted from a figure kindly provided by Paul Sharpe from *Mechanisms of Development*, volume 55, Shimeld, McKay, and Sharpe, The murine homeobox gene *Msx-3* shows highly restricted expression in the developing neural tube, pp. 201–210. Copyright © (1996) with permission from Elsevier Science.

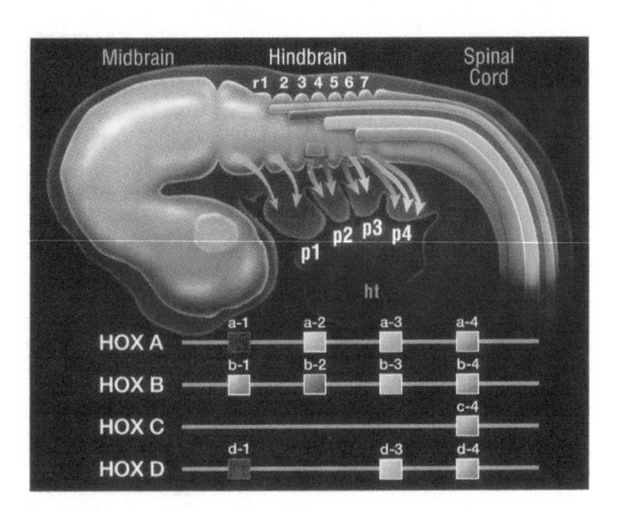

FIGURE 2.15. *Hox*-gene expression in the rhombomeres of the hindbrain (r1–r7) and in the pharyngeal arches (p1–p4) is shown in this reconstruction of a mouse embryo of 9.5 days of gestation. Colored bars in the neural tube and colored arrows in migrating neural crest cells represent expression domains of *HoxA–HoxD*, which are also shown in the panel at the bottom. Some genes, such as *Hoxa-2*, are expressed in the hindbrain but not in migrating neural crest cells. Reproduced from Manley and Capecchi (1995), *Development* 121:1989–2003, with the permission of Company of Biologists Ltd.

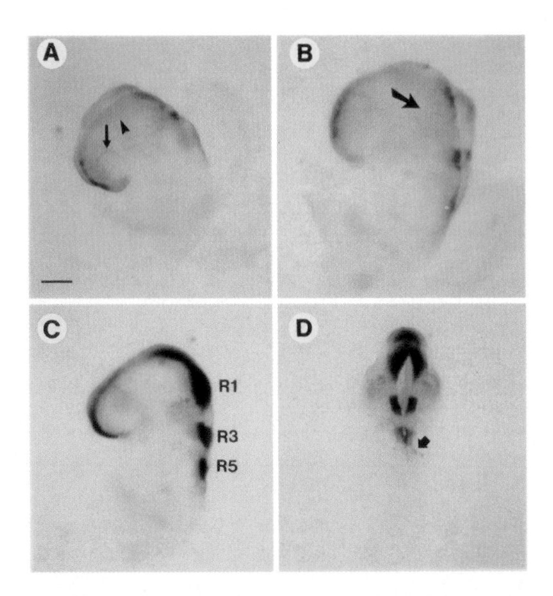

FIGURE 3.4. Expression of *Pax-7* in cranial neural crest as seen in lateral (A–C) or dorsal (D) views of whole mounts of murine embryos of 8.5 (A, B) and 9.5 (C, D) days of gestation. (A) Neural crest cells surrounding the optic vesicle (arrow) and at the level of the midbrain (arrowhead) are highlighted. (B) The arrow indicates neural crest cells migrating from the hindbrain. (C) Rhombomeres 1, 3, and 5 are identified and strongly express *Pax-7*. (D) This dorsal view highlights expression in alternating rhombomeres. R5 is marked by the arrow. Bar = 200 μm (A, C); 240 μm (B); and 160 μm (D). Reproduced from Mansouri et al. (1996), *Development* 122:834, with the permission of The Company of Biologists Ltd., from a figure kindly provided by A. Mansouri.

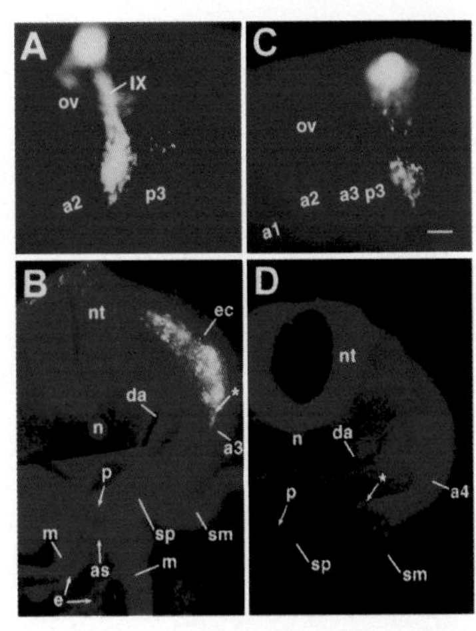

FIGURE 7.5. Migration of neural crest cells to visceral arches 3 and 4 is shown in these sections of chick embryos in which the neural crest has been labeled with DiI, shown as yellow fluorescence. (A) Labeled cells migrating from rhombomere 7 into arch 3 and forming the tract of the glossopharyngeal nerve IX. (B) In this transverse section of a chick embryo of H.H. stage 14, DiI-labeled neural crest cells (ec) that arose from rhombomere 7 can be seen migrating toward the third pharyngeal arch (a3). The leading edge of the neural crest cells (*) lies lateral to the pharynx (p). (C) Labeled neural crest cells from the level of somite 2 migrate toward the fourth visceral arch. (D) A cross section of C showing the ventral location (*) of the leading edge of the migrating cells. Other abbreviations are: a1–a4 and p3, pharyngeal arches; da, dorsal aorta; e, endocardium; m, myocardium; n, notochord; nt, neural tube; ov, otic vesicle; p, pharynx; sm, somatic mesoderm; sp, splanchnic mesoderm. Reproduced from Suzuki and Kirby (1997) from a figure kindly supplied by Margaret Kirby. Reprinted by permission of Academic Press, Inc.

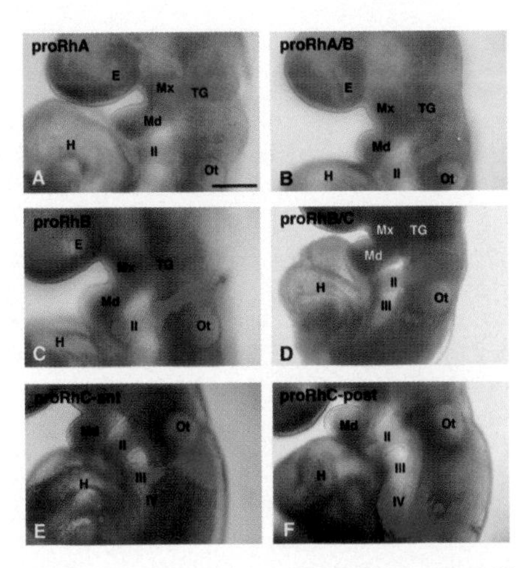

FIGURE 8.2. The segmental migration of neural crest cells (red) into individual pharyngeal arches as demonstrated following injection of DiI either into pro-rhombomeres A, B, or C of the mouse hindbrain (seen in A, C, E, F) or into the boundary between rhombomeres A/B or B/C (shown in B and D). II–IV, second to fourth pharyngeal arches; E, eye primordium; H, heart; Md, mandibular prominence; Mx maxillary prominence; Ot, otic vesicle; TG, trigeminal ganglion. Bar = 200 μm. Reproduced from Osumi-Yamashita et al. (1996), Rhombomere formation and hind-brain crest cell migration from prorhombomeric origins in mouse embryos. *Devel. Growth Differ.* 38:107-119, with the permission of Blackwell Science Pty. Ltd., from a figure kindly provided by N. Osumi-Yamashita.

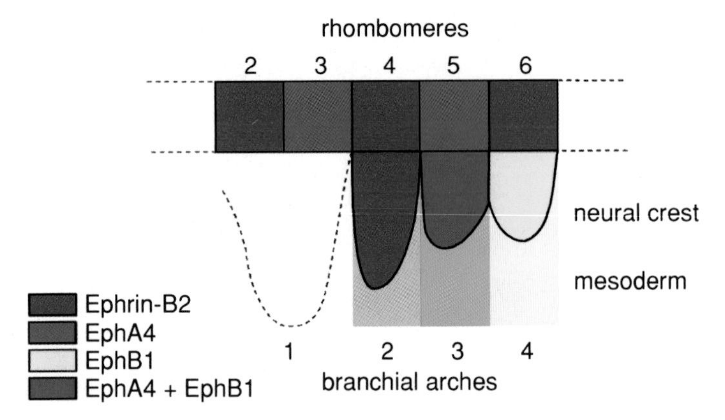

FIGURE 12.5. The expression of Ephrin-B2 (red), EphA (blue) and EphB1 (yellow) in rhombomeres 2 to 6, in migrating neural crest cells and in cranial mesoderm, is highly regionalized with respect to the four branchial arches. Ephrin-B2 is expressed in r4 and in the second branchial arch. EphB1 is expressed in r6 and the fourth branchial arch. The boundary of Ephrin-B2 is shown in red, the boundary of cells expressing EphA4 plus EphB1 in green. Receptor-ligand interactions at the boundary between branchial arches 2 and 3 restricts cells that are migrating from rhombomere 5 to the third branchial arch and prevents intermingling of second and third arch cells. Reproduced from A. Smith et al. (1997) with the permission of Current Biology Ltd., from a figure kindly supplied by David Wilkinson.

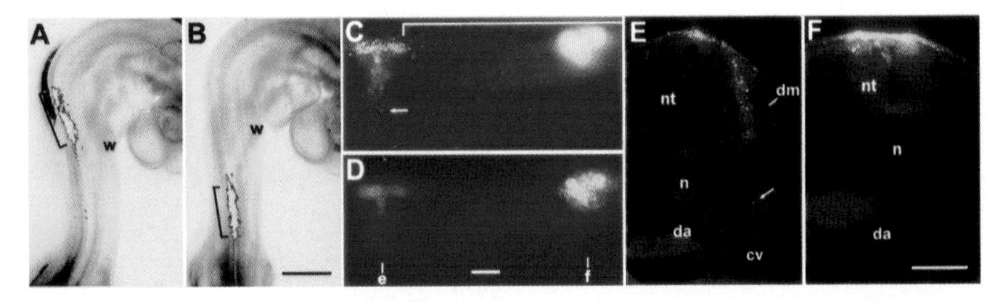

FIGURE 12.6. Ablation of the neural crest followed by injection of DiI demonstrates absence of regulation; no labeled cells are seen either in the normal migration pathway (illustrated in Figure 7.3) or at the final site. (A, B) Trunk neural crest was ablated (square bracket) either adjacent to the most caudal somites (somites 12–17 in A) or to the most cranial unsegmented mesoderm (square bracket in B) from 17- or 20-somite-stage embryos (A and B respectively). Multiple sites in the neural tube in the ablated region were then injected with DiI. In neither case can neural crest cells be seen emigrating from the ablated region, signifying lack of regulation. w, wing bud. Bar = 1 mm. (C, D) This 17-somite embryo is viewed in two focal planes. The bracket marks the ablated neural crest. The yellow fluorescence marks sites of DiI injection either immediately rostral to the ablated region (the left zone of fluorescence, seen in focus in C) or within the ablated region (the right zone of fluorescence, seen in focus in D). Neural crest cells are migrating from the rostral neural crest (arrow in C) but do not deviate into the ablated area. As expected, no cells migrate from the neural tube in the ablated region. e and f mark the position of the transverse sections shown in E and F. Bar = 100 μm. E. Transverse section rostral to the ablated region marked e in (D). DiI-labeled cells are migrating medial to the dermamyotome (dm) and in the primary sympathetic trunk (arrow). cv, cardinal vein; da, descending aorta; n, notochord; nt, neural tube. F. Transverse section through the ablated region marked as f in (D). DiI-labeled cells are concentrated adjacent to the dorsal neural tube and not seen in the normal migration pathway; compare with E. Abbreviations as in E. Bar = 100 μm for E and F. Reproduced from Suzuki and Kirby (1997) from a figure kindly supplied by Margaret Kirby. Reprinted by permission of Academic Press.

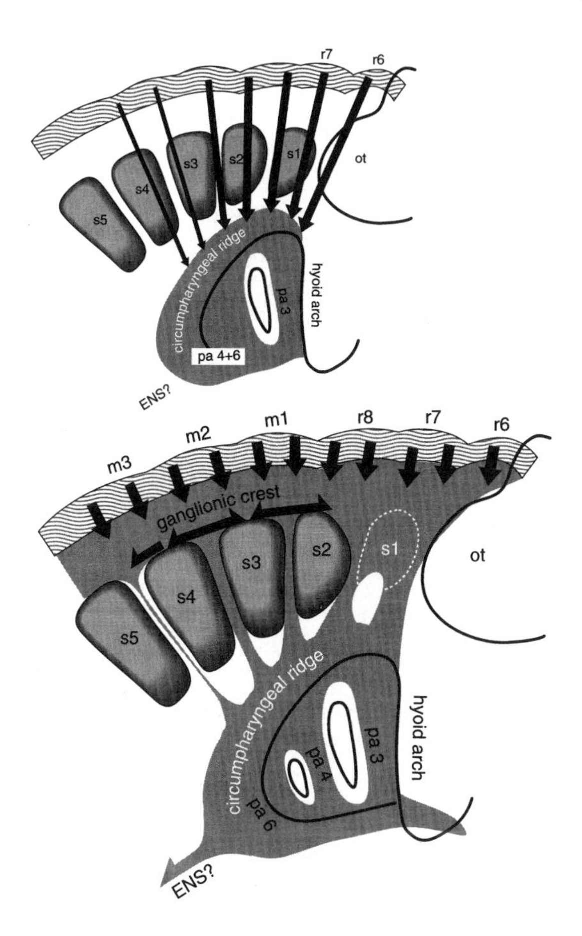

FIGURE 7.3. Patterns of early (top) and late (bottom) migration of the circumpharyngeal crest into pharyngeal arches 3, 4, and 6 (pa 3–pa 6) as deduced from DiI-labeling of premigratory neural crest cells. Arrows in the top figure show early migrating cells as a single population. Later migrating cells (short arrows in the bottom figure) form the ganglionic crest. Enteric neuroblasts (ENS?) originate from the postotic population of neural crest cells. m1–m3, myelomeres 1–3; ot, otocyst; r6–r8, rhombomeres 6–8; s1–s5, somites 1–5. Reproduced from Shigetani, Aizawa, and Kuratani (1995), Overlapping origins of pharyngeal arch crest cells on the postotic hind-brain. *Devel. Growth Differ.* 37:733–746, with the permission of Blackwell Scientific Pty. Ltd., from a figure kindly provided by Shigeru Kuratani.

there is overlap between populations contributing to each arch and that strict segmental identity is not maintained between neural tube and visceral arches. Like Noden, these authors demonstrated that the pattern of migration is intrinsic; it is preserved when cells are transplanted more caudally along the neural axis.

The cartilages and bones of the cranial and visceral skeletons of chick embryos are derived from neural crest, with the exception of the occipitals, sphenoid bones, and basal plate cartilage (which are all mesodermal) and the otic capsular cartilage and frontal bones (which receive contributions from both neural crest and mesoderm). Noden turned to quail–chick grafting to confirm and extend the fate maps of the craniofacial skeleton produced by Le Lièvre and Le Douarin, and to map the contribution of the neural crest to the trigeminal and ciliary ganglia (see Figs. 2.12 and 2.13). After neural crest cells have migrated, the ventral neural tube also contributes cells to the trigeminal ganglion, while neural tube–ectoderm interactions are required for placode formation. In mapping the origins of the ganglia that arise from the otic placode and associated neural crest in chick embryos, and the accompanying changes in proliferation, migration and modification of basal lamina, Hemond and Morest (1991) distinguished an "otic crest" adjacent to the otic placode and from which ganglia of the placode arise.[10]

Noden also determined the embryological origins of the cephalic and cervical musculature and blood vessels. None of the head dermis or craniofacial connective tissue arises from mesoderm. Dorsal iris muscles are of neural crest origin, but ventral muscles are not.

Some important attributes of neural crest cells have been established from Noden's studies:

- Mesenchyme derived from neural crest does not mix with mesodermally derived mesenchyme.
- Consequently, a neural crest/mesodermal junction can be identified clearly in the developing head.
- The final fate of neural crest cells is influenced by their position and by selective interaction with embryonic epithelia.

In emphasizing that serial evaginations of the pharynx, presence of a metameric paraxial mesoderm (somitomeres), and neural-crest-derived mesenchyme are unique to vertebrates and common to all vertebrates, Noden showed that neural crest cells impose patterns onto developing muscles, while Köntges and Lumsden (1996) discovered a previously unsuspected compartmentalization of rhombomeric neural crest cells such that muscles attach to neural crest cells with the same rhombomeric origin.[11]

Clonal culture of migrating cranial neural crest cells, or of cells from posterior visceral arches of avian embryos, allowed Bronner-Fraser (1987) to reveal an unexpected heterogeneity of cell lines. Visceral arch cell lineages are identified by whether they are HNK-1-positive or negative and by the cells types that arise from them in clonal cell culture; HNK-1 is both a marker for migrating neural crest cells (see Fig. 2.4) and is required for migration. Migration is blocked both in vivo and in vitro by perturbation of avian neural crest with an antibody against HNK-1. In vivo, neural crest cells accumulate beside the neural tube and ectopically within the lumen of the neural tube, migration having not only been blocked, but redirected medially. Addition of HNK-1 allows these cells to detach from the lumen.

Mesencephalic/metencephalic quail neural crest cells, cloned on a 3T3 feeder layer for some 7 to 10 days, display diverse developmental potentials ranging

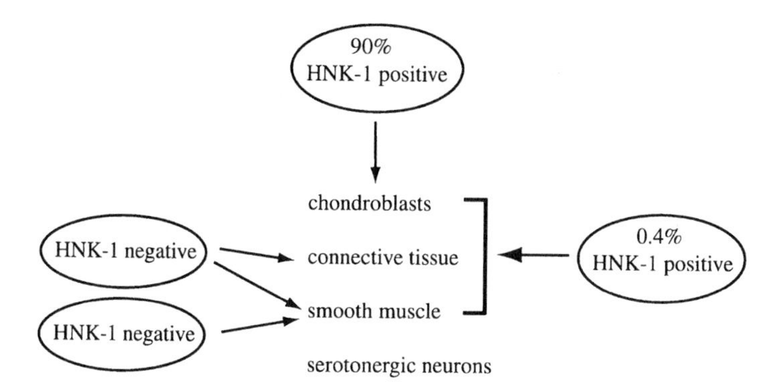

FIGURE 7.4. Four lineages of neural crest cells present in the mesencephalon/metencephalon of chick embryos are revealed following clonal cell culture. The percentages refer to percentages of cells that are HNK-1-positive in the two lineages containing HNK-1-positive cells. The more restricted lineages are HNK-1 negative. Based on data in Ito and Sieber-Blum (1993).

from pluripotency to committed neuronal cells. Ito and Sieber-Blum (1993) identified four clonal lineages (Fig. 7.4):

- A lineage, 90% of which is HNK-1-positive, that forms connective tissue, chondroblasts, smooth muscle, and serotonin-containing neurons. (Neural-crest-derived smooth muscles are found in the pharyngeal arteries and the muscles that activate feathers of the head and neck.)
- A lineage, only 0.4% of which is HNK-positive, which forms connective tissue, chondroblasts, and smooth muscle, but not neurons.
- An HNK-1-negative lineage that forms connective tissue and smooth muscle.
- Another HNK-1-negative lineage that forms only smooth muscle.[12]

Mapping the Trunk Neural Crest

Transplantation of quail neural crest into chick embryos has been used to identify the level of the neural crest from which particular portions of the peripheral nervous system and hormone-synthesizing cells arise.

- Parasympathetic (cholinergic) enteric ganglia of the gut are derivatives of the vagal neural crest, corresponding to neural crest at the levels of somites 1 to 7.
- Neural crest adjacent to somites 8 to 27 does not produce enteric ganglia but rather gives rise to the sympathetic (adrenergic) ganglia of the adrenal gland.
- The adrenomedullary cells themselves develop from neural crest that originates at the level of somites 18 to 24.
- Neural crest cells caudal to somite 28 (the sacral neural crest) produce enteric ganglia for the caudal (postumbilical) portion of the gut, including the ganglion of Remak of the ileum.[13]

- Some regions of the neural crest—those adjacent to somites 6 and 7 and caudal to somite 18—produce both cholinergic and adrenergic neurons. (Because of cell surface differences, HNK-1-positive adrenergic subpopulations can be isolated using fluorescence-activated cell sorting.[14])

Such precise cellular origins, a necessary prelude to investigation of mechanisms of differentiation, neoplasia and dysmorphogenesis (chapters 10 to 12), could not have been accomplished without the benefit of a label such as the quail nuclear marker.

Other approaches to the identification of subpopulations of trunk crest cells or that allow cranial crest to be distinguished from trunk crest are becoming available:

- A chick cytokeratin cDNA expressed extensively in trunk ectoderm but much less extensively in head ectoderm is a potential probe for trunk versus cranial derivatives.
- Isolation of cDNA libraries from vagal and thoracic neural crest cells as well as from the visceral arches of embryos of the Japanese quail will allow rapid advances in our knowledge of how the neural crest is specified.
- A further approach with considerable potential is differential display and comparison of cDNA fragments generated from cranial or trunk neural crest. One hundred bands expressed only in cranial neural crest have been identified. Sequencing of these cDNAs and probes generated against them have already been used to identify one cDNA expressed only in cranial crest caudal to rhombomere 2, and a second expressed only in migrating cranial neural crest cells.[15]

Mapping the Cardiac Neural Crest

The cranial neural crest has an indirect effect on heart and blood vessel development through the mesenchymal derivatives that arise from it and the connective tissue matrices they generate. Ablation of cranial neural crest, therefore, affects heart development not because cranial neural crest cells contribute to the heart directly, but because their derivatives alter the environment in which the heart develops by altering blood flow or the tissue milieu in which the heart develops. Although specific, such effects are indirect and independent of any direct contribution that crest cells may make to the heart.[16]

It has been known since the mid-1970s that neural crest cells also contribute directly to the smooth muscles of the aortic arches within the visceral arches of the embryonic chick. This region is now known as the *cardiac neural crest*. Indeed, to Fishman and Chien (1997), the vertebrate heart, with its cardiac neural crest cell contributions, is a new heart, part of a new cardiovascular system composed of modular units, only some of which existed in vertebrate ancestors.

Using transplantation of quail neural crest cells into chick embryos and DiI-labeling, Margaret Kirby and her colleagues and other workers established that neural crest cells extending rostrally from the mid-otic placode (rhombomeres 6 to 8) to the caudal end of the third somite, migrate through visceral arches 3, 4, and 6 coincident with the outgrowth of branches of cranial nerves IX, X, and XII (Fig. 7.5 [color plate]) and into the outflow region of the developing heart. In the heart, neural crest cells form cardiac ganglia, mesenchyme of the endocardial cushions, semilunar valves, and much of the aortico-pulmonary and conotruncal septa of the truncus arteriosus, but not the pulmonary veins. No smooth muscle α-actin is deposited where smooth muscle cells of the aortico-pulmonary septum should develop in embryos from which the cardiac neural crest has been removed. Hypobranchial muscles and Schwann cells for cranial nerve XII also arise from the cardiac neural crest. Neural crest cells adjacent to occipital somites 1 to 3 provide mesenchymal cells; neural crest cells adjacent to somites 1 to 7 provide cardiac ganglia. Deletions or deficiencies of these neural crest cells produce specific cardiac defects such as conotruncal anomalies, aortic coarctation, or bicuspid aortic valves, and provide an animal model for the DiGeorge syndrome, a human syndrome with heart involvement (see Chapter 11).[17]

The cardiac neural crest in avian embryos provides a unique set of proteins to the pharyngeal region. Using two-D gel electrophoresis, Abdulla, Slott, and Kirby (1993) showed that five proteins normally localized in migrating cardiac neural crest were eliminated in embryos from which the cardiac neural crest had been removed. Interestingly, several stages later, the normal protein profile had been restored, indicating regulation of cardiac neural crest. Other molecules can also be used as markers of migration of cardiac neural crest cells through the pharynx or their arrival in the cardiac primordia; Andrikopoulos et al. (1992) localized pro-α2 (V) collagen transcripts in craniofacial mesenchyme, skeletal precursors, and heart valves of murine embryos.

Clonal cell culture of quail neural crest shows that cardiac neural crest is made up of different cell types. Some are pluripotential and can form pigment cells, smooth muscle, connective tissue, chondrocytes, and sensory neurons. Others are more restricted. Ito and Sieber-Blum (1991) identified five clonal lineages:

- cells that form only pigment cells;
- mixed clones that form pigment and other cell types;
- clones that form all cell types other than pigment cells;
- clones that only form smooth muscle; and
- clones that form chondrocytes and sensory neurons.

More recently, Poelmann, Mikawa, and Gittenberger-de Groot (1998) identified a subpopulation of cardiac neural crest cells in the chick whose removal from the septum by apoptosis may play a role in initiating cardiac myogenesis. Such heterogeneity has implications for the regulation of cardiac neural crest, a topic discussed in Chapter 12.

The Thymic Neural Crest

The thymic epithelium, a derivative of the pharyngeal endoderm, is stimulated to divide by migrating neural crest cells. The neural crest also contributes connective tissue cells and elements of the blood vessels to thymus glands. Lymphoid stem cells are transported to the rudiment of the thymus via the developing vascular system. Under the influence of cells derived from the neural crest, thymic epithelial cells stimulate the proliferation of lymphoid stem cells and mediate their differentiation into thymocytes. The neural crest thus influences development of the thymus gland by contributing cells and by promoting thymocyte differentiation. Interestingly, neural crest cells from various rostro-caudal levels of the neural tube can participate in thymic development even if they normally do not. Absence or severe reduction of the thymus in humans is associated with cardiac defects in the DiGeorge syndrome. Such associations, recognized as neural crest defects or neurocristopathies, are the topic of Chapter 11.[18]

8

Mammals

Much of the interest in the neural crest over the past three decades was prompted by the knowledge that many craniofacial defects and inherited conditions in humans involve tissues known to be of neural crest origin in nonmammalian vertebrates. Defects involving pigment cells (albinism), the craniofacial skeleton (cleft lip and palate, asymmetrical facial growth, first arch syndromes), adrenal glands (medullary carcinoma), or sympathetic neurons of spinal ganglia (neuroblastomas) come into this category. Indeed, so linked are embryos, defects, and the genetics underlying defects that Pierce titled his 1985 paper on the topic, "Carcinoma is to embryology as mutation is to genetics."[1]

Would patterns of neural crest origin seen in other vertebrates be seen in mammals, especially in humans? Unfortunately, the scope for experimentation on mammalian embryos is much more restricted than it is for other vertebrates. Until recently, we had to rely on the interpretation of static evidence, usually from histological sections. After a long delay—caused in part by adherence to the germ-layer theory—knowledge of the derivatives of the mammalian neural crest increased rapidly from the late 1980s onwards. In the 1990s, there has been what amounts to an explosion of publications on the origin, migration, mapping, and differentiative capabilities of mammalian neural crest cells. No longer must we rely on extrapolation from the "lower" vertebrates. A synopsis of these mammalian studies is the subject of this chapter. I place special emphasis on the techniques that now allow mammalian neural crest to be studied.[2]

Fate Maps

Fate maps of the epiblast of murine embryos are now available and provide some insights into the location of prospective neural crest before neurulation. (There have of course been many studies mapping the brain but these are outside the scope of this book.) In 1991, Lawson and colleagues produced a sufficiently detailed fate map using clonal analysis that they could compare it with fate maps of the epiblast in chick and urodele embryos, while Tam and Quinlan's work provides an insightful comparison into what they term the "striking homology" between fate maps of representative fish, amphibian, avian, and mammalian embryos. The congruence of these fate maps extends to the location of presumptive neural crest between neural and epidermal ectoderm. One important finding, consistent with what is known from secondary neurulation (Chapter 2), is that clonal descendants within the epiblast are not confined to single germ layers and that germ layers are not fully segregated until gastrulation.[3]

An early regionalization of future cranial and trunk neural crest in murine embryos is suggested by the ^{3}H-thymidine-labeling study carried out by Rosa Beddington. Rostral or caudal ectoderm from embryos labeled immediately before the onset of neural crest cell migration (late primitive streak stage, 8 days of gestation) was inserted into the equivalent position in unlabeled embryos, which were maintained in whole embryo culture for three days. Rostral ectoderm formed cranial neuroepithelium while caudal ectoderm formed trunk neuroepithelium, suggesting that segregation into future cranial and trunk neural crest occurs before the incorporation of future neural ectoderm into the neural tube. Quinlan and colleagues mapped the neurectodermal fate of epiblast cells at the egg cylinder stage and demonstrated that neural primordia exhibit cranio-caudal patterning before neurulation. The most caudal neural crest cells in the mouse apparently consist of two populations, one derived from neuroectoderm (primary neural crest) and the second derived from the tailbud (secondary neural crest). I discussed this fundamental but little appreciated process of secondary neurulation and the fact that the tail region does not develop directly from primary germ layers in Chapter 1.[4]

Cruz and colleagues (1996) used DiI injection to map epiblast fate in the Australian dasyurid marsupial *Sminthopsis macroura*. Although they demonstrated that the neurectoderm (medullary plate) gives rise to epidermal and neural ectoderm, they did not map neural crest. Indeed, as noted in Chapter 1, there have been few investigations of the neural crest during marsupial development, though Hill and Watson, in studies published in 1958 but begun in 1911, documented neural crest cells and their contribution to cranial mesenchyme and ganglia in a number of Australian marsupials—the native cat (*Dasyurus*), bandicoot (*Perameles*), kangaroos and wallabies (*Didelphis*, *Macropus*)—and in *Petrogale*, the American opossum. Recently, Kathleen Smith (1997) laid the basis for a comprehensive analysis of the craniofacial development of four species of marsupials using specimens from the Hill collection. She compared marsupial craniofacial development with that of five eutherian mammals, especially with respect to the central nervous system and its relationship to skeletal and muscle development, and showed that development of the marsupial central nervous system is delayed relative to skeleto-muscular development, which is advanced. To Smith, neurogenesis and the short gestation period are the rate-limiting processes for marsupial development.

Labeling and Whole Embryo Culture

Grafting mammalian tissues into chick embryos was an early method used to attempt to follow the fate of the mammalian cells. Rawles (1940) grafted fragments of mouse neural folds intracoelomically into chick embryos and showed that murine melanoblasts are of neural crest origin. Fontaine (1979) grafted fragments of the thyroid rudiment, pharyngeal pouch, or pouch endoderm/mesenchyme in various combinations from 18 to 45 somite (9 to 11 day) mouse embryos onto the chorioallantoic membrane of embryonic chicks. Calcitonin-synthesizing cells first appear in pharyngeal pouch (fourth branchial arch) mesenchyme, then in the endoderm, and finally in the thyroid. Thus, C-cells are of neural crest origin in mice as they are in birds.

There had been few studies in which mammalian neural crest cells were labeled and traced until quite recently, attesting both to the unavailability of species-specific markers and to the general difficulty of working with mammalian embryos. Neural crest cells are especially sensitive to inhibition by Mitomycin C and take up and decarboxylate amine precursors. In the 1970s, Nozue and his collaborators utilized staining after administration of Mitomycin C, and autoradiography after administration of C^{14}-DOPA (they consider DOPA to be a specific marker for murine cells derived from neural crest), to identify tooth germs, vibrissae, head mesenchyme, thyroid, adrenal gland, pancreas, heart, spinal and sympathetic ganglia, and various portions of the digestive system as neural crest in origin.[5]

When trunk neural crest cells from 9-day-old embryonic mice are maintained in organ culture, a population of mesenchymal cells migrate away from the explant. These cells differentiate into melanocytes and adrenergic neurons, providing evidence for the neural crest origin of these cell types. Formation of melanophores from neural crest has also been demonstrated following culture of murine neural crest cells and their microinjection back into 9-day-old embryos. The injected cells migrated extensively, contributing pigment-forming cells to the hair and iris of the chimeric embryos.[6]

In the mid to late 1980s, however, effective procedures for labeling and following the migration and fate of rodent neural crest cells were developed and combined with whole embryo organ culture. Results from these studies are discussed below.

Tan and Morriss-Kay, and Smits-van Prooije and colleagues, labeled embryos with ^{3}H-thymidine and/or wheat germ agglutinin conjugated to gold particles, removed the neural folds, and microinjected them into the equivalent position in an unlabeled embryo, which was then allowed to develop in culture for up to 3 days (the maximum for in vitro survival of whole-rodent embryos of these ages). Using immunohistochemistry to visualize the wheat germ agglutinin-labeled cells, these authors describe the migration routes and final locations of cranial neural crest cells. Neural crest cells migrate well below the superficial ectoderm; neural crest cells in avian embryos primarily migrate subectodermally. Neural crest cells did not originate from the forebrain but did arise from midbrain caudally to the postotic caudal hindbrain, confirming the cytological studies of sectioned embryos discussed earlier.[7]

Because cultured whole embryos will not survive long enough to follow the subsequent fate of labeled cells, Tan and Morriss-Kay (1986) removed mandibular arches or other regions from cultured embryos and grafted them into the anterior chamber of the eyes of adult rats, where they continued to develop and differentiate. In this way, the effective life of labeled neural crest cells can be extended greatly. Labeled cartilage was found in such grafts, confirming its origin from neural crest cells.

Chan and Tam (1986) obtained neural plates or neural folds with their resident populations of neural crest cells from 0- to 4-somite (7.5- to 9.5-day-old) rat embryos and grafted them under the kidney capsules. Cartilage, bone, and hair follicles developed in these renal grafts. The capacity to form skeletal tissues was lost at the 5-somite stage, reflecting either prior emigration of the skeletogenic cells or lack of an appropriate epithelial environment to elicit differentiation. The latter is more likely, especially given observations about the dispersion of neural crest cells into the first visceral arch of developing mouse embryos. These neural crest cells migrate superficially, rather than more deeply, retaining an association with the overlying epithelium. Nichols (1986a) noted that this affinity of epithelial and neural crest cells immediately preceded the onset of the epithelial–mesenchymal interaction, which I showed in 1980 to be a prerequisite for the initiation of mandibular chondrogenesis and osteogenesis in mice.

Neural crest contributes mesenchyme throughout the head. In a further study, Chan and Tam (1988) followed the migration of mesencephalic neural crest from their initial emigration through breaks in the basement membrane. Midbrain cells were traced into craniofacial mesenchyme, trigeminal ganglia, and the visceral arches. Placodal cells were also observed in the trigeminal ganglion, confirming a mixed origin from placodal ectoderm and neural crest in the mouse, as had been shown in birds (see Figs. 2.12 and 2.13). Labeled neural crest cells were found in the trigeminal, vagal, and acousticofacial ganglia; maxillary, mandibular, and hyoid arches; periocular mesenchyme; anterior cardinal vein; and dorsal aorta. All are comparable locations to those seen in labeling studies in avian and amphibian embryos. A rostro-caudal regionalization of the cranial neural crest was very evident, confirming Beddington's (1985) finding of antero-posterior regionalization of the early ectoderm. Cells of the maxillary processes originate from midbrain and rostral hindbrain. Cells of the mandibular processes originate from rostral hindbrain and preotic caudal hindbrain, while cells of the hyoid arch arise from pre- and postotic caudal hindbrain.

Less effective as a cell-labeling system was the application of fluorescent latex microparticles to murine embryos by Fleming and George (1986). Although such microparticles are endocytosed, they are not incorporated into the cytosol and therefore only last through a few rounds of cell division. A different approach for labeling mouse neural crest cells, developed by Chan and Lee (1992), involves labeling neural crest cells with latex beads and then microinjecting the labeled crests into the amniotic cavities of 2- to 3-somite-stage mouse embryos. Chan and Lee confirmed that the beads were incorporated into the neural crest cells and could follow the migration of these labeled cells into lateral mesenchyme and the visceral arches.

Kuratani and Eto and their groups in Japan are using focal injections of DiI into defined regions of mouse neural tubes to great advantage in producing fine-grained maps of the neural crest. By injecting DiI into pro-rhombomeres of the hindbrain, both neural crest cell lineages and the segmental contribution of hindbrain lineages to the visceral arches have been identified (Figs. 8.1 and 8.2 [color plate]). Injecting DiI into pre- and postotic neural crest demonstrates that only even-numbered rhombomeres in the preotic crest produce neural crest cells, and that recognition of even- and odd-numbered rhombomeres is obscured in the postotic neural crest, the region that marks the head–trunk interface and the transition between cranial and trunk neural crest.

Injection of DiI into fore- or midbrain murine crest cells allowed Osumi-Yamashita *et al.* (1994) to demonstrate that the anterior edge of the prosencephalon gives rise to the head ectoderm, including the nasal placode, Rathke's pouch, and the oral epithelium, as it does in the chick (see Fig. 2.12). The lateral edges of the prosencephalon provide the neural-crest-derived mesenchyme of the frontonasal mass, while the mesencephalon provides neural crest mesenchyme of the first visceral arch. Culture of fragments of murine neural folds by Chareonvit *et al.* (1997) also demonstrates that midbrain neural crest cells primarily form mesenchyme. They do not form pigment cells.[8]

Osumi-Yamashita, Ninomiya, and Eto (1997) adapted their injection methods to analyze neural crest cell contributions to later stages of mammalian embryogenesis. They labeled specific populations of neural crest with fluorescent dyes and cultured whole embryos for as long as possible, after which regions such as

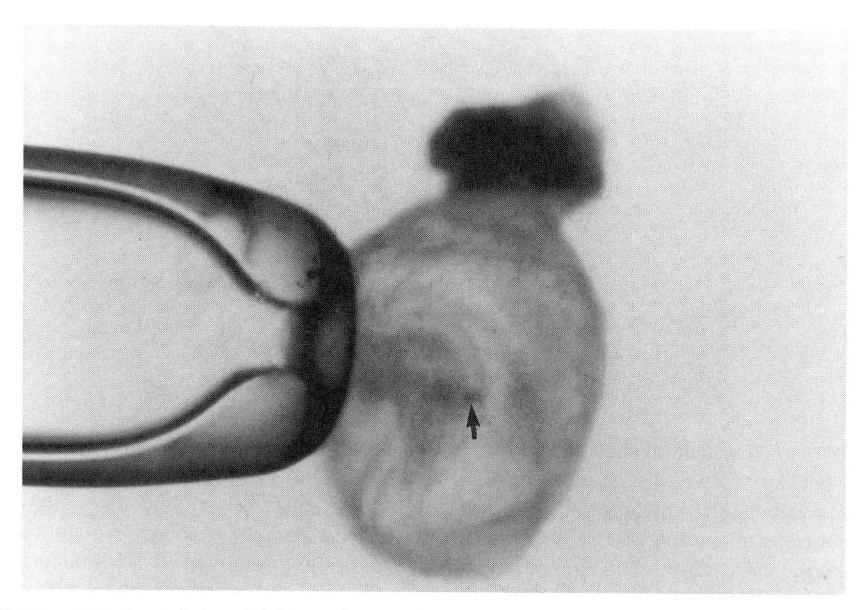

FIGURE 8.1. An 8.5-day-old (5-somite-stage) mouse embryo supported on the end of a micropipette. DiI has been injected into the rostral hindbrain (arrow). Reproduced from Trainor and Tam (1995), *Development* 121:2572, with the permission of The Company of Biologists Ltd., from a figure kindly provided by Patrick Tam.

the mandibular arches were established in organ culture. In this way they followed cranial neural crest into tooth development as late as the cap stage; see the following section.

Since the late 1980s, Tam and his colleagues have investigated whether neural crest and cranial paraxial mesoderm are codistributed in mouse embryos. They use a combination of neural crest cell grafting—labeling with either a fluorescent dye (DiI, DiO; Fig. 8.1) or with wheat germ agglutinin conjugated to gold particles, Fig. 8.3)—followed by 48 hours of in vitro culture. Their experiments show that mesoderm and neural crest that arise at the same axial levels have common destinations and share similar patterns of regionalization. Hence mesoderm from somitomeres I, III, IV, and VI contributes to the same craniofacial tissues as does neural crest adjacent to these somitomeres (forebrain, caudal midbrain, and rostral-caudal hindbrain), indicative of global segmental patterning (Figs. 8.3 and 8.4). The subsequent behavior of mesodermal and neural-crest-derived mesenchyme differs in craniofacial and visceral arch regions. Mesenchyme from both sources mixes extensively in the periocular, periotic, and cervical regions, but segregates within the visceral arches, the neural-crest-derived mesenchyme

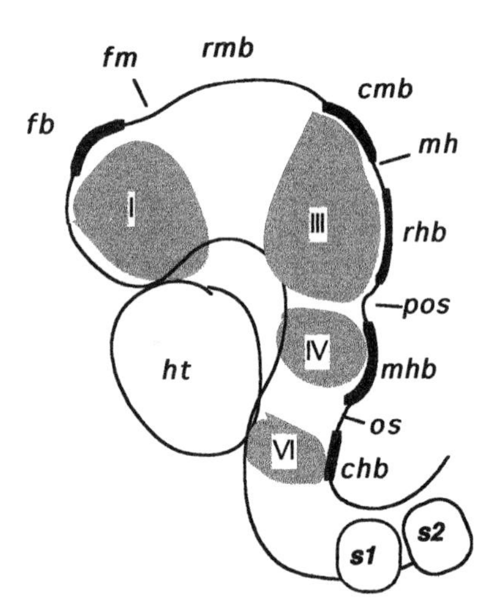

FIGURE 8.3. The cephalic region of an 8.5-day-old (5-somite-stage) mouse embryo showing the location of somitomeres I, III, IV, and VI (gray) and the regions of the developing brain that correspond in position to those somitomeres. Thick lines indicate labeled regions of neural crest whose fate is shown in Figure 8.4. The brain regions, from rostral to caudal, are fb, forebrain; fm, forebrain/midbrain junction; rmb, rostral midbrain; cmb, caudal midbrain; mh, midbrain/hindbrain junction; rhb, rostral hindbrain; pos, preotic sulcus; mhb, middle hindbrain; os, otic sulcus and chb, caudal hindbrain. ht, heart; s1 and s2, somites 1 and 2. Reproduced from Trainor and Tam (1995), *Development* 121:2573, with the permission of The Company of Biologists Ltd., from a figure kindly provided by Patrick Tam.

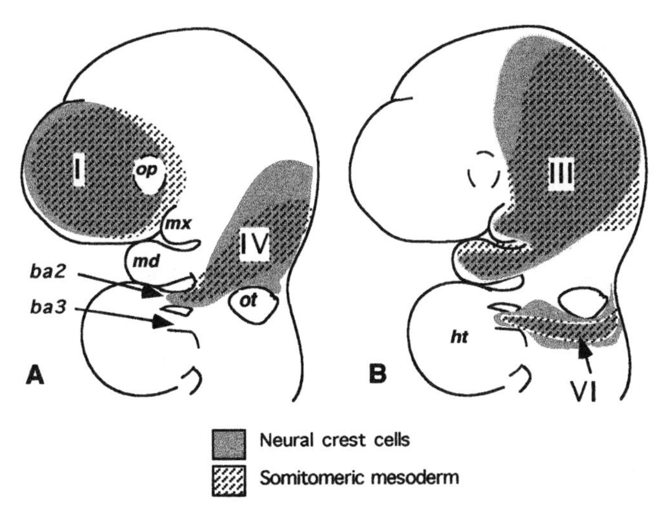

FIGURE 8.4. The distribution of cranial neural crest cells and somitomeric mesoderm as seen in a 10.5-day-old embryo labeled at 8.5 days (see Fig. 8.3) and cultured for two days. A. The distribution of neural crest cells from forebrain and middle midbrain correlates closely with the distribution of cells from somitomeres I and IV. B. The distribution of neural crest cells from caudal midbrain and from rostral and caudal hindbrain correlates closely with the distribution of cells from somitomeres III and VI. ba 2, ba 3, second and third branchial arches; ht, heart; md, mandibular process; max, maxillary prominence; op, optic vesicle; ot, otic vesicle. Reproduced from Trainor and Tam (1995), *Development* 121:2579, with the permission of The Company of Biologists Ltd., from a figure kindly provided by Patrick Tam.

toward the periphery, and the mesodermal mesenchyme toward the center (as was previously shown for avian embryos). In avian embryos, periocular mesenchyme functions differently from other mesenchyme, being able to interact with melanocytic neural crest cells and modify the type of pigment cell produced.[9]

Amongst other approaches that hold promise for future studies:

- Lysinated rhodamine dextran and DiI have been used with effect to label individual murine neural crest cells, follow their migration, and demonstrate that enteric ganglia are of neural crest origin (and arise from a sacral neural crest), and that single cells from trunk neural crest are multipotential, being capable of forming neurons of the neural tube and of contributing to the dorsal root ganglia, the sympathoadrenergic system, and Schwann and pigment cells.
- Antibodies that label murine neural crest cells and/or derivatives are also being developed. A rat anti-mouse monoclonal antibody (4Egr) against a vimentin-related molecule labels premigratory cranial and trunk neural crest cells, migrating cells, and mesenchyme of the visceral arches, but also labels rostral somitic mesoderm. Not all neural crest cells bind to this antibody, although 80% of cultured neural crest cells were labeled.

- m217c, a monoclonal antibody against a neural antigen and a Schwann cell marker, has been used to follow Schwann cell development from premigratory rat neural tube maintained in vitro for 1 or 2 days.
- A neural-crest-specific LacZ reporter, in combination with transplantation between murine and avian embryos, has also been used to analyze defects in neural crest or neural crest cell environment in *Splotch* mutant mice.
- A transgenic neuroanatomical marker also reveals cranial neural crest deficiencies, especially of cranial ganglia and cranial nerves, in *Splotch* mice (see Chapter 12 for *Splotch*).
- Another approach, with potential for following neural crest cells, is to use Y-chromosome-specific probes to monitor neural crest from male embryos transplanted into female embryos. Such a technique has already been used to follow the survival and migration of glial cells transplanted into the brains of adult female mice.[10]

The technical problems of survival in vitro may be mitigated by an experimental technique developed by Muneoka, Wanek, and Bryant (1986). Rodent embryos lying within the amniotic cavity are exteriorized onto the abdominal surface, treated either surgically or by injection, and then returned to the abdominal cavity, where they continue to develop *ex utero*, but normally, for the remainder of the gestation period. Serbedzija, Bronner-Fraser, and Fraser (1992) have used this technique in combination with DiI labeling and in vitro culture to determine the timing of cranial neural crest cell migration from murine neural tubes. Migration was initiated from the rostral hindbrain at the 11-somite stage, from mid- and caudal hindbrain at 14 somites and from forebrain at 16 somites. This pattern, whereby migration does not begin from the forebrain and spread caudally but rather begins more caudally, is typical of other vertebrates studied.

Teeth

Teeth are composite structures consisting of epithelially derived enamel deposited by ameloblasts and neural-crest-derived dentine and pulp deposited by odontoblasts and fibroblasts respectively. Teeth develop as a result of a complex series of epithelial–mesenchymal interactions that regulates division and differentiation, as in cartilage and bone formation, and that are most completely understood in mouse embryos. Pharyngeal endoderm, oral ectoderm, and neural crest mesenchyme must all interact for teeth to form. Epithelium is required for mesenchyme to form dental papillae and to differentiate as odontoblasts. Mesenchyme is required for oral epithelium to form enamel organs and to differentiate as ameloblasts. Subsequent interactions between ameloblasts and odontoblasts directs their further differentiation and the deposition of pre-enamel and pre-dentine, which then mineralize as enamel and dentine.

The essential epithelial–mesenchymal interactions in that cascade are:

- dental mesenchyme induces oral epithelium to proliferate;
- dental epithelium triggers dental mesenchyme to proliferate to form a dental papilla;
- The dental papilla induces dental epithelium to form an enamel organ containing preameloblasts;
- preameloblasts induce dental papilla cells to differentiate as preodontoblasts that transform into odontoblasts;
- odontoblasts induce preameloblasts of the enamel organ to differentiate into ameloblasts, synthesize, and deposit enamel.
- ameloblasts induce odontoblasts to synthesize and deposit dentine.[11]

Although it has long been known that amphibian teeth are derivatives of the neural crest, the first experimental analysis of the neural crest origin of mammalian teeth was not undertaken until the late 1980s (Lumsden 1987, 1988). The basic technique is to excise cranial or trunk neural crest from 6- to 12-somite-stage mouse embryos, recombine it with epithelium (either from mandibular arches or from limb buds of 9- and 10-day-old embryos) and graft these epithelial-neural crest recombinations into the anterior chamber of the eye. Cranial neural crest grafted alone forms cartilage. When grafted with either mandibular or limb bud epithelium, cranial neural crest cells form both perichondrial and membrane bone. When grafted with mandibular epithelium, cranial neural crest cells also form teeth. Both dentine and enamel and the cells that deposit them (odontoblasts and ameloblasts) were found in the teeth, indicative of differentiation of both neural crest and epithelial cells. The formation of cartilage, bone, and teeth in these grafts clearly demonstrates their neural crest origin and provides information on the role of epithelia in their differentiation. Patterning of the teeth—whether incisors (incisiform) or molars (molariform)—is specified, not by the mesenchyme derived from the neural crest as it is later in development, but by the epithelium, specifically from the site along the mandibular arch from which the epithelium is derived. Mandibular epithelium can evoke teeth from the most rostral murine trunk neural crest if the two tissues are brought into association; see Chapter 10.[12]

Teratomas

The teratomas that form when embryonic ectoderm is grafted under the capsule of the rat kidney contain such typical neural crest derivatives as cartilage, bone, and teeth. You will recall that amphibian neural crest cells only form cartilage and bone if they interact with an embryonic epithelium such as the pharyngeal endoderm. The cartilage in teratomas derived from rat neural crest cells usually lies adjacent to an epithelium. Indeed, grafts from older embryos will only form cartilage or bone if ectoderm or endoderm is included in the graft, providing presumptive evidence for epithelial-neural crest cell interactions in the rat. Direct experimental evidence for such interactions in chondrogenesis of murine neural crest cells is discussed by Hall (1980) and in Chapter 10.[13]

Part III

Mechanisms and Malformations

9

Mechanisms of Migration

Although other embryonic cells migrate—primordial germ cells from their site of origin to the embryonic gonadal ridges, sclerotomal cells from the somites to surround the notochord and spinal cord—the extraordinary migratory ability of neural crest cells sets them apart from all other embryonic cells. The loss of cell-to-cell attachments, basal translocation of cytoplasmic contents, movement of cells to a fenestrated basal lamina, penetration through the basal lamina, and subsequent migration along epithelial basal laminae or through extracellular matrices all play their roles in neural crest cell migration. In this chapter I address evidence of, and mechanisms that control this remarkable migratory behavior. I also ask how neural crest cells stop migrating when they reach their final site.[1]

Evidence of Migration

Early descriptions of neural crest cell migration were based on analyses of serially sectioned whole embryos in which investigators "saw" these cells move away from their superficial association with the developing dorsal neural tube to take up more peripheral positions as sensory ganglia. Cytology, and the colonization of increasingly more distant regions as development ensued, provided eloquent evidence of the massive cell movements associated with the relocation of neural crest cells.

In their cytological studies in the 1920s, Bartelmez and Holmdahl saw cells apparently migrating away from the neuroepithelium at the neural-plate stage of human and other mammalian embryos, a situation paralleling that seen in the other vertebrate groups. Bartelmez reported neural crest cells migrating away from the forebrain on either side of the optic primordium of 8-somite stage (day 22 of gestation) human embryos. These cells could be traced in older embryos into the trigeminal ganglion and as mesenchyme migrating around the developing pharynx to contribute to the mandibular and hyoid arches. Bartelmez also described neural crest cells in the neural folds of fore-, mid-, and hindbrain of 4- to 5-somite-stage rat embryos. Migration, which begins cranially, gradually extends caudally as in birds and amphibians; migration away from the forebrain begins at the 5-somite stage and is maximal at the 6 to 7 somite stage, when spurs of mesenchyme project into the cranial mesoderm and developing mandibular arches. Some 40 years after his first study, Bartelmez (1962) was still providing information on the neural crest in the rat, pig, *Hemicentetes* (an insectivore), primates and humans.[2]

Extirpation of neural crest from neural-fold-stage embryos resulted in embryos devoid of peripheral derivatives, indicating that a migratory cell population had been removed. Localized deficiencies after ablation of selected portions of the neural crest allowed conclusions to be drawn regarding the pathways and extent of neural crest cell migration. Extirpation is, however, a less desirable technique than those that allow neural crest cells to be removed and replaced with uniquely labeled cells, or that allow neural crest cells to be labeled in situ.

The classic technique of grafting radioisotopically labeled cells in place of the neural crest in an unlabeled host chick embryo, as undertaken in the pioneering studies of Weston and Johnston in the 1960s, was discussed in Chapter 7. Johnston mapped out the basic migration pathways as cranial neural crest cells moved around the developing eye in three basic streams to form the maxillary, mandibular, median, and lateral nasal processes. These cells went on to form connective tissue, cartilage, and bone. Weston identified two streams of migratory trunk neural crest cells—a dorsolateral and a ventral population—each with predictable pathways of migration:

- superficially, along the dorsal trunk epidermal ectoderm;
- more medially, along the lateral edges of the somites;
- between the somites;
- into the somites themselves along the dermatome/myotome boundary (subsequently shown to migrate through the rostral half of each somite) and into somitomeres, the equivalent of somites in the head; and
- the most medial population that migrates between spinal cord and somites.

These migrating cells, whose major pathways are shown in Figure 9.1, went on to form pigment cells, spinal ganglia, and sympathetic neurons. As discussed in Chapter 10, by exposing cells to different environments, differing pathways of migration influence the final fate of neural crest cells.[3]

Migration of trunk neural crest has been investigated using transplants between *Xenopus laevis* and *X. borealis* following injection of lysinated fluorescein dextran into *X. laevis* eggs. Migration is along the three pathways seen in the chick—a ventral pathway around the notochord and neural tube, a lateral pathway under the ectoderm, and a dorsal route across the caudal two thirds of each somite and into the dorsal fin. Pathways of migration (excluding intrasomitic) are broadly conserved across the vertebrates. For example, migration of cranial neural crest cells in amphibians with tadpoles, such as *Bombina*, *Xenopus*, and *Rana*, and in the direct-developing Puerto Rican frog *Eleutherodactylus coqui*, is in the rostral, rostral otic, and caudal otic streams typical of other vertebrates studied; patterns of early migration are not altered with loss of the tadpole stage (see Fig. 5.6). As discussed in Chapter 5, the only difference in coqui is some enhancement of the mandibular stream.

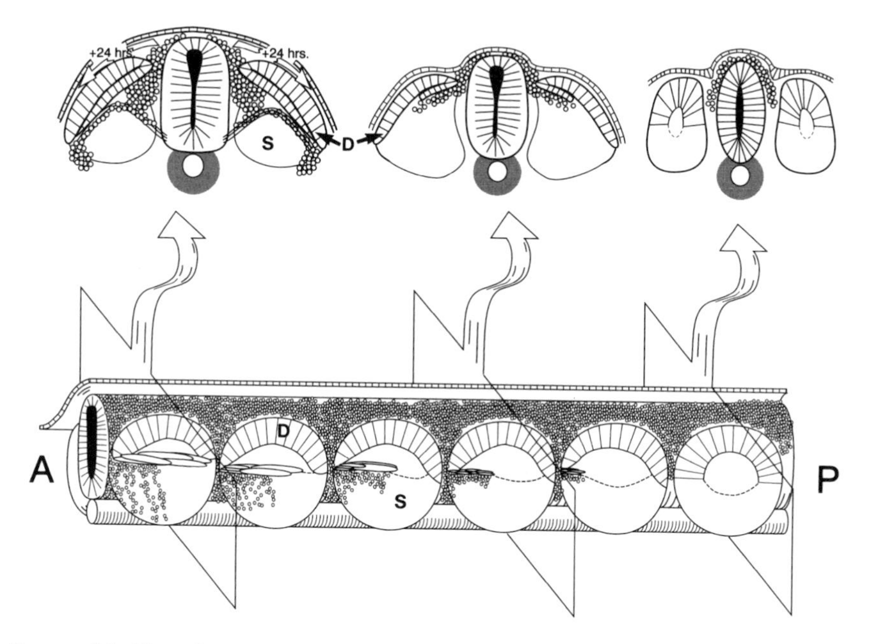

FIGURE 9.1. The migratory pathways followed by trunk neural crest cells are shown in cross sections of three levels of the neural tube (top) and in longitudinal section (bottom). A is anterior and P posterior. The progressive migration of cells between the neural tube and dermatome (D) and through the rostral and dorsal portions of the sclerotome (S) can be seen. Open arrows (top left) show migration of cells beneath the surface ectoderm. Reproduced from Erickson and Perris (1993) from a figure kindly supplied by Carol Erickson. Reprinted by permission of Academic Press, Inc.

The application of scanning electron microscopy has provided additional details of neural tube and neural fold formation in a number of mammalian embryos. The patterns of neural crest cell migration first documented by in the 1920s by Bartelmez (1922) and Adelmann (1925) and have now been mapped in much greater detail for mice and rats. Cranial neural crest cells begin to migrate while the neural tube is an open neural fold, unlike the situation in amphibians and birds, in which emigration is from a closed neural tube.[4]

In 1979, O'Rahilly and Gardner integrated an analysis of 16- to 37-day-old human embryos from the Carnegie collection with other published reports to identify three populations of cranial neural crest cells (rostral, facial, and postotic) migrating between 20 and 24 days of gestation. Subsequently, in a series of elegant and important studies detailing the development of the human brain (published as a series of papers in the 1980s and as an atlas of developmental stages in 1994), Müller and O'Rahilly documented:

- the earliest appearance of neural crest cells in association with the initial development of the major divisions of the brain at stage 9;

- the absence of neural crest arising from the future forebrain and the presence of prominent populations of mesencephalic and rhombencephalic crest cells at stage 10;
- the migration of neural crest from both open and closed regions of the neural tube at stage 11;
- the maximal extent of the optic neural crest at stage 12, which is also when the caudal neuropore closes and secondary neurulation begins; and
- the emergence of trunk neural crest and placodal development at stage 13, by which stage the neural tube is fully closed.

Utilizing staining with toluidine blue after cetylpyridinium chloride fixation, Nichols (1987) differentiated early migrating neural crest cells from the adjacent neuroepithelium and ectoderm in mouse embryos, and visualized neural crest cells leaving the midbrain and rostral portion of the hindbrain at the 3- to 4-somite stage. By the 5- to 7-somite stage, neural crest cells are moving away from the ectoderm at fore-, mid-, and hindbrain levels, a migration that continues until the 16-somite stage. Migration is facilitated by discontinuities in what was previously a continuous basal lamina underlying the neuroepithelium. These cells do not immediately break through the basement membrane[5] of the neuroepithelium, but rather accumulate above it to lie beneath the basal surface of the ectoderm. Nichols described the following sequence in mice:

- Neural crest cells elongate and reposition their organelles to the basal region of the epithelium.
- Apical cell-to-cell contacts are lost, leading to the development of a stratified epithelium with basal free elongated cells.
- Processes from the basal cells penetrate the basal lamina, which is then degraded or disrupted, allowing the basal cells to break free of the neuroepithelium.
- The apical cells form a new basal lamina.

It has been assumed that escaping neural crest cells create the focal breaks in the basal lamina, perhaps by selective enzymatic degradation, but Nichols' study indicates that this may not be so. Studies on neural crest emigration in the long-tailed macaque, *Macaca fascicularis*, indicate that gaps in the basement membrane, which appear at stages 11 to 12, are repaired by stage 13. N-CAM labeling of migrating hindbrain neural crest cells in this species revealed that they migrate under the ectoderm and into arches 1 to 3 using laminin and collagen type IV of the basement membrane as the substratum. Initial migration of rat neural crest cells is also mediated via laminin and collagen type IV, both of which remain attached to neural crest cells as the cells migrate. The molecules that control migration are discussed more fully later in this chapter.[6]

Extracellular Space

It is important for neural crest cells, especially trunk crest, to have sufficient cell-free extracellular space so they can migrate normally. In comparison with forces exerted by fibroblasts, neural crest cells exert relatively weak traction forces on the substrate, an attribute that allows them to migrate on or through what appear to be relatively weak extracellular matrices. Hyaluronan (hyaluronic acid, an alternating polymer of glucuronic acid and *N*-acetylglucosamine joined by ß1–3 linkage) is involved in the initial separation of neural crest cells from the neural tube and the opening up of spaces through which they migrate. Earlier in development, hyaluronan is involved in elevation and closure of the neural folds, the extracellular matrix around the neural folds being rich in hyaluronan.

Neural crest cells produce proteases such as plasminogen activator that help create the spaces through which they migrate. Indeed, neural crest cells from murine embryos synthesize both urokinase- and tissue-type plasminogen activators from as early as 8.5 days of gestation. A small population of neural crest cells normally fails to migrate but remains attached to the neuroepithelium. These cells do not produce plasminogen activator. Inhibitors of protein kinases can initiate the epithelial to mesenchymal transformation associated with initiation of neural crest cell migration.[7]

Intrinsic or Imposed?

Is specification of the migratory route a property of neural crest cells themselves or of the extracellular environment through which the cells migrate?

Although migratory routes are highly predictable from individual to individual within a given species, pathways of migration can vary from breed to breed, and indeed from species to species. Thus superficial, pigment-forming neural crest cells of the Japanese quail migrate along the epidermal–dermal junction when grafted into embryos of the white leghorn breed of domestic fowl, but migrate below the epidermis when grafted into embryos of the silkie breed. Furthermore, quail neural crest cells do not respond to positional signals within the dorsal trunk feathers of chick embryos, a finding that argues for pr-patterns in the dorsal body wall that precede neural crest invasion. On the other hand, species-specific local cues do exist; melanoblasts from one individual can "read" local patterning signals in feather papillae of another.[8]

When neural crest cells are transplanted to more rostral or caudal regions within the neural folds in order to follow their pathways of migration, we see for the most part, that directionality of migration is not an intrinsic property of neural crest cells. Neural crest cells that are grafted into a different position along the neural tube do not seek out their original path, but rather migrate along paths typical of their new location. For example, rostral midbrain-level neural crest cells, which would normally migrate rostral to the eye, migrate caudally to the eye when they are grafted more caudally along the neural tube.[9]

Interestingly, although migratory patterns of such broad regions of neural crest cells are not intrinsic, specific patterns emerge in cells that arise from a single rostro-caudal level of the neural crest but that are destined to form several different cell types. Thus, in the embryonic chick:

- precursors of dorsal root ganglia remain at the same rostro-caudal level of the neural tube from which they emerge so that only cells adjacent to a single somite contribute to each dorsal root ganglion;
- neurons that make up an individual sympathetic ganglion migrate two somite lengths rostrally or three lengths caudally from their point of emergence, so that neural crest cells adjacent to six somites contribute to a single sympathetic ganglion; while
- melanocyte precursors migrate extensively along the rostro-caudal embryonic axis.[10]

It is tempting to suspect that these specific migratory behaviors reflect some property associated with each determined subpopulation of neural crest cells, rather than variations in the microenvironments that cells encounter as they leave the neural tube. In an especially informative study, Erickson and Goins (1995) labeled cells in vitro with fluoro-gold and grafted them into the neural crest cell migration pathway. Migration of the labeled cells was then followed with high-resolution microscopy. Migration along the dorso-lateral pathway was restricted to cells already specified as melanocytes, other neural crest cells being excluded from that pathway. Migration per se, however, does not evoke differentiation of this subpopulation, which still requires specific environmental signals (such as the growth factor endothelin-3) to differentiate.[11]

In *Xenopus laevis,* the initiation of migration of neural crest cells, onset of segregation of paraxial mesoderm, and initiation of chondrogenesis are all associated with production and deposition of the extracellular matrix protein cytotactin and cytotactin-binding proteoglycan, implicating extracellular matrix control over onset of migration and differentiation. Extracellular matrix products from both *Ambystoma mexicanum* and avian embryos have been adsorbed onto Nuclepore filters and the filters implanted in vivo (Fig. 9.2) or used as substrates on which neural crest cells are cultured. In *A. mexicanum* large proteoglycan complexes produced by the ectoderm are involved in neural crest cell migration. Extracellular matrix associated with the superficial ectoderm preferentially promotes migration of the superficial population of neural crest cells. The basal lamina of the dorsal epidermal ectoderm, along which superficial cells migrate, promotes adherence between neural crest cells and basal lamina so that superficial cells effectively migrate as a sheet. In the pericellular matrix immediately surrounding neural crest cells, enzymes such as galactosyltransferases bind to sugar residues such as N-acetylglucosamine in the extracellular matrix and on basal laminae. Such binding affects cell adhesion and de-adhesion and therefore cell motility, and could control the movement of neural crest cells in vivo.[12]

Ectoderm in the white axolotl mutant produces an inhibitor of neural crest cell migration, differentiation, and survival, which along with intrinsic deficiencies in the neural crest cells and in the extracellular matrix through which they migrate,

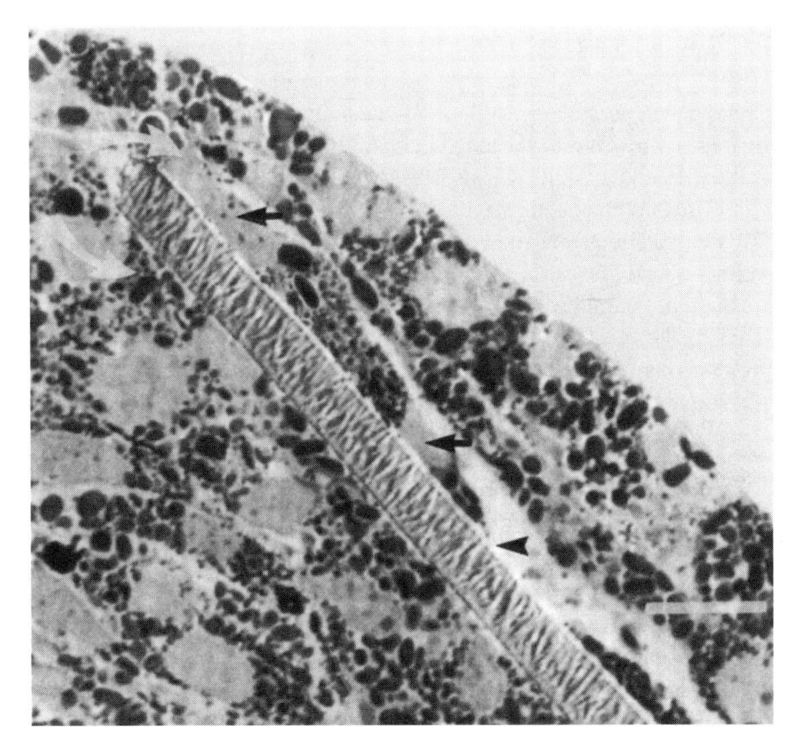

FIGURE 9.2. A Nuclepore filter (arrowhead) coated with fibronectin and implanted beneath the ectoderm of a white mutant axolotl embryo, stimulates migration of neural crest cells (arrows). No migration is seen on the contralateral (control) side (not shown). Bar = 20 μm. Modified from Olsson, Svensson, and Perris (1996).

renders the animals albino. While organization of the ectodermal basement membrane and collagen I and IV are all normal, collagen II and chondroitin-6-sulfate are abnormal in albino axolotls.[13]

Patterns of migration and dispersal of pigment cells in the newts *Taricha rivularis* and *T. torosa* appear to be intrinsic to the cells. As determined both in vitro and following heterospecific grafts of neural crest cells in vivo, *T. rivularis* cells disperse, producing even pigmentation patterns, while *T. torosa* cells aggregate, producing pigmented bands.[14]

In murine embryos (and perhaps in other vertebrates as well), each pigmented stripe is the product of a clone of cells derived from a single melanoblast. This was demonstrated by retroviral marking of single cells in neurulating albino embryos to produce pigmented stripes in otherwise albino animals. Culture of neural crest cells from mutant embryos followed by injection of these cells in vivo demonstrated that the host genotype modified coat pattern, presumably by the type of extracellular-matrix-mediated interactions demonstrated to operate in amphibians.[15]

Extracellular Matrix

As just indicated, neural crest cells use molecular cues from their environment to govern migration. This has been demonstrated by substantial analysis of the extracellular matrix through which neural crest cells migrate, and the migratory behavior of neural crest cells cultured on or in various extracellular matrix products. This extracellular matrix exists as a fibrillar meshwork that is structurally altered immediately before neural crest cells enter it and modified biochemically by the transit of neural crest cells through it. The matrix is rich in the glycosaminoglycans hyaluronan and chondroitin sulfate, but also contains chondroitin sulfate proteoglycan (aggregan, versican), type I and type II collagens, tenascin, laminin, and fibronectin. To summarize the functions of these molecules:

- Aggregan and versican inhibit migration.
- Fibronectin and collagen types I and IV promote migration.
- Collagen types II, V, and IX, which are found in regions from which neural crest cells are excluded, inhibit or deflect migrating cells.[16]

These comments on molecular control of migration apply primarily to trunk neural crest cells. Cranial and trunk cells use different mechanisms of attachment to extracellular matrices. Trunk neural crest cells attach to fibronectin, laminin, and collagen types I and IV. Cranial neural crest cells do not attach to the two collagens although they do attach to basal laminae. Attachment of trunk neural crest cells to fibronectin is Ca^{++}-dependent, attachment of cranial crest cells is not.[17]

Cell surface and extracellular matrix products such as fibronectin, laminin, and entactin play important roles in neural crest cell migration in avian and amphibian embryos, as they do in other situations in which active cell migration is important. Laminin provides a scaffold for migrating cells during primitive streak formation in avian embryos; antibodies that bind laminin alter migration to such an extent that no normal primitive streak forms. Indeed no embryonic axis forms. Analyses of the composition of the cell surface and pericellular matrix of the neural tubes of rat and mouse embryos at the time of initiation of neural crest cell migration (using antibodies against fibronectin, laminin, and entactin) identified these extracellular matrix components in the rat, but did not find any site-specific localization that could be related to cell migration.

Lectins, which are plant proteins that bind to specific sugar residues in membrane glycoproteins of animal cells, have been used to localize D-mannose, D-glucose, and N-acetyl-D-glucosamine, especially in the zone of fusion of the neural folds in the mouse and rat. Some lectins are also mitogenic for animal cells. Developmentally regulated lectins also influence pigment patterns in axolotls; an ectodermal defect and resulting low lectin levels in albinos prevents melanophores from colonizing the skin. Otherwise, melanophores from albino axolotls are capable of normal migration and colonization.[18]

In the following section I use fibronectin to illustrate how extracellular matricial components influence neural crest migration.

Fibronectin

Fibronectin is a complex, large (400,000 Da) glycoprotein with structurally and functionally distinct cell-, collagen-, heparin-, and hyaluronan-binding domains that constitute some two thirds of the molecule. The cell-binding domain is the largest, weighing in at 120,000 Da.

Although glycosaminoglycans modify and perhaps even inhibit migration of neural crest cells, fibronectin is the major extracellular matrix molecule used by neural crest cells to control migration. Fibronectin is found in especially high concentrations in epithelial basement membranes, including those of the superficial dorsal ectoderm and somites along which neural crest cells migrate. The localization of neural crest cells to the anterior half of each somite at the boundary between dermatome (future dermal connective tissue) and myotome (future muscle) as shown in Figure 9.1, may be facilitated because migrating neural crest cells use fibronectin in the somitic basal lamina as a substratum. Migration requires that mesoderm be at a stage of differentiation equivalent to the sclerotome; neural crest cells do not migrate through segmental plate mesoderm.[19]

Fibronectin binds both cells and other molecules. Neural crest cells preferentially bind to fibronectin-coated substrates when presented with a choice of extracellular matrix products. Such direct in vitro evidence for the binding of fibronectin to neural crest cells, coupled with colocalization of fibronectin and migrating neural crest cells in vivo and expression of fibronectin receptors on neural crest cells, argues strongly for a role for fibronectin in neural crest cell guidance in vivo.[20]

As might be expected, neural crest cells attach to the cell-binding domain of fibronectin; this need not have been the case, for fibronectin could, in theory, attach to components of the pericellular matrix via the collagen or hyaluronan-binding domains. Neural crest cells fail to migrate on fibronectin to which an antibody raised against the cell-binding domain has been bound. They also fail to migrate on a synthetic decapeptide (with the amino acid sequence Arg-Gly-Asp-Ser-Pro-Ala-Ser-Ser-Lys-Pro) that contains the recognition site for the cell-binding domain. Latex beads can be coated with cell-, collagen-, or heparin-binding portions of fibronectin and implanted into the pathway taken by migrating avian neural crest cells *in ovo* to assess which portions of fibronectin inhibit bead translocation. Only beads coated with the cell-binding domain failed to translocate, further supporting the role of the cell-binding domain in mediating crest cell migration. Antibodies against fibronectin can also be used as immunoselective agents to isolate subpopulations of neural crest.[21]

Matrix Components Interact

Using a novel bead aggregation assay, Turley and colleagues (1989) demonstrated interactions among matrix extracellular molecules encountered by migrating neural crest cells. For example, fibronectin has multiple binding sites for chondroitin sulfate (binding that is promoted by cations) but does not bind to hyaluronan. Chondroitin sulfate proteoglycan binds to collagen type I and to aggregan but not to fibronectin, collagen type IV, or laminin, while integrins (which regulate cell adhesion and migration) bind to fibronectin and laminin. Integrin antisense oligonucleotides block integrin attachment to fibronectin or laminin and block migration of neural crest cells. Integrins play other roles; apoptosis of cranial neural crest cells is increased in $\alpha 5$-integrin-null mice.[22]

These interactions between neural crest cells and matrix molecules are regulated in time as well as spatially. In rat embryos, for example, chondroitin sulfate proteoglycans, which retard migration, are at low levels on day 9 but higher levels on day 10, and so are associated with migrating but not postmigratory cells. Treatment of the extracellular sheath that surrounds the avian notochord with chondroitinase removes chondroitin sulfate proteoglycan and permits neural crest cells to invade this matrix, into which they (unlike sclerotomal mesenchymal cells) normally cannot penetrate. The inhibition of neural crest cell migration by chondroitin sulfate proteoglycans is mediated by the hyaluronan-binding region through cell-surface interactions.[23]

Neural Crest Cells Contribute to Extracellular Matrices

As already noted, neural crest cells produce proteases and plasminogen activator to create a path through which they migrate. At least some matrix components are synthesized by neural crest cells themselves. Avian neural crest cells synthesize tenascin, which they appear to use to create an environment through which to migrate; tenascin binds to chondroitin sulfate proteoglycan but not to fibronectin. Again in avian embryos, ascorbic acid released from migrating trunk neural crest cells in levels as high as 1.5 μg/mg protein increases collagen synthesis by somites or muscles between 2.5 and 6 times.[24]

Novel classes of adhesion (and de-adhesion) molecules have also been localized in neural crest cells at different phases of migration. ADAM 13, a member of the ADAM family of membrane-anchoring proteins (named from **a d**isintegrin **a**nd **m**etalloprotease domain, the former serving an adhesion, the latter an antiadhesion function), has been cloned and mRNA and protein localized by Alfandari et al. (1997) to cranial neural crest and somitic mesoderm during *Xenopus* embryogenesis (Fig. 9.3).

In summary, extracellular glycosaminoglycans and proteoglycans, pericellular galactosyltransferases, but especially extracellular fibronectin, are the major candidates controlling neural crest cell migration through inhibition or promotion of cell-to-cell and cell-to-matrix adhesion. An extracellular environment with adhesive molecules such as hyaluronan and fibronectin on the migration pathway and nonadhesive molecules such as chondroitin sulfate on the nonmigratory pathway, seems to be sufficient to determine where neural crest cells will migrate. Direc-

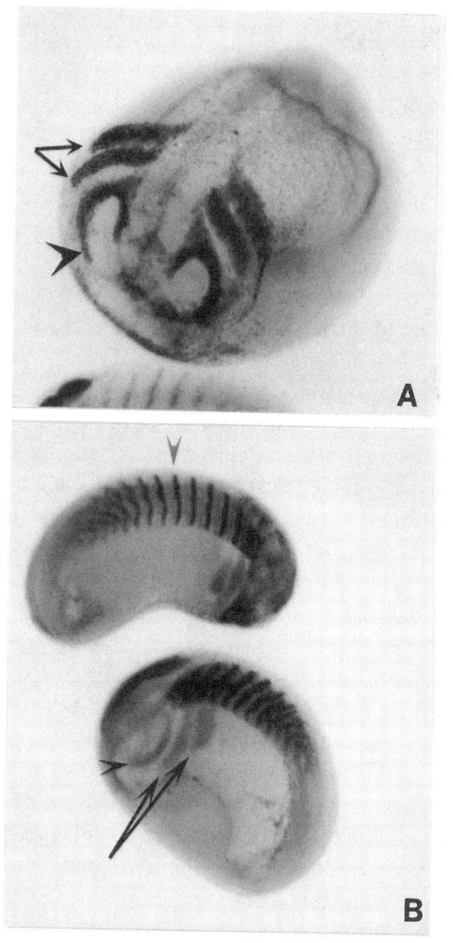

FIGURE 9.3. (A) ADAM 13 mRNA is expressed in streams of migrating neural crest cells (arrows) in embryos of *Xenopus laevis*. The most rostral stream in this stage-20 embryo subdivides into pre- and postoptic streams around the optic vesicle (arrow head). (B) Lateral views of an early tailbud (stage 22) embryo (top) and a late neurula (below) to show expression of ADAM 13 protein in somitic mesoderm (arrowhead in the stage-22 embryo) and in migrating neural crest cells in the late neurula (arrows). Reproduced with permission from Alfandari et al. (1997) from a figure kindly supplied by Douglas DeSimone.

tionality of migration is less well understood. It appears not to come from these matrix components but may involve processes such as contact guidance, contact inhibition, chemotaxis (orientation along chemical gradients), haptotaxis (movement against an adhesive gradient), and galvanotaxis (movement guided by electrical or electromagnetic fields).[25]

Cessation of Migration

There should be a brief comment about how neural crest cells stop migrating and settle down at their final sites. Neural crest cells could merely "run into" barriers and accumulate. Such physical barriers—basal laminae, blood vessels, or other cells (such as mesodermally derived mesenchyme, somitomeres in the head, somites in the trunk, mesenchyme surrounding the notochord)—could direct migration of neural crest cells and/or govern cessation of migration and cell accumulation. The embryonic locations of neural crest cells, however, are much too precise to be explained by such a sloppy mechanism, unless those physical barriers also carried specific molecular information.

For example, neural crest cells grafted into the lumen of the neural tube cannot penetrate the basal lamina and so accumulate against it. Such membranes could act as nonspecific barriers that direct cells passively, or they could function with greater specificity because there are specific biochemical components in their cell membranes or within their pericellular matrices to which migrating populations or subpopulations of neural crest cells respond selectively. Blood vessels in developing Japanese quail are associated with a fibronectin-rich extracellular matrix that provides the substratum for migration. Other apparent "physical" barriers may have similar molecular bases. Indeed, cessation of migration and localized accumulation of specific subpopulations of neural crest cells is controlled in as complex a manner as is initiation and directionality of migration, sometimes by factors intrinsic to neural crest cells, sometimes by the extracellular environments they encounter.[26]

One population of sensory neurons, thought to be of neural crest origin, defies the normal patterns of neural crest migration and remains within the midbrain. These, the largest sensory neurons in the central nervous system, are the cells that form the mesencephalic nucleus of the trigeminal nerve; they transmit mechanosensory input from muscles of the jaws and from the extraocular muscles. Evidence for their neural crest origin is equivocal, with different experimental approaches (extirpation, transplantation) in different vertebrates (birds, frogs) giving conflicting results; see M. Jacobson (1991) for a discussion. Whether or not they are of neural crest origin, these nuclei, their relationship to the neural crest and placodal cells that form the trigeminal ganglion, and their pattern of migration all deserve further study.

The hyaluronan-, chondroitin sulfate-, and aggregan-rich extracellular matrix encountered by deeply migrating neural crest cells inhibits the formation of cell-to-cell attachments so that neural crest cells migrate as single cells. Consequently, chondroitin sulfate and other glycoconjugates provide barriers to crest cell migration. Microinjection of chondroitin sulfate or of a xyloside into rhombomeres of the chick hindbrain at H.H. stage 9 (immediately pre-migration; Fig. 9.4) inhibits neural crest cell migration or results in cells moving into the neural epithelium instead of migrating laterally. Injection of retinoic acid has similar effects (Chapter 12). Extracellular matrix also effectively directs speed of migra-

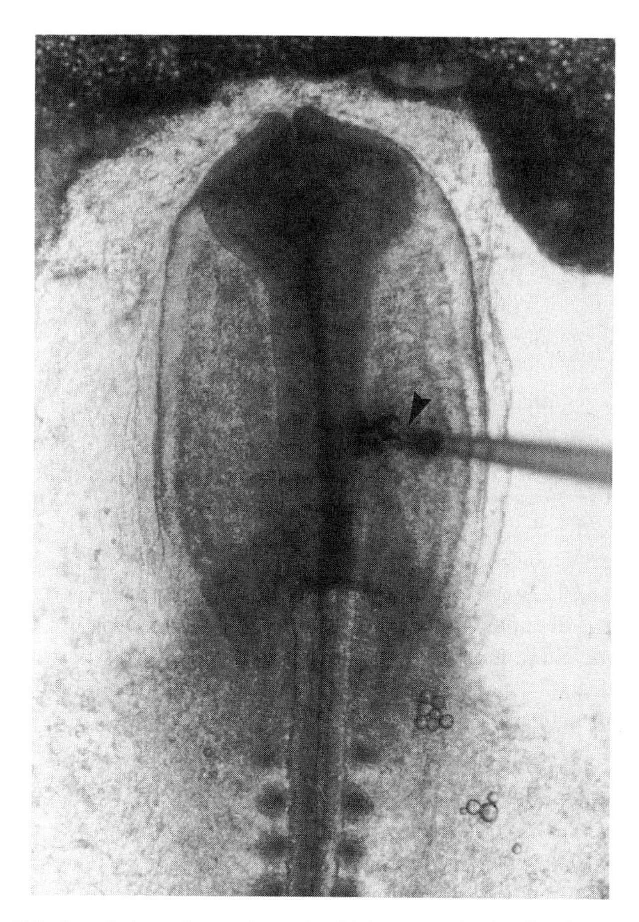

FIGURE 9.4. This dorsal view of an embryonic chick shows the tip of a pipette (arrowhead) through which chondroitin sulfate was delivered to the rostral rhombencephalon before onset of neural crest cell migration. Reproduced from Moro-Balbás et al. (1998) from a figure kindly supplied by Jose Moro-Balbás. Reprinted by permission of the author and the Servicio Editorial of the International Journal of Developmental Biology.

tion by decreasing adhesion of neural crest cells to substances known to enhance neural crest cell migration in vitro. By placing neural crest cells into the interstices of hydrated collagen gels containing extracellular matrix components in various combinations and concentrations, Tucker and Erickson (1984) elegantly demonstrated that chondroitin sulfate and chondroitin sulfate proteoglycan (aggrecan) increase, but collagen decreases migration rate, while hyaluronan inhibits migration.[27]

Avian and mammalian neural crest cells (and axons of cranial nerves) migrate only through the dorso-rostral sclerotome, specifically avoiding ventro-caudal sclerotome and peri-notochordal mesenchyme. Anterior sclerotome is rich in tenascin and cytotactin, molecules that facilitate intrasclerotomal migration. The notochord produces a trypsin- and chondroitinase-labile molecule that inhibits neural crest cell migration and thus prevents neural crest cells from colonizing the peri-notochordal sheath. This is true even if the notochord is implanted ectopically in situ or if notochord and neural tube are extirpated or inverted dorso-ventrally and reimplanted, in which case neural crest emigrates from the new ventral (non-notochord-associated) surface of the neural tube. Aggrecan is one molecule whose removal with hyaluronidase or chondroitinase permits neural crest cells to invade the peri-notochordal matrix.[28]

Other subpopulations—e.g., those forming the tissues of the mandibular arch of the embryonic chick or the trabecular and visceral arch cartilages in amphibians—possess regional specificity before leaving the neural tube. Thus, even when forced by an ingenious experimenter (Drew Noden) to migrate into the second or third visceral arches, these cells still form first arch structures, as evidenced by development of an extra external auditory meatus and ectopic mandibular skeleton in the hyoid arch. The localization of such cells during normal development is controlled very precisely to avoid abnormal development. Indeed, as discussed in chapters 11 and 12, major craniofacial defects can occur when migration of neural crest cells is disrupted.[29]

The microenvironment of the final site occupied by neural crest cells plays an important role in determining where and when they cease migrating, accumulate, and differentiate. Five examples are outlined briefly

- Versican (a large chondroitin sulfate proteoglycan with a hyaluronan-binding domain, a lectin and two epidermal growth factor [EGF] repeats) is expressed selectively in tissues that act as barriers to neural crest cell migration or axonal outgrowth and is absent from tissues that are invaded by neural crest cells or axons. Thus, versican is found in caudal sclerotome, in the extracellular matrix surrounding the notochord, and beneath ectodermal cells. Versican inhibits migration because it inhibits molecules such as fibronectin, laminin, and type I collagen that are required for migration. (Tenascin and cytotactin, which facilitate migration, are found in the cranial sclerotome.) Expression of versican is reduced in mutants such as the white mutant axolotl in which neural crest cell migration is abnormal.
- *Lethal-spotting* (*Ls*/*Ls*) mutant mice lack ganglia in the terminal two millimeters of the bowel. Mesenchyme of the presumptive aganglionic bowel prevents migrating neural crest cells from colonizing the bowel, resulting in an abnormal, aganglionic bowel segment. A close study of the basal lamina of the gut of *Ls*/*Ls* mutant embryos reveals abnormalities in laminin, collagen type IV, and proteoglycans from day 11 onwards. Excess matrix components lead to formation of an abnormally thick basal lamina, which blocks colonization, but only in the future aganglionic or hypoganglionic regions of the bowel.
- When back-transplanted between somites and neural tube, neural crest cells from the developing quail bowel migrate again and colonize the neural tube, spinal ganglia, peripheral and sympathetic nervous systems, and adrenal

glands, but not the bowel. Timing of entry to the gut may be critical; back-transplantation of the embryonic gut wall adjacent to the neural tube enhances proliferation of neural tube but not ganglion cells, whose ability to respond is temporally restricted.

- Abnormal basal laminae around the eye rudiment in the *Eyeless* mutant of the Mexican salamander disrupts neural crest cell migration.
- Neural crest cells normally enter the cranial half of the occipital somites, where they form axons but not dorsal root ganglia. If grafted to the trunk, however, occipital-level neural crest cells take a different pathway of migration and subsequently differentiate into dorsal root ganglia; in unoperated embryos, these cells are prevented from doing so by the environment at their usual site.[30]

Migration and Patterning

Migrating neural crest cells influence the patterning of other cell types, such as the neuromasts deposited by lateral line primordia as they migrate along the body. Removal of segments of trunk neural crest from the axolotl, *Ambystoma mexicanum*, demonstrates that while neuromast precursors can migrate through neural-crest-free zones, they fail to deposit neuromast precursors in those regions, resuming deposition of precursors once in a region populated by cells derived from neural crest. The lateral line sensory system also influences patterning of melanocytes in larval salamanders; melanophore-free zones are maintained by an influence(s) from lateral line. The first studies on *Hox* gene expression in axolotl lateral lines and neuromasts (both of which express *Msx-2* and *Dlx-3*) appeared in 1997, providing a preview of molecular markers that could be used in future studies of effects of neural crest on neuromast origins, differentiation, and morphogenesis.[31]

Type II Collagen

Factors controlling localization and accumulation of specific subpopulations of neural crest cells were investigated by Peter Thorogood, who demonstrated that neural crest cells destined to form the cartilaginous cranium of the embryonic chick accumulate specifically along the neuroepithelium of the developing brain at sites rich in type II collagen. Subsequent epithelial–mesenchymal interactions at these sites elicit chondrogenesis from the accumulated mesenchymal cells. Based on the pattern in chick embryos, Thorogood developed a "fly-paper model" and suggested that morphogenesis of the developing chondrocranium may be the result of a combination of neural crest cell migration and neuroepithelial folding, the latter determining where type II collagen will be deposited and therefore where neural crest cells will become trapped. In murine embryos, type II collagen is found near the basal lamina on 10–15 nm fibrils around the optic and otic vesicle, ventral brain, pharyngeal endoderm, and visceral arch ectoderm. In guinea pig embryos, type II is found in developing ear cartilages but also in many other epithelial structures of the ear. In the zebrafish, type II is found around the notochord, otic capsule, visceral arches,and head mesenchyme. Clearly, type II collagen, once thought to be the exclusive product of cartilage, is much more ubiquitous; in *Xenopus laevis* embryos it is present on the ventral brain, sensory vesicles, and notochord. As type II collagen is expressed after neural crest cell migration (expression is first seen at stage 21 and is most intense between stages 33 and 36), it appears not to be involved in migration or cessation of migration in this highly derived amphibian.[32]

10

Mechanisms of Differentiation

Neural crest cells do not self-differentiate. Instead they diversify into particular cell types following interactions with other cells, extracellular matrices, growth factors, and/or hormones. These differentiative signals may be encountered in the premigratory environment in the neural tube, while neural crest cells are migrating, or after they arrive at their final site. I have already touched on how pharyngeal endoderm, either alone or with dorsal mesoderm and/or ectoderm, elicits chondrogenesis from visceral arch mesenchyme in amphibians and possibly also in lampreys (chapters 4 and 5), and how oral epithelium interacts with dental mesenchyme to initiate odontoblast differentiation and deposition of dentine during tooth development in amphibians and mammals (Chapter 8). When do neural crest cells become committed for particular cell fates? In this chapter I consider whether either pre- or postmigratory cranial or trunk neural crest cells are comprised of subpopulations of cells, and whether commitment occurs early during embryonic development or only at the time of interaction(s) with the environmental cues that evoke differentiation. I then provide overviews of the evidence supporting inductive specification of neural crest cells, discuss inductive interactions and epigenetic cascades in craniofacial development, and discuss neural crest cells as inhibitors of inductions involving non-neural-crest cells.[1]

Subpopulations of Neural Crest Cells

Are neural crest cells:

- a homogenous multipotent population that is selected for or switched into particular pathways of differentiation; or
- a heterogeneous assembly of subpopulations, each behaving as lineage-restricted cells with perhaps only one or at the most two or three possible cell fates?[2]

We know that the entire neural crest is not a single homogenous population of cells. Cranial neural crest cells have different differentiative properties from cardiac and trunk neural crest cells. Indeed, cranial, trunk, and cardiac crests are themselves heterogeneous. This is seen in the regionalized ability of amphibian and avian cranial neural crest cells to produce specific elements of the craniofacial skeleton or of trunk neural crest to form different neuronal derivatives. It is seen in the segregation of cells for particular skeletal tissues such as cartilage, bone and teeth. It is also seen in the segregation of particular cell lines (such as neuronal cells) within premigratory neural crest (as evidenced using quail/chick chimeras and cell culture techniques and as discussed in the next section) or during and after neural crest cell migration (as demonstrated using monoclonal antibodies against specific neural crest cell types and discussed in the subsequent section).[3]

To state that the neural crest consists of subpopulations is not to say that those subpopulations can only express one possible cell fate, i.e., that they are unipotential. Identification of a cell lineage or cell population is not a demonstration of a single state of determination for that lineage. Of course, cells only ever do express one differentiative phenotype, but many such subpopulations are at least bipotential and may remain bipotential even after differentiating along one pathway.[4]

Dedifferentiation

Can an alternate cell fate be expressed after initial differentiation along one pathway? Such an event would require dedifferentiation of one cell type and subsequent redifferentiation into another, or that the two differentiated cells shared part of a common pathway from a shared stem cell. The phorbol ester 12-0-tetradecanoylphorbol-13-acetate transforms neuronal cells of the dorsal root ganglion into pigmented cells, a transformation consistent with at least some neural crest cells being able to dedifferentiate and then redifferentiate into a second cell type. Exposure of dorsal root ganglia to growth factors also produces changes in cell fate. Exposure to bFGF (but not to PDGF, TGF-α, TGF-ß1, or NGF) causes some 20% of DRG cells to transform into pigment cells, a transformation that is blocked by TGF-ß1. Consequently, cells that have dedifferentiated can redifferentiate along another pathway. Indeed, some current studies regard the neural crest as comprised of multipotential, self-renewing populations of stem cells.[5]

Restricted Premigratory Populations

A classic way to demonstrate bipotentiality is to determine whether clonal cell cultures can differentiate into more than one cell type. In 1975, Cohen and Konigsberg established clonal cultures from trunk neural crest cells of the Japanese quail. A small proportion of the clones differentiated into two distinct cell types: pigmented cells (distinguished by their production of melanin) and adrenergic neurons (distinguished by their production of catecholamines). The cells that differentiate into adrenergic neurons *in ovo* are those that migrate across the neural tube and somites, from which they receive a diffusible signal permitting adrenergic neuronal differentiation. The differentiation of adrenergic neurons both *in ovo*, and from clones shown from in vitro studies to consist of pigmented and undifferentiated cells, demonstrates the bipotentiality of these neural crest cells. Adrenergic neuronal differentiation can also be elicited from cloned neural crest cells not expressing a neuronal phenotype if they are exposed to extracellular products secreted by somitic cells or grafted in association with somites.[6]

Boisseau and Simonneau (1989) used the neural tube culture technique developed by Cohen and Konigsberg to establish murine neural tubes in a chemically defined medium and to follow the emigration of neural crest cells and neuronal differentiation in vitro. Further clonal analyses undertaken by Ito and Sieber-Blum in the early 1990s establish heterogeneity within a diversity of neural crest cells in both murine and avian embryos:

- Murine neural crest consists of heterogeneous populations of pluripotential and restricted cells at the onset of migration.
- Quail cardiac neural crest contains five clonal cell lines: (1) pigment cells, (2) five types of mixed cells that can form pigment or other cell types, (3) unpigmented clones that form a variety of cell types but not pigment cells, (4) a clonal line that produces smooth muscle, and (5) a line that forms chondrocytes and sensory neurons.
- Posterior visceral arches of avian embryos contain four clones of neural crest cells (see Fig. 7.4).[7]

Substantial evidence, primarily from studies of avian embryos, supports the existence of restricted lineages capable of forming disparate cell types within premigratory neural crest. Thus in 9- to 13-somite-stage quail embryos, the cranial neural crest consists of precursors capable of producing neurons, glia, and chondrocytes and a small number of precursors from which neurons, glia, chondrocytes, and pigment cells arise (Fig. 10.1). Injection of the proto-oncogene *V-myc* into trunk neural crest of 10.5-day-old rat embryos immortalizes a clonal cell line that then gives rise to glial progenitors and sympathoadrenal cells.[8]

A classic series of studies, performed by Bronner-Fraser and Fraser in the late 1980s, consisted of injecting single cells of the trunk neural crest of avian embryos with lysinated rhodamine dextran to monitor migration and cell fate. Single trunk neural crest cells gave rise to sensory neurons, pigment cells, ganglionic support cells (Schwann cells), adrenomedullary cells (sympathetic neurons), and neurons of the central nervous system (Fig. 10.1). These results are fascinating. They reveal multipotentiality of single neural crest cells, and that neural crest and central nervous system neurons share a common lineage. The latter also speaks directly to the developmental origin of the neural crest and to its close connection to neural ectoderm, discussed in Chapter 2. As discussed later in the present chapter, interactions with environmental factors are required to select, elicit, and/or direct differentiation of these multipotential cells.[9]

Clearly, subpopulations exist in the premigratory neural crest. How does such heterogeneity arise? This problem may be part and parcel of understanding how the neural crest itself arises. Early regionalization may be a component of the primary inductive events that segregate presumptive ectoderm into neural and epidermal ectoderm and neural crest and that then regionalize that neural ectoderm into fore-, mid-, and hindbrain and spinal cord. In this connection the two studies by Couly and Le Douarin (1985, 1987), which are discussed in Chapter 7, may be

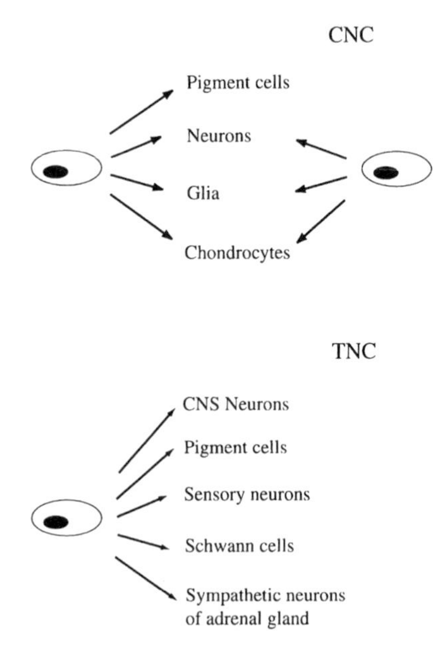

Figure 10.1. Premigratory cranial neural crest (CNC) of quail embryos consists of precursors that can produce neurons, glia and chondrocytes and a small number of precursors that give rise to pigment cells, neurons, glia, and chondrocytes. Premigratory trunk neural crest (TNC) of avian embryos contains individual precursor cells which give rise to neurons of the central nervous system (CNS), pigment cells, sensory neurons, Schwann cells, and sympathetic neurons. See text for details.

especially important. They demonstrated that the superficial ectoderm of the roof of the mouth, olfactory cavities, head and face in avian embryos arise from the neural folds and neural plate of the prosencephalon, i.e., from cells immediately rostral to the most rostral neural crest (see Fig. 2.12). Neural ectoderm, neural crest, and facial ectoderm may be patterned during primary embryonic induction. As discussed in Chapter 2, a code of *Hox* genes establishes these patterns.

Restriction During Migration

The potentiality of neural crest cells could be restricted during their migration. Monoclonal antibodies raised by Heath, Wild, and Thorogood (1992) against premigratory avian neural crest cells, however, reveal potential heterogeneity, but only after cells have been cultured for some 15 hours, at which point subpopulations diverge from a less heterogeneous premigratory population. Several possible mechanisms can be entertained:

- Neural crest cells might be restricted by some intrinsic mechanism according to the time they emigrate from the neural tube.

- Restriction might occur during migration because cells with particular potentials take particular migration routes.
- Restriction might occur during migration because cells encounter and respond to different environmental cues along different pathways.

Perhaps not surprisingly, neural crest cells at different levels of the neural tube and/or from different species use a variety of mechanisms to limit their potentialities for differentiation. In the sections below, I briefly consider such differences as seen in trunk and cranial neural crest cells.

Trunk Neural Crest

Avian trunk neural crest cells that follow a ventral pathway of migration become neuronal and contribute to the dorsal root ganglia. Cells that take a more dorso-lateral pathway become pigment cells. Late migrating cells preferentially take the dorso-lateral pathway and have a more restricted set of developmental options than those cells that migrate early and can take either the ventral or the dorso-lateral pathway. The restriction of early migrating cells as ganglionic also varies along the rostro-caudal axis of the trunk neural crest. Interestingly, lampreys, which lack sympathetic ganglia, lack a ventrally migrating population of neural crest cells.[10]

A monoclonal antibody designated E/C 8 and directed against neural-crest-derived dorsal root ganglia of the embryonic chick identifies a subpopulation of postmigratory cells that form neurons but not pigment cells. Separate lineages of sensory and autonomic neurons are established early in neural crest cell migration from cells that are pluripotential and can form sensory and autonomic neurons or melanocytes. Clonal analysis of cells derived from dorsal root or sympathetic ganglia of quail embryos demonstrates both (a) tripotential cells capable of forming pigment cells, sensory neurons, and adrenal medullary cells, and (b) cells that are bipotential, either for sensory neuronal/adrenal medullary cells, or for pigment cells/sensory neurons (Fig. 10.2). At least some migrating trunk neural crest cells in avian embryos are bipotential when they migrate. Fraser and Bronner-Fraser (1991) labeled migrating cells (previously, only cells within the neural tube had been labeled), and found that some cells formed sensory and sympathetic neurons and others formed Schwann and non-neuronal cells.[11]

Cranial Neural Crest

What applies to trunk may not apply to cranial or cardiac neural crest cells, alerting us to be cautious when extrapolating mechanisms from one region of the neural crest to another. Early- and late-migrating cranial neural crest cells of chick embryos appear to have equivalent developmental potential; exchange of early- and late-migrating cell populations does not produce the deficiencies that would

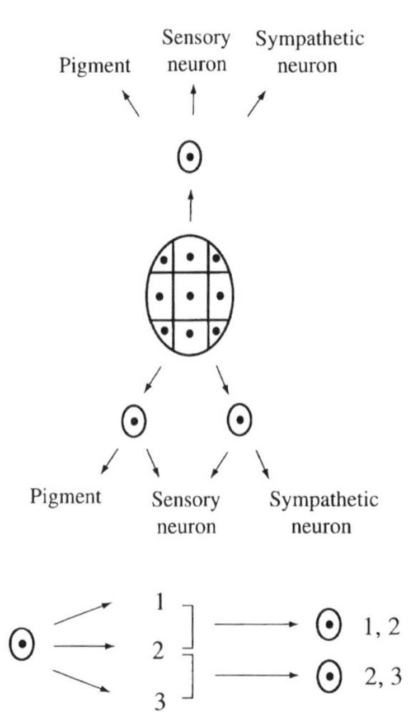

FIGURE 10.2. Quail ganglia (center) contain cells capable of forming pigment cells, sensory and sympathetic neurons (top), and bipotential cells that can form either pigment cells or sensory neurons, or sensory or sympathetic neurons. These bi- and tripotential cell lineages are shown diagrammatically in the flow chart at the bottom. See text for further details.

be expected if cell fate were fixed early. Early-migrating cells are defined as those migrating from embryos with 8 pairs of somites, late-migrating cells those from embryos with 12 pairs of somites. Most late-migrating cells migrate dorsally and therefore normally contribute less to skeletal tissues than do early-migrating cells. Interpretations for these results are that early-migrating cells migrate more medially and/or that they occupy spaces that late-migrating cells fail to fill.[12]

Differentiation of Bipotential Cells: A Role for Growth Factors

Extracellular matrix modulates the differentiation of bipotential cells; trunk neural crest cells differentiate into pigment cells if grown on plastic but into catecholamine-containing adrenergic neurons if cultured in media conditioned by somitic cells. During normal development of the axolotl, components of the extracellular matrix along the dorsolateral migratory route elicit differentiation of pigment cells, specifically melanophores and xanthophores, which, in *Xenopus*, can arise from separate lineages or from a common cell lineage (as determined by clonal cell culture). One such factor is melanocyte-stimulating hormone, which regulates melanocyte differentiation in a dose-dependent way by acting through a cAMP-mediated pathway.[13]

Even if specified as melanocytes before migration, neural crest cells labeled in vitro and grafted back into the migration pathway still require specific environmental signals, such as the growth factor endothelin-3, to promote differentiation. Such restriction is reflected in the decline in the proportion of pluripotential neural crest cells in the skin as embryos age; the fate of pluripotential cells is determined late and in response to local environmental signals. In the Japanese quail, for example, 20% of colonies established from the skin of embryos of H.H. stage 21 form mixed populations of cells (melanocytes, sympathoadrenal cells, sensory neurons) but only melanocyte colonies arise from cells isolated from embryos of H.H. stage 30. Sieber-Blum (1990) has argued from her studies on neuronal and melanocytic differentiation that two classes of signals are required: positive signals for commitment of cell fate and negative signals to restrict fate.[14]

Growth factors play important roles in selecting the pathways of differentiation expressed by neural crest cells, as discussed later in the chapter for the skeletogenic neural crest. Indeed, TGF-ß is distributed in neural crest and other mesenchymal derivatives in the mouse in positions and at times when epithelial–mesenchymal interactions known to play a role in initiation of differentiation are taking place. Neural crest cells themselves are sources of some of the growth factors. Cranial neural crest cells synthesize and secrete a latent TGF-ß that is activated by proteolysis of neural crest cells. Hormones also enhance and stabilize chosen pathways of differentiation.[15]

The growth factor neurotrophin-3 (NT-3), produced in the central nervous system, is present in the neural tube before neural crest cell migration. NT-3 is mitogenic for neural crest cells and influences survival and/or differentiation of postmitotic neuronal precursors. Brain-derived neurotrophic factor (BDNF) and nerve growth factor (NGF) promote the differentiation of sensory neurons from pluripotental neural crest cells maintained in clonal culture. These pluripotential cells, which can give rise to sensory or autonomic neurons or melanocytes, are also sensitive to other signals such as neurotransmitters: norepinephrine uptake inhibitors such as lidocaine and chlorpromazine inhibit expression of the adrenergic phenotype, implicating norepinephrine as a signaling molecule in adrenergic neurogenesis.[16]

Other switches confirm both the bipotentiality of neural crest cells and the role of environmental products in switching cells from one pathway into another. They include a switch:

- from adrenergic to cholinergic neuronal differentiation in response to a soluble factor released by heart cells or as a consequence of grafting neural crest cells to different locations along the neural axis, thereby exposing cells that would normally form adrenergic neurons to environments that switch them into cholinergic neurogenesis;

- from medullary cells of the adrenal gland to sympathetic neurons in response to NGF;
- of mammalian neural crest cells to glia, smooth muscle, or autonomic neurons under the influence of glial growth factor (GGF), TGF-ß, and BMP-2/4, respectively.
- of avian periocular mesenchymal cells from chondrogenesis to neuronal differentiation if associated with embryonic hind gut, a tissue that they would normally not "see" *in ovo*, but to which they can respond; and
- of periosteal cells on avian membrane bones from osteogenesis to chondrogenesis in such situations as articulations, where ligaments or muscles insert or attach, or in fracture repair, where mechanical conditions favor cartilage formation by up- or down-regulating N-CAM.[17]

These examples show that neural crest cells can express more than one cell fate in vivo and can be "made" to express cell fates in vitro that they normally do not express in vivo. Potential for differentiation is greater than that normally expressed. Quite commonly, the signal for differentiation is provided by epithelial cells encountered by neural crest cells before, during, or after their migration. One example is the differentiation of cartilage.

Epithelial Evocation of Cartilage

Hörstadius and Sellman took fate-mapping of the neural crest as far as available techniques allowed them in the mid-1940s. They also took the mechanistic approach characteristic of Hörstadius' work and investigated factors controlling migration of neural crest cells. *En passant*, they obtained presumptive evidence for inductive tissue interactions required for the initiation of various head cartilages, including chondrification of the otic capsule (a cartilage that develops from head mesoderm under the influence of the otic vesicle) and differentiation of neural-crest-derived visceral arch cartilages under the influence of pharyngeal endoderm. They extended Raven's studies on differences between cranial and trunk neural crest by demonstrating that cranial neural crest grafted into the trunk fails to chondrify unless pharyngeal endoderm is included in the graft. Gill-like openings and associated visceral cartilages develop in such composite grafts as organized site-specific structures are formed. Hörstadius concluded "...that the crest-cells in the neural ridges are not yet determined to form cartilage, their potency in this respect has to be achieved" (1950, p. 66). Important conclusions from this and other early studies include:

- Pharyngeal endoderm, over which amphibian neural crest cells migrate as they populate the visceral arches, is necessary for mesenchyme to differentiate into visceral arch cartilage.
- Ectoderm cannot substitute for pharyngeal endoderm, indicating specificity of the inductive component of the interaction.

- The epithelial otic vesicle is necessary for differentiation of the mesodermally derived otic capsule, confirming the pioneering studies of Lewis, Luther, and Kaan on *Ambystoma punctatum*, *Rana sylvatica*, and *R. esculenta*, respectively. This is not a species-specific interaction: frog otic vesicle will induce salamander otic capsule formation.
- Holtfreter demonstrated that the epithelial otic vesicle could not induce neural-crest-derived mesenchyme to chondrify, indicating specificity of mesenchymal ability to respond to induction and that epithelial ectoderm does not play a role in initiating otic capsular cartilages.[18]

Neural crest cells neither preferentially migrate toward pharyngeal endoderm in vitro nor differentiate as chondroblasts when neural crest and pharyngeal endoderm are placed near but not in contact with one another. Evidently, pharyngeal endoderm does not exert its influence by a diffusible factor; direct contact with neural crest cells is required to initiate chondrogenesis. Once chondrogenesis is initiated, however, neural crest cells can transmit the ability to chondrify to other neural crest cells over distances of some 300 μm, apparently without direct cell-to-cell contact. As discussed in Chapter 2, such a self-generating transfer of cell fate, known as homoiogenetic induction, is also responsible for the spread of neural induction. This epigenetic view of differentiation of cartilage that develops from neural crest cells has been repeatedly confirmed and extended to other neural crest cell types and has dominated subsequent work on the neural crest.[19]

We now have evidence for inductive interaction between pharyngeal endoderm and visceral-arch-destined mesenchyme for more species of amphibians than any other vertebrate group. Much of this evidence was obtained using hanging-drop cultures of tissues from *Triturus alpestris*, *Pleurodeles waltl*, *A. mexicanum*, *A. maculatum*, *Discoglossus pictus*, and *Xenopus laevis*. The story of cartilage induction is more complex than a simple pharyngeal endodermal induction, however. While only pharyngeal endoderm is inductive in *T. alpestris*, both pharyngeal endoderm and stomodaeal ectoderm induce in *A. maculatum* and pharyngeal endoderm and dorsal mesoderm induce in *P. waltl*. For birds and mammals, experimental evidence is available for the domestic chick, Japanese quail, and common mouse. For fish and reptiles, evidence for inductive interactions is either indirect or extrapolated from other vertebrate classes.[20]

Mediation of Differentiation by the Extracellular Matrix

The finding that visceral arch chondrogenesis requires interaction between amphibian pharyngeal endoderm and neural crest mesenchyme has been augmented by a considerable body of knowledge demonstrating that all craniofacial cartilages and bones begin to differentiate (i.e., chondroblasts or osteoblasts first appear) as a consequence of one or more epithelial–mesenchymal interactions. We have perhaps the greatest knowledge for the embryonic chick, for which we can identify the epithelia that evoke the differentiation of all the major cartilages

TABLE 10.1. Epithelial-mesenchymal interactions that initiate the differentiation of cartilages and bones of the avian craniofacial skeleton.[a]

Skeletal element	Epithelium	Timing of the interaction (days of incubation)
Cartilages		
Meckel's	dorsal cranial ectoderm	1.5
scleral	pigmented retinal epithelium	2 to 3
otic capsule	otic vesicle	3 to 5.5
Bones		
basisphenoid	rhombencephalon, notochord	2 to 2.5
parasphenoid	notochord	2 to 2.5
squamosal	mesencephalon	2 to 2.5
occipital	rhombencephalon	2 to 2.5
parietal	mesencephalon, rhombencephalon	2 to 2.5
frontal	prosencephalon, mesencephalon	2 to 2.5
	cranial ectoderm	3.5 to 7
maxilla	maxillary	3 to 4
mandible	mandibular	3 to 4.5
palate	palatal	? to 5
scleral ossicles	scleral	7 to 10

a. The interactions are arranged in the temporal sequence in which they occur in the embryo. Note that two epithelia are involved in the differentiation of the basisphenoid and that two separate epithelial-mesenchymal interactions are involved in the differentiation of the frontal. See Hall (1987a) for details and literature.

and bones of the craniofacial skeleton (Table 10.1). As also indicated in Table 10.1, the timing of each of these interactions is tightly regulated.[21]

Epithelia transmit differentiative signals to adjacent mesenchymal cells by one of three mechanisms:

- The epithelium produces a diffusible ion or molecule that acts upon mesenchymal cells that may be as much as 300 μm away. Such interactions are classified as long-range, diffusion-mediated epithelial–mesenchymal interactions (Fig. 10.3a).
- The epithelium—perhaps even mesenchyme in some situations[22]—deposits a product(s) into the epithelial basal lamina or basement membrane. To encounter this matrix product mesenchymal cells must lie within 20 to 40 μm of the basal lamina. The matrix product could be a diffusible ion or a molecule such as a growth factor bound to the basal lamina. Such interactions are known as

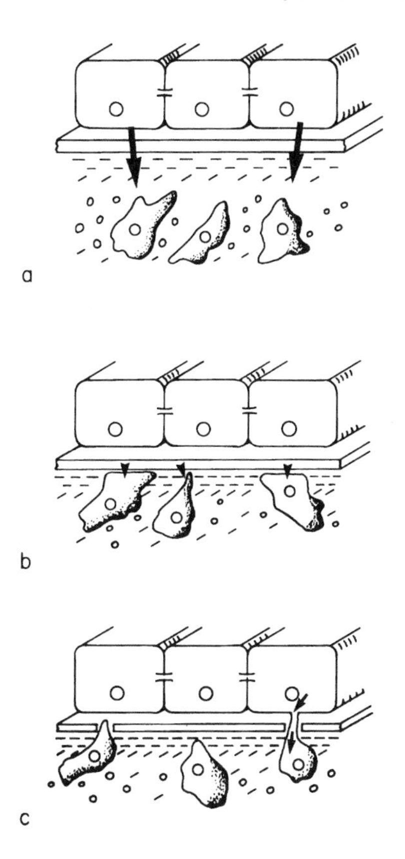

FIGURE 10.3. Epithelia (shown as the connected row of three cells) may transmit signals to mesenchyme (shown as the three isolated cells) by one of three mechanisms: (a) long-range, diffusion-mediated interaction (arrows) affecting cells located some distance from the basal lamina; (b) short-range, matrix-mediated interaction affecting cells that approximate the basal lamina (arrow heads); or (c) short-range, contact mediated interaction affecting cells that have penetrated the basal lamina (arrows). See text for detail.

short-range, matrix-mediated, epithelial–mesenchymal interactions (Fig. 10.3b).[23]

- Epithelial and mesenchymal cells must come into direct contact for interaction to occur. A direct interaction requires selective degradation of the extracellular matrices (epithelial basal lamina and the pericellular and extracellular matrices around mesenchymal cells). Perforation of the basal lamina allows cell membranes to come into contact and epithelial and mesenchymal cells to communicate via gap junctional connections, which allow molecules of some 1300 to 1500 Da to move between cells. Ionic coupling allows the transfer of much smaller constituents. Such interactions are short-range, cell-contact-mediated, epithelial–mesenchymal interactions (Fig. 10.3c).

Matrix mediation is the most common basis for those interactions that initiate skeletogenesis from neural crest cells. The evocation of scleral cartilage by pigmented retinal epithelium and of mandibular membrane bone by mandibular epithelium in the chick are both short-range, matrix-mediated interactions. Induction of mandibular bone is discussed in detail in the next section. Some of the major steps in the induction of teeth, in which mesenchyme derived from neural crest cells interacts with mandibular and maxillary epithelia in a series of epithelial–mesenchymal interactions (Chapter 8), are also matrix-mediated, although the initial interaction that specifies enamel organs and dental papillae appears to be mediated by direct contact between epithelial and mesenchymal cells.[24]

Mandibular Membrane Bone

Neither premigratory neural crest cells nor mesenchyme migrating toward the mandibular arches in chick embryos can form bone if maintained in isolation; such cells have yet to undergo the necessary epithelial–mesenchymal interaction required to differentiate, although they may already be determined subpopulations of osteogenic cells. Signaling takes place in the mandibular arches between 2 and 4 days of incubation; bone appears at 7 to 7.5 days. On the other hand, avian mandibular chondrogenic cells that form Meckel's cartilage receive their signals from epithelial ectoderm while in or soon after leaving the neural tube (Table 10.1). Amphibians receive the equivalent signals from pharyngeal endoderm.

Avian mandibular mesenchyme and epithelium can be separated using enzyme solutions that degrade the basal lamina. Mesenchyme isolated before 4 days of incubation does not form bone; mesenchyme isolated after 4 days does. Mesenchyme from the younger embryos can form bone, however, if recombined with mandibular epithelium, thus demonstrating that an epithelial–mesenchymal interaction is a prerequisite for membrane bone formation. This is a matrix-mediated interaction, as demonstrated by the following independent sets of evidence:

- A detailed ultrastructural examination of the epithelial–mesenchymal interface during interaction fails to demonstrate any perforations of mesenchymal cell processes through the epithelial basal lamina and fails to demonstrate any direct cell-to-cell contact between epithelial and mesenchymal cells (Fig. 10.4). A continuous intact basal lamina rules out direct cell-to-cell contact as the means of epithelial–mesenchymal interaction.
- Mesenchymal cells flatten out along the basal lamina so that some 60% to 70% of the lamina is associated with mesenchymal cells or cell processes.
- Recombination of mandibular mesenchyme and epithelium across Millipore or Nuclepore filters of various thicknesses and porosities demonstrates that bone only forms when the filters are less than 10 μm in thickness and have pores small enough to allow cell processes to penetrate. These data eliminate long-range, diffusion-mediated interaction and support short-range, matrix- or cell-to-cell-mediated interaction. In combination with the ultrastructural evi-

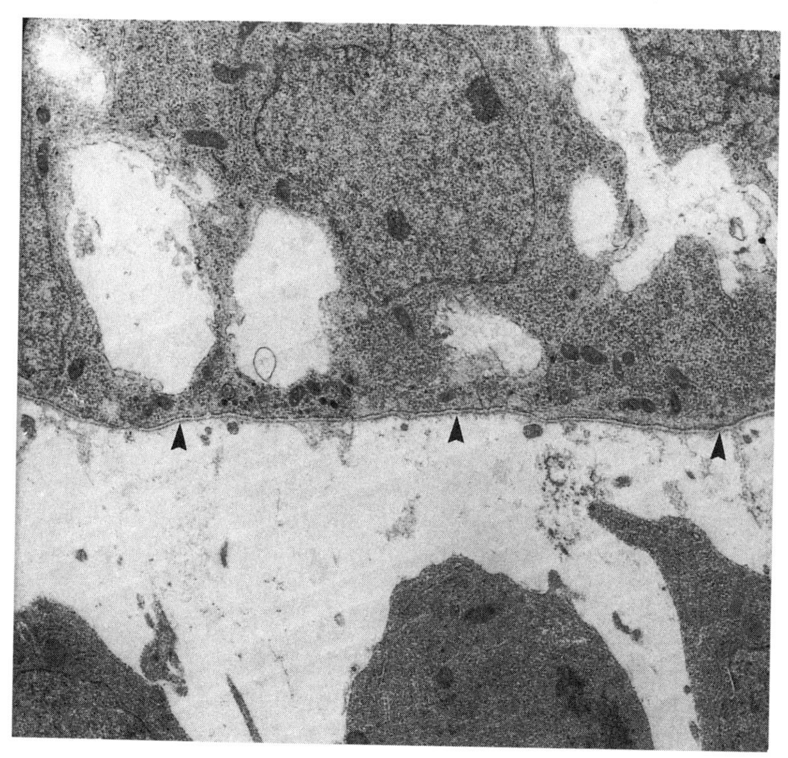

FIGURE 10.4. This electron micrograph shows the presence of an intact basal lamina (arrowheads) underlying mandibular epithelial cells (top), that lie adjacent to mandibular mesenchyme, seen at the bottom.

dence of an intact basal lamina, these data suggest (but do not prove) a matrix-mediated interaction.

Two further independent sets of evidence confirm the interaction as matrix-mediated:

- When cultured in isolation, mandibular epithelium deposits an extracellular basal lamina-like extracellular matrix onto the filter substrate. Mandibular mesenchymal cells differentiate into osteoblasts and deposit bone when cultured on this extracellular matrix.
- The use of a chelating agent to separate mesenchyme from epithelium does not degrade the basal lamina (as occurs when the separation is effected using enzyme solutions) but cleaves the links between basal lamina and epithelial cells, leaving the basal lamina intact on the mesenchyme (Fig. 10.5). When cultured in isolation from epithelium, mandibular mesenchymal cells that are covered with basal lamina deposit bone, i.e., the basal lamina is sufficient to provide the epithelial signal.

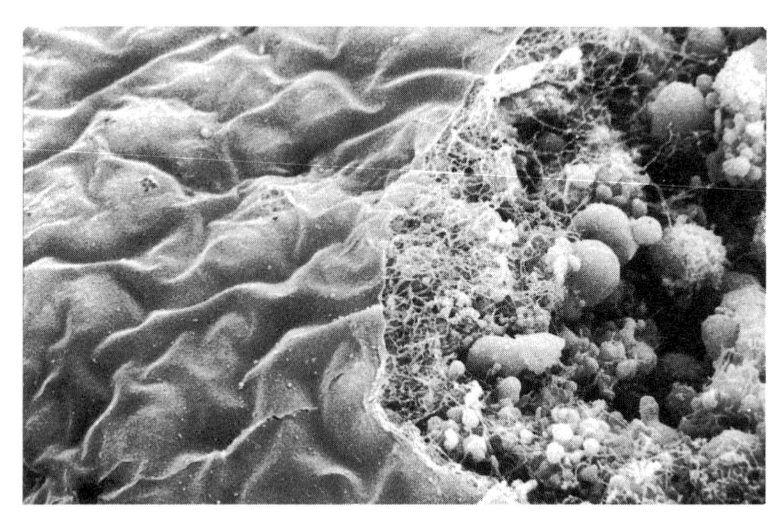

FIGURE 10.5. The sheet-like basal lamina (left) overlying mesenchyme of the mandibular arch of an embryonic chick remains intact after the epithelium has been removed using a chelating agent.

In combination, these data unequivocally demonstrate that the epithelial–mesenchymal interaction is matrix-mediated.[25]

In theory, there is no reason why the epithelial signal could not be one or more steps removed from the mesenchymal differentiative response. We may be seeking two signals; one a mitogen to accumulate a subpopulation of cells, the second a differentiation signal. One could imagine a scenario in which a specific subset of mesenchymal cells accumulates against an epithelium (being trapped by type II collagen, as suggested by Peter Thorogood and his colleagues and discussed in Chapter 9) and that the epithelium then triggers increased proliferation of these mesenchymal cells, allowing them to accumulate to a density that permits interactions that trigger cytodifferentiation. This is the well-known phenomenon of *condensation formation*, the density-dependent enhancement of synthesis of cell-specific products and subsequent differentiation of skeletogenic and other cells. The signaling molecules involved in condensation of prechondrogenic cells and the switch from condensation to overt chondrogenesis are illustrated in Figure 10.6. Enhancement of mesenchymal proliferation and condensation formation beneath and not between epithelial scleral papillae are the initial steps in the epithelial–mesenchymal interaction that triggers differentiation of mandibular bone and of the scleral bones that surround the eyes of embryonic chicks (Fyfe and Hall 1983; Dunlop and Hall 1995).[26]

Matrigel is a complex reconstituted basement membrane derived from the mouse EHS tumor. It contains laminin, type IV collagen, heparan sulfate, fibronectin, entactin, and many growth factors, including EGF, IGF-1, bFGF, PDGF, and TGF-ß1. Cultured quail trunk neural crest cells respond to Matrigel

CONDENSATION

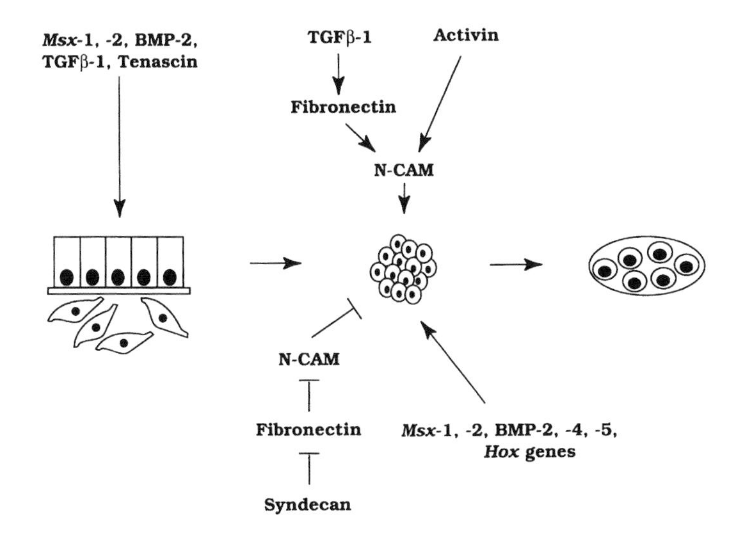

DIFFERENTIATION

FIGURE 10.6 A summary of the major classes of signaling molecules associated with epithelial–mesenchymal interaction (left), condensation (center), and differentiation (right) of prechondrogenic cells. Signals involved in condensation are shown above the three phases of chondrogenesis; signals involved in differentiation are shown below. Epithelial–mesenchymal interactions, which are regulated by *Msx-1*, *Msx-2*, BMP-2, TGF-ß-1, and tenascin, are responsible for formation of a condensation of prechondrogenic cells. Condensation is enhanced by up-regulation of N-CAM, either directly through activin or indirectly through TGFß-1 and fibronectin. The switch from condensation to overt differentiation of chondroblasts is controlled by two pathways. Up-regulation of syndecan blocks fibronectin and N-CAM and so blocks condensation. Up-regulation of *Msx-1*, *Msx-2*, BMP-2, -4, and -5, and *Hox* genes, such as genes of the *Pax* family, provides positive signals for differentiation of chondroblasts. For further details, see Hall and Miyake (1995b).

by differentiating as adrenergic catecholamine-containing cells (which increase by some 50-fold), while melanocytes decrease by 50-fold. I have shown that avian mandibular mesenchyme responds to Matrigel by differentiating into bone cells. Because of the heterogeneity of Matrigel, however, mechanisms underlying the response of neural crest cells must be interpreted with care. That said, some components of Matrigel can be eliminated from consideration; the response of trunk neural crest cells cannot be elicited using collagen type I, collagen type IV, or laminin. BMP-7 (osteogenic protein-1), which binds to type IV collagen of basement membranes, does increase the number of catecholamine-positive cells, and so may be an active component in the interaction. TGF-ß, which also binds to type IV collagen in basement membranes, inhibits this transformation.[27]

Growth factors bound to the basal lamina or structural components of basal laminae, some of which have EGF repeats and so could function as mitogens, are candidates as signaling molecules. BMP, a mitogen known to act as an inducer of cartilage and ultimately of bone from mesenchymal cells, is localized at the epithelial–mesenchymal interface when avian mandibular bone is being elicited. Barlow and Francis-West (1997) and Ekanayake and Hall (1997) have both demonstrated that BMP-2 and BMP-4 are involved in morphogenesis of skeletal tissues of the mandibular and other facial processes. BMPs may exert their control over differentiation by up-regulating *Hox* genes; see below.

The presence of PDGFR-α in *Xenopus* pharyngeal endoderm and pre- and postmigratory neural crest suggests that PDGF may be involved in signaling associated with cartilage differentiation in that species. In mice, the PDGF-α receptor subunit is expressed in both mesodermal and neural crest mesenchyme as well as in such ectodermal derivatives as the lens and choroid plexus; the mutation *Patch*, a deletion in this receptor, results in both mesodermal and neural crest deficiencies. PDGF-α and its receptor are expressed in separate but adjacent cell layers—PDGF in epithelial cells, the receptor in mesenchymal cells—in organs that depend on epithelial–mesenchymal interactions for their initiation, such as branchial arches, the otic vesicle, sclerotome, hair, and mammary glands.[28]

Hox genes also play a role. In avian embryos, *Msx-1* (previously *Hox-7* or *Hox-7.1*) expression is widespread, including the primitive streak, neural tube and neural crest (mesenchymal but not neuronal or pigment cells), otocysts, limb mesenchyme, heart valves, and the superficial ectoderm of facial processes and visceral arches. *QMsx-1* is expressed medio-ventrally in quail mandibular processes early in development and then in preosteogenic cells. *QMsx-1* expression is required for the epithelial–mesenchymal interaction that initiates differentiation of mandibular membrane bone, and *Msx* is regulated by BMP. *Msx-1* is also regulated by BMP-4 during murine tooth development; indeed BMP-4 can substitute for the dental epithelium and up-regulate *Msx-1* in dental mesenchyme.[29]

Murine mandibular bone is also induced by signals from mandibular epithelium, as documented using similar approaches to those used for chick embryos, viz., tissue separation and recombination. So too is murine Meckel's cartilage, in contrast to the chick, in which Meckel's cartilage is induced by cranial ectoderm before neural crest cell migration. *Msx-1* may be involved in mice as well. *Msx-1* is expressed in neuroepithelium, Rathke's pouch, limb bud mesenchyme, and in neural crest and neural-crest-derived craniofacial mesenchyme at sites of known epithelial–mesenchymal interactions for bone and tooth development.[30]

Mice that are null mutants for the transcription factor AP-2 have defective midline fusion, underdeveloped mandibular skeletons, and abnormal cranial ganglia, and lack the malleus and incus of the middle ear. Apoptosis is also increased in brain and proximal arch mesenchyme at 9 to 9.5 days postconception, often resulting in formation of a single maxillary-mandibular element. Even earlier in the development of these mutants the neural folds fail to close, although migration of neural crest cells is normal. AP-2 is expressed in neural ectoderm, neural crest, and facial ectoderm, and may be a useful marker for lineage-related epidermal derivatives. AP-2 is also expressed in ganglia and facial, limb, and kidney mesenchyme, so its function is not unique to ectoderm or ectodermal derivatives. Nevertheless, with care, it could be used as a marker for early ectodermal derivatives. Inhibition of mandibular skeletogenesis following normal neural crest cell migration, and involvement of AP-2 in determination of basement membrane components, are both consistent with the mutant gene acting on the epithelial–mesenchymal interactions responsible for differentiation of murine mandibular bone and cartilage. FGF-3 is expressed in the ectoderm of the second and third visceral arches of avian and murine embryos and may be a marker for ectoderm of these arches and/or be involved in the interactions.[31]

Specificity of Interactions

It is important to understand how an extracellular matrix such as an epithelial basal lamina can be required for one population of mesenchymal cells (avian osteogenic) to differentiate while not being required for chondrogenic or fibroblastic populations; recall that chondrogenesis of avian mandibular mesenchyme is independent of the mandibular epithelium, but requires an earlier epithelial–mesenchymal interaction (Table 10.1). Are there highly specific molecules or sets of molecules within individual basal laminae, each responsible for activating a particular differentiative program within mesenchymal cells? Or is the epithelial signal a relatively nonspecific one, perhaps based on the three-dimensional configuration of a number of interacting basal lamina components—type IV collagen, laminin, fibronectin? In this situation, specificity would not be provided by a specific epithelial signal but by a subpopulation of mesenchymal cells of neural crest origin. Or are there several signals: an initial general signal that might be shared by several systems, and subsequent more specific signals, as indeed is the case in the development of epidermal appendages such as scales, feather, and hair. The mouse mandible, in which mandibular epithelium evokes both cartilage and bone (Hall 1980), is an excellent model system in which to evaluate epithelial or mesenchymal specificity and the existence of subpopulations of cells (Atchley and Hall 1991).

Cascades of Interactions in Craniofacial Development

There is evidence from both anurans and urodeles that interactions involving the neural crest and adjacent embryonic layers are part of a cascade of interactions leading to the integrated formation of associated tissues and organs to form functional units.

The European frog *Discoglossus pictus* was used in a series of experiments by Fagone and Cusimano-Carollo to investigate the role played by the neural crest in the formation of the tissues and structures associated with the developing mouth. The experimental procedure used was to associate neural crest with one or more adjacent regions of neural-fold-stage embryos, and then to graft them either into the flanks of similarly aged embryos or into the blastocoele of younger embryos. Supernumerary mouths formed in these ectopic sites. Formation of cartilage was a prerequisite for induction of the mouth, oral papillae, horny beak, and teeth. Cusimano-Carollo interpreted her results as evidence for a cascade of interactions emanating from the pharyngeal endoderm as follows:

- Initially, the endoderm from regions equivalent to the 0°–90° sectors mapped by Chibon (see Fig. 5.4d) acts inductively on rostral neural crest cells to initiate differentiation of the supra- and infrarostral cartilages, which are the cartilages associated with the larval mouth.
- At the same time, pharyngeal endoderm acts inductively on future stomodaeal ectoderm.
- Once differentiated, the cartilages act upon the ectoderm, enabling it to invaginate to form the horny teeth, horny beak, and oral papillae of the developing larval mouth.

Formation of the mouth therefore requires interaction between stomodaeal endoderm and ectoderm, neural folds, and neural crest cells.[32]

Whether the taste buds of the tongue develop from or are induced by neural crest cells was unclear until the recent studies by Barlow and Northcutt on salamanders and the axolotl. Using a combination of transplantation, evocation of ectopic taste buds, and in vitro culture of taste buds, Barlow and Northcutt demonstrated that:

- taste buds are the last of the sensory receptors to arise during development;
- taste buds arise from local pharyngeal endoderm and not from neural crest or placodal ectoderm;
- neither neural crest nor paraxial mesoderm induces taste buds; and
- taste buds do not require innervation to develop.[33]

Some anurans such as *Xenopus laevis* lack supra- and infrarostral cartilages and keratinized mouth parts. Urodeles also lack these cartilages. Consequently, the cascade seen in *Discoglossus pictus* cannot be representative of all anurans. The only experimental studies on urodele embryos that parallel those on *Discoglossus* are the transplantation studies on *Pleurodeles waltl* and *Triturus alpestris* performed by Cassin and Capuron and by Corsin. All structures associated with the mouth only formed when neural crest, prechordal mesoderm, lateral head mesoderm, stomodaeal endoderm, and stomodaeal ectoderm were all included in the grafts or organ cultures. As Corsin emphasized, the earliest ossifications are dermal and associated with tooth-bearing bones; endochondral bones are poorly developed in urodeles and develop later, immediately before metamorphosis. Formation of such bones as the vomers, palatine, dentary, and splenial depends on prior differentiation of cartilage: differentiation of dentary and splenial depends on Meckel's cartilage; differentiation of the vomers and palatine on the cartilaginous trabeculae.[34]

Of interest, in terms of epigenetic cascades, is the inductive role of the annular tympanic cartilage, which both induces and maintains the epithelial tympanic membrane in frogs. Contact between epithelium and cartilage during larval life in *R. pipiens* results in epithelial dedifferentiation and transformation into the tympanic membrane; continued contact with the cartilage after metamorphosis maintains the epithelium as a membrane. The quadrate and suprascapular were also shown to be able to induce a tympanic membrane in *Rana palustris*.[35]

Skull morphogenesis is controlled in part by inductive interactions and in part by influences such as the local mechanical environment. Exchange of nasal vesicles between *Ambystoma tigrinum* and *A. punctatum* produces nasal capsules intermediate in size between those typical of the two species, indicating a role for induction in organ size. Earlier studies demonstrated that the otic capsule of frogs and salamanders fails to develop if the otic vesicle is removed early in development. When the nasal placode is removed from embryos of *Rana sylvatica*, *R. pipiens*, or *Triturus alpestris*, all elements of the skull form but skull shape is abnormal and adjacent elements are misplaced. Removal of the olfactory sac from embryos of *Hynobius keyserlingii* results in absence of the nasal bones and ascending process of the premaxilla. Removal of the otic sac results in anomalies of the frontal-parietal as it does in ambystomatids. A classic series of studies is Medvedeva's experiments on inductive interactions between the nasolacrimal duct and skull bones such as the prefrontal, lachrymal, and septomaxilla that form late in larval life or during metamorphosis in urodeles. These bones fail to form or are small or incomplete following removal of the duct or its placodal precursor.[36]

These dependent associations continue into adult life; regeneration of the bone of attachment of the teeth during jaw regeneration in *Rana pipiens* depends on prior formation of the dental lamina. The roots and bone of attachment are intrinsic parts of the tooth primordium. Because components of the craniofacial region are so tightly integrated, alteration in a single key early process can affect the entire region. The neurocristopathies discussed in Chapter 11 are examples of the types of defects that can arise when integration breaks down.[37]

Migration is a key event in neural crest cell development. Anomalous migration can lead to failure of organ formation, fused cartilages, or the development of teeth in epithelial tissue. Levy, Detwiler, and Copenhaver (1956) demonstrated this in a study with an interesting history. Based on a reexamination of sections from a study initially carried out in 1940 by two of the authors, Detwiler and Copenhaver, it illustrates the utility of not throwing out the slides once the study is done.[38]

Mutations also affect epigenetic cascades. The *premature death* (*p*) mutation in *Ambystoma mexicanum* affects the subpopulation of neural crest cells that produce the craniofacial cartilages and contribute to the heart. Other neural crest derivatives form normally but neural crest cartilages fail to form. The site of

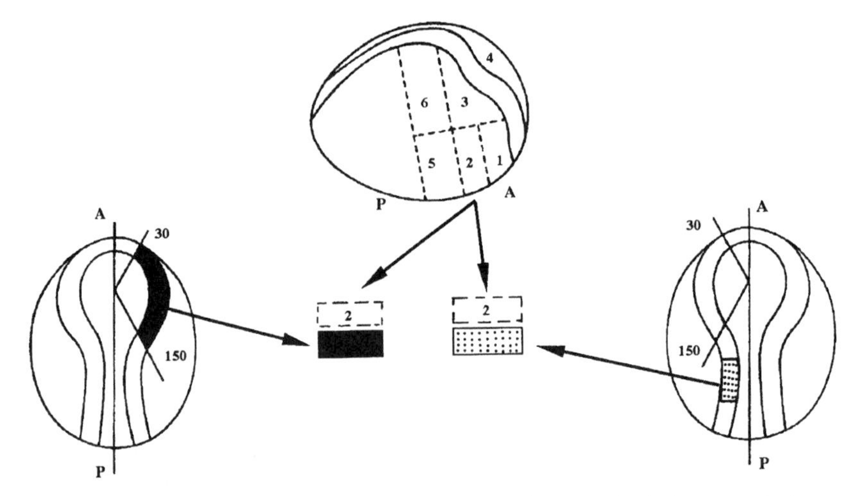

FIGURE 10.7. Cranial neural crest (black) or trunk neural crest (stippled) from wild-type or mutant amphibian embryos (shown here at the neurula stage, viewed from the dorsal surface) can be recombined with an epithelium (shown as region 2 of 6 regions of inductively active endoderm in the lateral view of a neurula) and maintained in organ culture to test both inductive activity of the epithelium and responsiveness of particular populations of neural crest cells. The region bracketed between 30° and 150° represents the chondrogenic neural crest (see also Fig. 4.4d). A, anterior; P, posterior. Modified from Hall (1998a).

action of mutant genes such as *p* can be identified by combining neural crest cells with known inductive epithelia and maintaining the tissue recombinants in vitro. (Fig. 10.7). In such recombinants, *p* endoderm induces heart and cartilage development from wild-type embryos but *p* mesoderm cannot respond to inductive influences from wild-type pharyngeal endoderm. Competence of the chondrogenic subpopulation of neural crest cells to respond to induction is defective in *p* mutants.[39]

Teeth from Trunk Neural Crest

We are accustomed from amphibian and avian studies to regard the trunk neural crest as nonskeletogenic (not capable of forming cartilage and bone), and have assumed that this regionalization applies to odontogenic capabilities as well. Premigratory avian trunk neural crest cells form "mesectodermal" derivatives such as connective tissue, dermis, and muscle, but neither cartilage nor bone, when they are grafted to the level of the cranial neural crest and allowed to mix with cranial neural crest cells. This ability is found only in very young, premigratory cells and requires some as yet unspecified cranial environmental influences to elicit the mesenchymal expression.

Surprisingly, Lumsden (1987, 1988) found that cells derived from the most rostral murine trunk neural crest could participate in tooth formation if combined with mandibular epithelium. The teeth developed in association with a follicle, periodontal ligament, and alveolar bone of attachment, although no other bone and no cartilage developed in these grafts. Lumsden concluded that bone of attachment develops from the dental papilla, not from osteogenic mesenchyme, a finding supported by studies in which tooth germs from older embryos are allowed to develop in ectopic sites.[40]

Lumsden's demonstration that the most rostral trunk neural crest of mouse embryos can form teeth if challenged with the appropriate epithelium has now been extended to the axolotl. By combining axolotl trunk neural crest with endoderm (which is the inducer of teeth in the axolotl), as illustrated in Figure 10.7, Graveson, Smith, and Hall (1987) demonstrated that neural crest, which would be defined as trunk on the basis of the map of the chondrogenic neural crest, can form form teeth but not cartilage (Fig. 10.8). While it remains true for both mice and the axolotl that trunk neural crest is not chondrogenic, the most rostral trunk neural crest is odontogenic. Teeth are remnants of a once extensive exoskeleton.

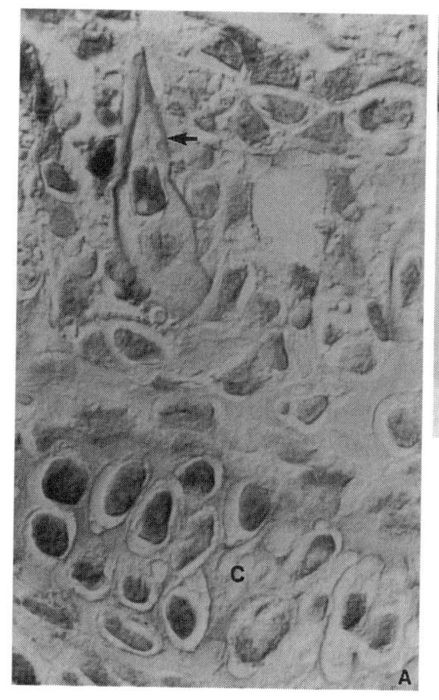

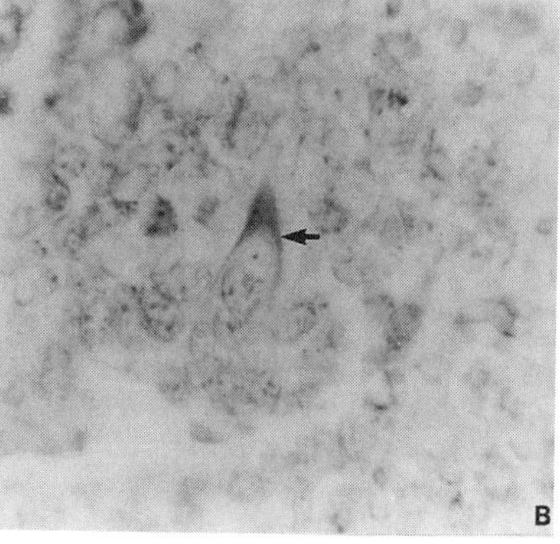

FIGURE 10.8. (A) A tooth (arrow) is shown adjacent to Meckel's cartilage (c) from the lower jaw of the axolotl, *Ambystoma mexicanum*, in a figure kindly provided by Moya Smith. Both the tooth and Meckel's cartilage are derived from cranial neural crest. (B) When the most rostral trunk neural crest is cultured in association with embryonic endoderm (as depicted in Figure 10.7), teeth develop (arrow) but cartilage does not. See text for details.

We might therefore expect that in vertebrates that have retained a more extensive exoskeleton (the dermal denticles of sharks and skates; the bony skin ossicles of some teleost fishes), the odontogenic neural crest might extend even more caudally. Indeed, it may well be that the entire trunk neural crest was capable of forming exoskeleton in those early agnathans that possessed an extensive dermal armor.[41]

Neural Crest as Inhibitor

A little-studied aspect of inductive interactions, but one that may be of considerable importance in integrating tissues during development, is inhibition of other cell types by neural crest cells. Neural crest cells inhibit induction of the heart, kidney, and lens of the eye, apparently by producing diffusible inhibitory factors. Although suggestions of such inhibition first appeared in the early 1960s, the concept of neural crest cells inhibiting induction, indeed, the general concept of inhibition as part and parcel of induction, is not well appreciated.

Woellwarth (1961) demonstrated that neural crest inhibits lens induction in *Triton taeniatus*. Forebrain, or forebrain plus neural folds containing neural crest, was removed from embryos which were then reared at the optimal temperature for lens induction, 14°C. Lenses formed in 26.5% of embryos from which forebrain had been removed, but in 95% of embryos lacking forebrain and neural crest. Henry and Grainger (1987) confirm this inhibitory action. They argue that by coming into contact with surface ectoderm, neural-crest-derived mesenchyme suppresses the lens field except where the optic cup contacts the overlying ectoderm. This contact prevents mesenchyme from invading that region. Effectively, neural crest mesenchyme positions the lens over the center of the developing optic cup.

The neural crest also inhibits heart development in *Taricha torosa*, cranial crest having a greater suppression than trunk crest when precardiac mesoderm is exposed to homogenates of neural crest in hanging drop cultures. Other studies from the same laboratory demonstrated that both premigratory neural crest and mesenchyme derived from neural crest suppressed kidney formation in *Taricha* and *Xenopus laevis*. Such inhibitory or delaying actions on kidney development may not be limited to amphibians. Kidney development is aberrant in *congenital hydrocephalus* (*ch*) mutant mice, which die at birth. These mutants display bulging cerebral hemispheres from early in development (a consequence of retained fluid within the brain), multiple skeletal defects, fewer than normal coeliac ganglia, and additional mesonephric kidney tubules. The latter defect has been attributed to neural crest cells inhibiting kidney development during the ninth or early tenth days of gestation when preganglionic crest cells are migrating along mesonephric mesoderm.[42]

Some intriguing, previously published studies involving the neural crest merit reexamination in light of the neural crest as inhibitor. For example, injection of a combination of histamine and EDTA into newborn mice kills neural crest cells. Mast cells move into the site previously occupied by the neural crest cells. Nozue (1988a) interpreted this as indicating a neural crest origin for mast cells. An alternative—and more likely explanation, given what is known of mast cell origins—is that neural crest cells normally inhibit invasion of mast cells into the area. Injection of EDTA alone produced what Nozue (1988b) regarded as neural crest tumors arising from undifferentiated stem cells, but no markers were available to reliably identify the stem cells as neural crest in origin.

In other situations neural crest cells promote differentiation. Thus, anuran primordial germ cells, which normally migrate through mesoderm to make their way to the gonad rudiments, have an enhanced attraction for mesoderm if neural crest is present. Even from the limited studies carried out so far, it is clear that much activity of the neural crest, especially in inductive interactions and epigenetic cascades, remains to be discovered.[43]

The picture of neural crest cell differentiation that emerges is of subpopulations of cells,

- some of which are present in the premigratory neural crest of the neural tube;
- others of which appear during or after neural crest cell migration;
- many if not all of which are at least bipotential; and
- with the particular pathway expressed depending on interactions with extracellular matrix and cell products such as growth factors, hormones, and neurotransmitters encountered before, during, or after cell migration.

11

Neurocristopathies

Interest, indeed fascination, with developmental abnormalities stretches back at least to the Neolithic period. A sculpture of a two-headed individual dated 6500 BC was excavated from southern Turkey, while cuneiform characters on seventh century BC clay tablets discovered in the Royal Library at Nineveh record 62 different human malformations known to the Babylonians. Syndromes such as Treacher Collins were depicted in Mexican statues as early as AD 2. Given such extended interest, we might expect that all possible syndromes would have been identified long ago. But syndromology is not dead; one or more new syndromes are described somewhere in the world each week. Syndromes involving neural tube defects (Table 11.1) are second only to heart defects as a source of perinatal mortality in humans, accounting for some 15% of perinatal deaths in the United Kingdom alone. Fortunately, folic acid has proven to be very effective in preventing as many as 70% of human neural tube defects, such as spina bifida, to which defective neural crest contributes. Because of the relationship between ultraviolet light and folic acid, melanin may also protect against neural tube defects.[1]

Given that abnormalities are often expressed as a syndrome displaying defective development of more than one cell type or tissue and that so many different cell types arise from the neural crest, it was inevitable that neural crest origin would be seen as the common "explanation" for syndromes comprised of some combination of nerve, pigment, craniofacial skeletal, heart, adrenomedullary, or other cell types derived from the neural crest. Bolande introduced the term *neurocristopathy* in 1974 for syndromes, tumors, and/or dysmorphologies involving neural crest cells. Commonality of embryonic origin—the rationale for neurocristopathies—brings us back full circle to germ layers and the germ-layer theory discussed earlier.

Does grouping neural crest defects as neurocristopathies merely classify them on the basis of an outmoded germ-layer theory? Or does such a classification help us to understand the etiology of the defects and the mechanisms involved when development goes astray? Because neural crest cells share so many characteristics—commonality of origin, bipotentiality, migratory capability, and activation by epigenetic factors—we can invoke the explanatory powers of neural crest origin, just as pathologists and developmental geneticists in the past correlated developmental effects/defects on the basis of commonality of origin from a single germ layer. The pervasiveness and utility of germ-layer thinking is amply illustrated by the willingness of clinicians to classify malformations on the basis of

TABLE 11.1. A survey of the major neurocristopathies

Tumors

Pheochromocytoma[1]: A tumor of the chromaffin tissue of the adrenal medulla that can arise as early as 5 months of age in humans. High concentrations of vanillinemandelic acid are characteristic.[2] Synonym: Endocrine neoplasia III, Multiple.

Neuroblastoma: A common, malignant embryoma involving the adrenal medulla and ganglia of the autonomic nervous system. Displays a very high incidence of spontaneous regression. [The autonomic nervous system can also function abnormally in defects other than neuroblastomas. Prader-Willi syndrome is one example (Di Mario et al. 1994).]

Medullary carcinoma of the thyroid[1,3]: A tumor of parafollicular calcitonin-forming "C"cells of the thyroid, representing 7% of all thyroid tumors. Synonyms: endocrine neoplasia II, Multiple, hypercalcitoninemia.

Carcinoid tumors: Tumors of the bowel and gastrointestinal tract involving enterochromaffin cells.

Nonchromaffin paraganglioma (chemodectoma) of the middle ear[4]: A tumor involving ganglia of the neck and the middle ear. Over 90% of patients suffer hearing loss. Associated with erosion of bone.

Hirschsprung disease[5]: Absence or reduction of ganglia of the colon. 1:5000 to 1:8000 live births; 80% of cases are in males. Synonyms: aganglionic megacolon, colon aganglionosis.

Neuroectodermal pigmented tumor[6]: A rare tumor of the maxilla consisting of clusters of neuroblast-like cells, with or without melanin. Can arise as early as the first year of life. High urinary excretion of vanillinemandelic acid. Results in destruction of bone and displacement of teeth. Synonym: Melanotic progonoma.

Clear cell sarcoma[7]: A sarcoma of tendons and aponeuroses.

Tumor Syndromes

von Recklinghausen neurofibromatosis[8]: Multiple neuroid tumors of the skin and abnormal pigmentation. An autosomal dominant; 1:3000 live births. Six or more café-au-lait spots (areas of pigmentation) greater than 1.5 cm diameter are diagnostic for the syndrome. Neural neoplasia occurs throughout the nervous system. Skeletal abnormalities and scoliosis may be associated.

Sipple syndrome[9]: An association of pheochromocytoma and medullary thyroid carcinoma. An autosomal dominant, displaying elevated levels of calcitonin, catecholamine in excess, and ectopic production of ACTH. Synonyms: Medullary thyroid carcinoma and pheochromocytoma syndrome.

Multiple mucosal neuroma syndrome[10]: Mucosal tumors (neurofibromas) of the tongue, lips and eyelids; medullary carcinoma of the thyroid, sometimes with neural involvement.

Wermer-Zollinger-Ellison syndrome[11]: Neoplasia of endocrine glands (pituitary, adrenal, parathyroid, pancreas, thyroid). A rare autosomal dominant. Synonym: Multiple endocrine adenomatosis.

Neurocutaneous melanosis: Giant pigmented nevi (birth marks) of the skin.

Malformations

Mandibulofacial dysostosis[10]: Hypoplastic zygomatic arch, orbital and supraorbital ridges, dysplastic ears, micrognathia, defective ear ossicles; high incidence of cleft palate. An autosomal dominant, largely treatable with corrective surgery. Synonym: Treacher Collins syndrome.

TABLE 11.1. (Continued) A survey of the major neurocristopathies

Otocephaly[11]: Absence of lower jaw, low-set ears, which may be fused in the midline; microstomia; ear ossicles, temporal bone, palate and maxilla may be deformed. Absence of mesenchyme derived from the neural crest. Associated with heart, visceral and limb anomalies. Synonyms: Agnathia, microstomia and synotia.

CHARGE association[12]: **c**oloboma, **h**eart disease, **a**tresia of choanae, **r**etardation of physical and mental development, **g**enital hypoplasia in males, and **e**ar (CHARGE) anomalies and/or deafness

Others

Albinism[13]: A very heterogeneous group of conditions. Absence of skin pigmentation because of lack of melanoblasts. Synthesis of tyrosinase inhibited. An autosomal dominant. Synonym: Piebaldness.

Waardenburg syndrome[14]: An autosomal dominant exonic mutation in the HuP2 paired domain as an intragenic deletion in *Pax-3*. Maps to the distal arm of chromosome 3. Incidence 1:4000 live births. Pigment defects of the hair (white forelock) and skin. May be associated with deafness. Increased incidence of cleft palate. Ocular hypertelorism. Due to abnormal migration or survival of melanoblasts.

1. Bergsma (1979), syndrome # 732; 2. Bertani-Dziedzik et al. (1979); 3. Bergsma (1979), syndrome #351; 4. ibid., syndrome # 145; 5. ibid., syndrome # 192; 6. ibid., syndrome # 711; 7. Kindblom, Lodding, and Angervall (1983); 8. Riccardi and Eichner (1986); 9. Bergsma (1979), syndrome # 351; 10. Gorlin, Cohen, and Levin (1988); 11. Bergsma (1979), syndrome # 350; 12. Pagon et al. (1981), Siebert, Graham, and MacDonald (1985); 13. Bergsma (1979), syndrome # 31; 14. Gorlin, Cohen, and Levin (1988), Baldwin et al. (1992), Moase and Trasler (1992), Tassabehji et al. (1992), Mansouri et al. (1996), and Dahl, Koseki, and Balling (1997). Waardenburg syndrome is named after Petrus Johannes Waardenburg (1886-1979), the Dutch ophthalmologist and pioneer in the application of genetics to ophthalmology. In 1932, the year he wrote his classic textbook, *The Human Eye and Its Genetic Disorders,* Waardenburg was one of the first to suggest that Down syndrome might be a chromosomal aberration.

germ layer of origin. We use the terms *embryopathy* for defective development involving the entire embryo and *neurocristopathy* for defective development of the neural crest or its derivatives. It is less usual to use the terms *mesoderm-*, *ectoderm-* or *endodermopathy*, but particular syndromes are classified on the basis that derivatives from only one germ layer are involved. Thus osteochondromas affect mesodermal tissue, ectodermal dysplasias affect ectodermal derivatives, and juvenile colonic polyposis affects cells of endodermal origin. Identifying commonality of origin from a single germ layer does mean that the likely time, location, and sometimes the cause of the defective development can be identified.[2]

Types of Neurocristopathies

Neurocristopathies may involve the adrenal medulla, endocrine organs of neural crest origin, pigment cells, ganglia or Schwann cells of the autonomic nervous system, the heart, craniofacial skeletal, and/or mesenchymal neural crest cells. A typical neurocristopathy, described almost 40 years ago, is a malignant embryoma, with pulmonary metastases, removed from the chest of a 61-year-old male. Ganglia, connective tissue, cartilage, neurolemma, and pigment cells (all neural crest derivatives) were all found. Some well-studied neurocristopathies, with the affected neural-crest-derived tissues in parentheses, are:

- neuroblastoma (adrenal medulla, autonomic ganglia);
- mandibulofacial dysostosis, first arch syndrome, Treacher Collins syndrome (facial skeleton, ears); and
- otocephaly (absence of lower jaws, ear defects, heart anomalies).

Because Schwann cells accompany axons of peripheral nerves into the limbs, limb abnormalities may accompany neurocristopathies.[3]

To Bolande, a neurocristopathy could arise following a defect at any stage of neural crest cell development—migration, proliferation, cell-to-cell interaction, differentiation, or growth. He divided neurocristopathies into:

- *simple*, each of which represents a single, usually localized, pathological condition; and
- *complex*, usually multifocal syndromes for associations of several simple neurocristopathies.

The diversity of tissues involved in a neurocristopathy is illustrated very nicely by the CHARGE association (Table 11.1), in which heart, nasal and oral cavities, brain, reproductive organs, ear, and overall growth are affected. I have grouped neurocristopathies in Table 11.1 into:

- *tumors*, in which only one tissue is affected;
- *tumor syndromes*, where several tissues are involved;
- *malformations* (mandibulofacial dysostosis, otocephaly, and the CHARGE association); and
- a *miscellaneous* group of others.

M. C. Jones (1990) developed an alternative classification based on disruption of two basic mechanisms (migration and proliferation) affecting neural crest cells. Table 11.2 illustrates how the neurocristopathies in Table 11.1 fit into these two categories. In Figure 11.1, I have grouped the neurocristopathies according to both the rostro-caudal level of the neural crest involved and the type of neural crest cell from which they arise.[4]

Syndromes can be genetically complex and dependent on the genetic background in which they are expressed. The *first arch* (*far*) locus in the mouse, which produces a classic first arch syndrome of neural crest defects, is partly dominant in ICR/Bc mice, but recessive in BaLB/c mice. Such a shift in dominance allows two syndromes to arise from the same mutation.[5]

TABLE 11.2. A summary of the neurocristopathies according to whether they are based in migration or proliferation defects of neural crest cells[a]

Migration defects	Proliferation defects
Defects of the anterior chamber of the eye	Pheochromocytomas
Cardiac aorticopulmonary septation defects	Neuroblastoma
CHARGE association	Medullary carcinoma of the thyroid
Cleft lip ± cleft palate	Carcinoid tumor
Cleft palate	Melanotic progonoma
DiGeorge syndrome	Neurofibromatosis
Frontonasal dysplasia	Wermer (Sipple) syndrome
Goldenhar syndrome	Neurocutaneous melanosis
Hirschsprung disease	
Waardenburg syndrome	
Isotretinoin embryopathy	

a. Migration defects produce malformations, proliferation defects produce dysplasias. This arrangement follows M. C. Jones (1990), who may be consulted for supporting literature.

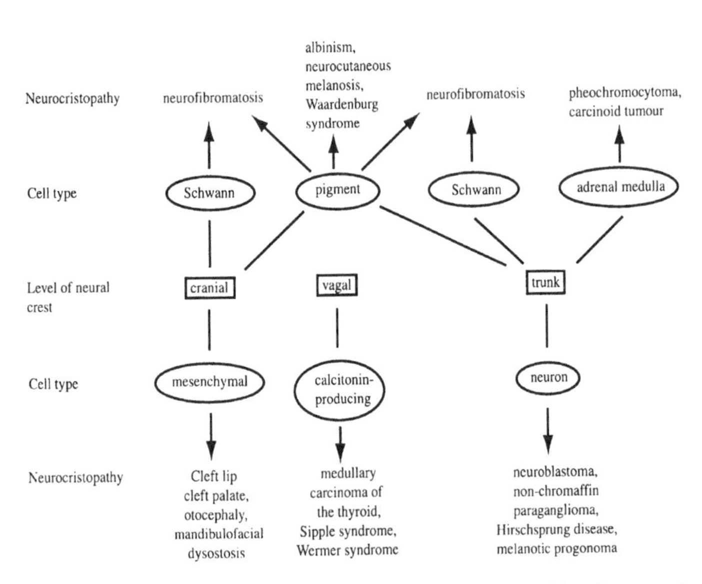

FIGURE 11.1. This flowchart illustrates the origin of neurocristopathies from particular cell types (circled) arising from cranial, vagal, or trunk regions of the neural crest. For example, abnormalities in Schwann cells from cranial or trunk crest are associated with production of neurofibromatosis; abnormalities of mesenchymal cells from cranial neural crest are associated with cleft lip, cleft palate, otocephaly and mandibulofacial dysostosis, and so forth.

Tumors

Malignant neoplasias are rare in humans, the estimated occurrence being 1:30,000 live births. Half the malignant neoplasias in human fetuses or newborns are primary leukemias or Wilm's tumor of the kidney. The other half are neuroblastomas, which are tumors of the adrenal medulla and autonomic nervous system (Table 11.1).

Neuroblastomas

Malignant tumors of neural crest origin are somewhat more common in children than in adults. Beckwith and Perrin (1963) reported a very high incidence (40 times expected) of microscopic neuroblastomas in the adrenal glands of infants under 3 months of age. They concluded that the vast majority of potential neuroblastomas never appear clinically and so are not diagnosed. Such tumors could remain dormant, regress, or transform into frank neuroblastomas that could, in turn, become malignant ganglioneuromas, ganglioneuroblastomas, or pheochromocytomas.

Neuroblastomas are neurocristopathies because their neoplastic cells are exclusively of neural crest origin. Not that neuroblatomas are all identical; cell lines derived from different neuroblastomas respond differently to TGF-ß with respect to proliferation, morphology and synthesis of extracellular matrix products. A survey of 90,000 surgical biopsies of malignant tumors undertaken in the mid-1970s revealed ten skin tumors, each of which contained two or three neoplastic tissues whose normal counterparts are derived from the neural crest. These were (with the normal counterpart in brackets):

- melanomas (melanocytes);
- gangliomas (neurons);
- Schwannomas (Schwann cells); and
- chondromas (chondroblasts).

These tumors, which had been classified as malignant melanomas or malignant Schwannomas, were identified by Wahlström and Saxén as of neural crest origin because of the association of several neural-crest-derived tissues within a single tumor. Similarly, a fourth ventricular Schwannoma was interpreted as a neurocristopathy because of the tissues involved. Transmission electron microscopic and/or enzymatic analysis can also provide evidence for neural crest origin, as it did for a human soft tissue sarcoma that displayed both Schwann cell (myelin sheaths and myelinated axons) and fibroblast differentiation (cross-banded collagen), or a clear cell sarcoma of tendons and aponeuroses.[6]

In vitro analysis of cell lines can also provide strong evidence for the neural crest origin of neuroblastomas and the ability of cells derived from neuroblastomas to modulate their phenotype. The human neuroblastoma cell line

La-N-5, which has been analyzed for neurotransmitters and extracellular matrix proteins after treatment with dibutryl cAMP and retinoic acid, undergoes cholinergic differentiation when challenged with retinoic acid. Indeed, because neuronal, Schwann, and melanocytic cells developed in vitro, Tsokos and colleagues saw these cell lines as recapitulating neural crest cell development. Reversal to normal neural crest cells is not complete; the Schwann cells produced laminin, fibronectin, and type IV collagen and the melanocytes secreted fibronectin, but the neuronal cells did not produce any extracellular matrix proteins.[7]

The Neoplastic State

Neuroblastomas often consist of cells that are morphologically identical to migrating neural crest cells, i.e., they appear to be mesenchymal cells that have failed to differentiate. Such cells often undergo spontaneous regression, not to normal neurons or chromaffin cells but to nonmalignant cells, forming a ganglioneuroma or neurofibroma. An intriguing question is obviously why they do not (cannot?) transform to normal cells? The answer lies partly in the neoplastic state as a stable differentiated state that differs from both the embryonic and the normal fully differentiated cell states. Neoplastic cells are identified by the production of tumor-specific cell products. They are neither embryonic cells expressing some abnormal products nor normal cells expressing some embryonic products (although neoplastic cells do express fetal antigens). In the transformation to a neuroblastoma, a neural crest cell expresses a new pathway of differentiation that is only partially reversible; the cell can go back along the neoplastic pathway but not as far back as the stage when the decision to become neoplastic was made.[8]

Diagnosis

The techniques of nuclear medicine facilitate detection and treatment of neuroblastomas, e.g., iodine^{135} metaiodobenzyl-guanidine can be used to detect neuroblastomas and pheochromocytomas but not other neural crest tumors. Diagnosis of tumors such as pheochromocytomas or melanotic neuroblastoma (Table 11.1), in which synthesis of catecholamine is abnormal, can be greatly facilitated by measuring urinary vanillinemandelic acid (4-hydroxy-3-methoxymandelic acid [VMA], a metabolite of catecholamine). Mean urinary VMA levels of 2.68 μg/mg creatine (range 0.5–4.8 μg) for 14 control patients and 15.7 μg/mg (range 4.5–50) for 12 patients with pheochromocytoma were reported.[9]

Some tumors, such as human melanoma, display the 7S NGF receptor complex. This receptor is lost from neural crest cells as they differentiate; a monoclonal antibody against the NGF receptor can be used to isolate multipotential neural crest stem cells that form neurons or Schwann cells, but not to isolate neurons or Schwann cells. Reappearance of the receptor on tumor cells is taken as evidence for the neural crest origin of the tumor. Further evidence comes from monoclonal antibodies generated against neuroblastomas. These antibodies, which cross-react with such normal neural crest cells as ganglia, satellite cells, and chromaffin cells of the adrenal gland, are being used to screen for neuroblastomas.[10]

Model Systems

Model systems are available for such simple neurocristopathies as neural crest cell tumors. When maintained in vivo or as a cell line in vitro, the C-1300 mouse neuroblastoma continues to produce axonlike extensions (typical of neurons) and to synthesize catecholamines (typical of adrenergic neurons). Neural crest tumor cell lines contain lectin receptors and respond to lectins by increased proliferation, i.e., the balance between proliferation and cytodifferentiation, which is so important in neoplasia, can be manipulated in vitro. Direct experimental evidence for responsiveness to tumor-promoting agents of neural crest cells or of tumors derived from neural crest cells comes from in vitro studies using trunk neural crest cells or cell lines established from tumor with a neural crest origin such as a neuroblastoma, melanoma, or pheochromocytoma. Such treated neural crest cells divide more rapidly, delay pigmentation, and block the adrenergic phenotype, showing that tumor promoters can redirect the differentiation of neural crest cells.[11]

von Recklinghausen Neurofibromatosis

Neurofibromatosis is one of the most common, if not the most common tumor involving the nervous system and pigment cells. Given an occurrence of 1:2500 to 1:3000 live births (Table 11.1) and 3.6 million births/year in the United States, 1200 to 1400 new cases of neurofibromatosis appear each year, giving an estimated 80,000 affected individuals in the United States alone. The inauguration of a National Neurofibromastosis Foundation in the United States in 1978, launching of a journal, and publication of the proceedings of two workshops and of a comprehensive treatise on neurofibromatosis in the mid-1980s focused scientific interest on the disease.[12]

Included in the proceedings of the 1981 conference edited by Riccardi and Mulvihill is a translation of the classic account of neurofibromatosis from the 1882 treatise on neuromas written by von Recklinghausen,[13] who provided a clinical description of two individuals with neurofibromatosis and the first clear indication of this new syndrome involving the nervous system. Although he identified neurofibromatosis as a distinct entity, von Recklinghausen neither described the first cases nor named the condition, which is now known as von Recklinghausen neurofibromatosis or type 1 neurofibromatosis (NF-1). Robert Smith, the first professor of surgery at the medical school in Dublin, published the first compilation of information on 77 cases, two of his own patients, and 75 cases from the literature (R. W. Smith 1849).

An estimated mutation rate of 1/10,000 gametes/generation places von Recklinghausen neurofibromatosis alongside Duchenne muscular dystrophy with the highest mutation rate of any human gene. The high mutation rate in the muscular dystrophy gene may relate directly to its very large size—it spans at least 33 kb of genomic DNA. The high mutation rate of neurofibromatosis is probably partly artifact, reflecting the heterogeneity of the neurofibromatoses, which are categorized into at least four types—central, Schwannoma, segmental, and other variant neurofibromatoses. Individual syndromes involving more than one set of genes are often pooled in single analyses, while other syndromes may have been misidentified as von Recklinghausen neurofibromatosis. The lack of heterogeneity in individuals reliably diagnosed with von Recklinghausen neurofibromatosis suggests that single locus mutations produce the majority of the new cases. The NF-1 gene was mapped to the long arm (pericentromeric region) of chromosome 17 in 1987 and cloned in 1990. The gene is complex; embedded within its introns are several other genes transcribed in the opposite orientation. Targeted disruption of the NF-1 gene produces such typical syndromes as heart malformations and hyperplasia of sympathetic ganglia. Other neurofibromatoses are genetically distinct: central or bilateral acoustic neurofibromatosis (NF-2) maps to chromosome 22.[14]

Involvement of Non-Neural-Crest Cells

Non-neural-crest cells are also affected in neurofibromatosis; there is both a secondary involvement of mesodermal derivatives—scoliosis in 53% of 47 patients and tibial pseudarthroses in 19% of the same patients—and associated cardiac malformations. Malignancies not of neural crest origin (such as leukemia, stomach cancer, and Wilm's tumor of the kidney) are also associated with neurofibromatosis. An osteoma of the skull consequent to an adrenal medullary tumor was reported in Gardner syndrome. Bone in other tumors, as in Ewing sarcoma (which contains a mix of mesenchymal and neural derivatives), is regarded as neural crest in origin; some of the differences between neural crest and mesodermal bone are discussed in Box 3.1.[15]

Investigators have gone to considerable lengths to explain involvement of non-neural-crest derivatives in neurocristopathies. Explanations include:

- genetic pleiotropy—the same gene is affected in more than one tissue—the most likely explanation;
- displaced migration and metaplasia—neural crest cells migrate into tissues that normally lack them, and differentiate ectopically;
- epigenetic—the microenvironment surrounding more than one cell type is affected; or
- production of stimulatory factors which act on an adjacent second cell type.

Joseph Merrick—The "Elephant Man"

von Recklinghausen neurofibromatosis has received much publicity because of a single case—Joseph Merrick (1862–1890), the "Elephant Man," the subject of magazine articles, a Broadway play, a motion picture, and at least four popular books. Joseph Merrick was diagnosed as having the most severe expression of von Recklinghausen neurofibromatosis, characterized by grossly abnormal skin (hence the name Elephant Man) and severe skeletal abnormalities, including overgrowth and exostoses of the calvarium (known to be of neural crest origin) and hemihypertrophy of such mesodermally derived long bones as the right femur and radius. The skeletal defects may be secondary to a primary defect in the neural crest; neurofibromas produce a bone-matrix stimulating factor(s) that could account for the skeletal overgrowth and hemihypertrophy.[16]

The difficulty with the diagnosis of von Recklinghausen neurofibromatosis in this case is the absence of café-au-lait spots, which are diagnostic for the syndrome (Table 11.1), and the severity of the skeletal overgrowth and hypertrophy, especially since the majority of the affected skeleton is mesodermal and not of neural crest origin, and when no other individual diagnosed with von Recklinghausen neurofibromatosis has such skeletal involvement. The solution to this dilemma turns out to be devilishly simple. Joseph Merrick did not have von Recklinghausen neurofibromatosis. Rather he exhibited what was thought in the 1980s to be an even rarer condition, Proteus syndrome, only 14 cases of which had been reported by 1987. There are now 120 reported cases with a further 30 to 40 cases diagnosed in the past year but not yet reported.[17]

Proteus syndrome is characterized by macrocephaly, hyperostosis of the skull, hypertrophy of the long bones, and thickening of the skin and dermis, including plantar hyperplasia and lipomas. Proteus syndrome is not a neurocristopathy. Joseph Merrick still represents the most severe case of the syndrome, perhaps because he was the longest-lived individual so far diagnosed, dying at age 28. So, 110 years after Merrick was "exhibited" to the Pathological Society of London in 1888, his condition has been correctly diagnosed. This is important for the psychological well-being of those individuals and for the parents of those children diagnosed as having von Recklinghausen neurofibromatosis. Such is the publicity associated with the Elephant Man that fears of automatic progression of neurofibromatosis to such a disfiguring condition can easily be raised. Such fears are groundless.

Animal Models and Mutations

Part of the past difficulty in determining the mechanism of pathogenesis of neurofibromatosis was that no mammalian animal models were available until the late 1980s, a surprising situation given the plethora of mutants that affect neural crest derivatives. Neurofibromas are occasionally seen in other species—cattle, horses, dogs, and birds—but there they arise following viral infections rather than as inherited defects. Furthermore, these neurofibromas are associated neither with pigment nor with skeletal defects.

Transgenic mouse models are now available in which Schwannomas and facial bone tumors can be investigated. One is associated with osteogenic sarcomas of the facial bones with extension into the periodontal ligament—both structures of neural crest origin—making this a good model for neurocristopathies. In 1987, Hinrichs and colleagues described a transgenic mouse model utilizing tumors induced by the *tat* gene of human T-lymphotropic virus type 1, tumors that closely resemble those produced in von Recklinghausen neurofibromatosis. Some mouse mutants exhibit pleiotropic effects on several neural crest derivatives such as pigment cells, nerve cells and cells of the craniofacial skeleton, but none mimic neurofibromatosis. One such mutant, *Patch*, a deletion of the gene encoding the α subunit of the PDGR receptor, is evidenced by abnormalities of the glycosaminoglycans of the extracellular matrix, resulting in abnormal pigmentation, craniofacial defects, and defects to the thymus, heart, and teeth.[18]

The same phenotype can be expressed by two genes acting in quite different ways. The mutations *Dominant Spotting* and *Steel* both result in similar defects in melanocytes. *Dominant Spotting* affects neural crest cells directly while *Steel* affects the extracellular matrix, altering both collagen and glycosaminoglycans to suppress melanocyte development. Steel factor (stem cell factor), which is a growth factor and the ligand for the *c-kit* protein kinase, is required for survival but not for differentiation of melanocyte precursors. Both soluble and cell-bound forms of steel factor are known: soluble steel disrupts the migration of melanocyte precursors, while membrane steel disrupts their survival. In contrast to situations in which mutations in different genes produce the same phenotype, interactions between genes can produce syndromes not seen with either gene alone. Thus, mice that are double mutant for *Patch* and *Undulated* display an extreme form of spina bifida; single mutants show no evidence of even mild spina bifida.[19]

The avian neurofibromatosis-1 gene (*a*NF-1), which Kavka and Barald (1994) cloned as a 432-bp cDNA, has 82% similarity to the human NF-1 gene at the nucleic acid level and 93% similarity to the protein sequence. A probe to the 12.6 kb transcript of the chick gene is expressed from H.H. stage 11 through to adult life. While mRNA is expressed ubiquitously, the protein is most highly expressed in a subset of migrating neural crest cells and subsequently in a subset of neural crest derivatives.

In Vitro Studies

To date, no in vitro study has provided a complete explanation for neurofibromatosis. Although fibroblasts from patients with neurofibromatosis have been cultured and shown to have slowed growth, diminished ability to bind and/or respond to EGF, and abnormal tyrosine and tryptophan metabolism, the fibroblasts are usually derived from skin, are not therefore presumably of neural crest origin and so may not shed light on the primary neural crest defect in the syndrome. A defect in EGF is plausible, however; neural crest cells from the Japanese quail possess receptors for EGF (10^5 receptors/neural crest cell) while EGF stimulates release of proteoglycans and hyaluronan and incorporation of ^{3}H-thymidine into neural crest cells. Erickson and Turley (1987) postulate that EGF may play a dual role in neural crest cells, promoting proliferation and repressing differentiation.[20]

All the neurofibromas removed by Riopelle and Ricardi (1987) from patients with von Recklinghausen neurofibromatosis and established in culture released NGF, implicating local production of NGF in the formation of neuronal elements within the tumors. However neither EGF, nor indeed any other growth or humoral factor, modulates, let alone alleviates, neurofibromatosis. The evidence for primary humoral imbalance in neurofibromatosis is tenuous.

Neurofibromas were reported in the goldfish, *Carassius auratus*, but the most promising animal model, the bicolor damselfish, *Pomacentrus partitus*, which inhabits the coral reefs off Florida and commonly displays tumors such as neurofibrosarcomas, neurofibromas, and Schwannomas associated with abnormal scale pigmentation. These phenotypes and the ability to maintain the tumors in vitro makes this a potential model system. The mechanism of transmission may differ from that typical of mammalian tumors; subcutaneous or intraperitoneal innoculation of tumor cells into tumor-free bicolor damselfish evoked Schwann cell tumors in 84% of fish inoculated. The tumors arose by transformation of host nerve cells and not from the tumor cells inoculated, prompting Schmale and Hensley (1988) to conclude that the tumor cells produced an infectious, transmissible factor such as a virus, which induced the tumors.[21]

APUDomas

The acronym APUD, for cells that exhibit **a**mine **p**recursor **u**ptake and **d**ecarboxylation, was coined by Pearse in 1969. Such cells include polypeptide-hormone-synthesizing cells of the gut, calcitonin-producing cells of the ultimobranchial gland, and cells of the pancreas, adrenal gland, and carotid body, which is a chemoreceptor. Pearse argued that cells that share this characteristic are all derivatives of the neural crest. Like the neurocristopathies, the APUD concept seeks to explain a suite of characters on the basis of commonality of developmental origin.

APUD cell types—there may be as many as 40—decarboxylate the catecholamine precursor DOPA (3, 4-dihydroxyphenylalanine) or the serotonin precursor 5-hydroxytryptophan. Either amine can be visualized using a fluorescent or radioactive label, facilitating cytochemical visualization of APUD cells. The amines and/or peptides function as hormones or as neurotransmitters for what is termed a diffuse neuroendocrine system involved in regulating the autonomic nervous system, the somatic division of the nervous system, and non-endocrine cells, including other APUD cells. Indeed, injection of a neurotransmitter such as serotonin (Box 11.1) into newborn mice leads to the production of tumors involving APUD cells (APUdomas), while exposure of newborn mice to adenosine causes dysplasia of neural crest and multiple neural crest tumors, including tumors containing cartilage and bone surrounded by melanocytes.[22]

Publication of the APUD concept led Nicole Le Douarin and her collaborators to search for experimental evidence for the possible neural crest origin of APUD cells. Approximately half of the 40 APUD cell types are of neural crest or neurectodermal origin, including calcitonin-synthesizing cells and cells of the carotid body, adrenal gland, and aortic paraganglia. APUD cells that are not neural crest in origin include the endocrine islet cells of the pancreas and the enterochromaffin cells of the gut and respiratory tract.[23]

The developing pancreas of rat embryos is colonized by two waves of neural crest precursors. An initial wave colonizes the bowel. These cells then acquire the ability to migrate as a second wave into the pancreas. Although murine pancreatic cells can produce neurite outgrowths and although the neurites contain neurofilament protein, they are endodermal in origin. Endoderm and neural crest therefore share the ability to form neurite outgrowths, just as mesoderm and neural crest share the ability to generate mesenchyme. Neither commonality of cell type nor

---

Box 11.1
Serotonin and the Migration of Neural Crest Cells

Serotonin may play a role in regulating neural crest cell migration and morphogenesis. Serotonin inhibits the migration of murine neural crest cells in Boyden chambers, and neural crest cells have the 5-HT serotonin receptor on their surfaces. Serotonin is expressed transiently in murine embryonic facial epithelia (palatal, tongue, nasal septum, maxillary, mandibular) at times (especially days 10 and 11 postconception) that correlate with morphogenesis. Shuey, Sadler, and Lauder (1992) propose that serotonin is a regulator of craniofacial morphogenesis.

Site-specific malformations occur following exposure of mouse embryos to serotonin-uptake inhibitors. The mechanisms involve decreased proliferation and increased cell death of mesenchyme located five or six cell diameters away from the epithelium, and increased proliferation in subepithelial mesenchyme. Shuey and colleagues conclude that serotonin is one of the agents regulating the epithelial–mesenchymal interactions that initiate craniofacial differentiation and morphogenesis. Injection of serotonin into newborn mice is followed by the death of neural crest cells and production of APUDomas, again implicating serotonin and serotonin receptors on neural crest cells in development and dysmorphogenesis.

Serotonin also affects morphogenesis of the cardiac cushion in murine embryos, perhaps because of its action on the neural crest derivatives in the cushion. Other neurotransmitters are also involved in neural crest cell differentiation; inhibition of norepinephrine with uptake inhibitors inhibits the adrenergic phenotype in clonal cultures of pluripotential trunk neural crest cells.[24]

---

commonality of metabolism need imply commonality of developmental origin, a caution to keep in mind when invoking the "explanatory" powers of germ layer of origin.[25]

Other evidence used to argue for commonality or origin comes from the associations of APUD cells in tumors. Bosman and Louwerens (1981) argued that APUD cells in ovarian teratomas or testicular teratocarcinomas are not of neural crest origin because they are associated with intestinal and respiratory epithelia, rather than with neuroepithelia. The migratory capabilities of neural crest cells renders suspect conclusions based on such associations or lack of associations. Furthermore, cell lines derived from teratocarcinomas produce fibroblasts and pigment cells and at least one cell line (AT805 murine teratocarcinoma cells) also forms cartilage in response to 10 μg/ml insulin, the association of pigment and chondrogenic cells being typical of a neural crest cell origin. Neoplastic changes can be experimentally produced in APUD cells of neural crest origin by administration of steroid hormones or endotoxin, providing an approach toward developing an animal model for neoplastic changes affecting these neural crest cells.[26]

Cardiac Defects and the DiGeorge Syndrome

Neural crest cells provide neuronal cells (parasympathetic cardiac ganglia), mesenchymal cells, and cells of the aorticopulmonary conotruncal septa, valves, and major vessels to the developing heart in the embryonic chick (see Chapter 7). Using a combination of DiI labeling and embryo culture, Fukiishi and Morriss-Kay (1992) demonstrated a cardiac neural crest that contributes cells to the heart after migrating through the visceral arches in rat embryos.

The hearts of many vertebrates contain cartilages. Depending on the species, these may either be occasional, apparently ectopic nodules or constitutive elements present in all individuals. Habeck (1990) observed islands of cartilage or bone at the margins of the carotid bodies in 0.5% of 1395 rats. He thought (as it turns out, correctly) that these arose from visceral arch mesenchyme, though Habeck thought they were of mesodermal origin. A neural crest origin is more likely.

Deletion of cardiac neural crest, i.e., cells that lie at the level of occipital somites 1 to 3, produces:

- persistent truncus arteriosus;
- malformations of the aorticopulmonary septum;
- transposition of the major heart vessels;
- high ventricular septal defects;
- single outflow vessels emerging from the right ventricle or over the ventricular septum; and
- abnormal outflow of blood from the heart, including increased velocity of blood flow through the dorsal aorta, and lower blood pressure.

Specific defects can be induced by controlling the time of neural crest cell deletion or by deleting cells adjacent to only one of somites 1 to 3. Deletion of more caudal neural crest cells (adjacent to somites 10 to 20) removes sympathetic cardiac innervation, producing sympathetically aneural hearts. Unexpectedly, deletion of the cardiac neural crest results in intrinsic myocardial defects, evidenced as defective contraction of heart muscle accompanied by ventricular dilatation, implicating a role for neural crest cells in regulating the functioning of heart muscle. More recently, a subpopulation of cardiac neural crest cells, whose removal from the septum by apoptosis may play a role in initiating cardiac myogenesis, has been identified in the chick.[27]

Heart defects arising from the deletion of cardiac neural crest are associated with underdevelopment of the thymus and parathyroid glands, both of which organs also receive contributions from neural crest (see Table 2.1). One significance of these defects and of the association of thymic and conotruncal cardiac abnormalities in the embryonic chick, is their resemblance to the DiGeorge syndrome of thymic agenesis in humans, in which heart, thymus, and parathyroid glands either fail to develop or are severely underdeveloped. Of 161 individuals with DiGeorge syndrome examined by van Mierop and colleagues in the mid-1980s , 97% had heart defects. Aortic arch development is abnormal and craniofacial malformations such as micrognathia, palatal, and ear defects characterize DiGeorge syndrome. This association in the one syndrome of three regions (thymus, cardiac, craniofacial region) containing cells derived from neural crest, coupled with the experimental evidence from embryonic chicks, implicates a defective neural crest in the etiology of DiGeorge syndrome. A major gene involved has been mapped to the 22q11.2 region in humans; the syndrome results from deletion of a 250 kb region of this gene. The phenotype of individuals with DiGeorge syndrome overlaps the phenotypes of individuals with velo-cardio-facial syndrome or with conotruncal anomaly face. Of individuals with these three syndromes, 80% or more have deletions of one allele of chromosome 22q11.2. Consequently, the three syndromes have been grouped as the CATCH-22 syndrome for **c**ardiac defects, **a**bnormal facies, **t**hymic hypoplastia, **c**left palate, **h**ypocalcemia, associated with chromosome 22 microdeletion.[28]

Cell communication is also altered in association with heart defects. The gap junction protein connexin 43 is expressed in the dorsal neural tube and in subpopulations of neural crest cells, including those of the cardiac neural crest. Transgenic mice that overexpress the Cx43 gene for connexin 43 display conotruncal heart defects (Ewart et al. 1977).

Whether the DiGeorge syndrome should be regarded as a neurocristopathy is perhaps a matter of semantics. I have not included it in Table 11.1 only because:

- it is clearly not *only* a neural crest defect—much of the heart is not of neural crest origin;
- many structures that arise from the neural crest are not involved; and
- the involvement of the neural crest may be secondary to a primary defect elsewhere in the embryo.

Nevertheless, our ability to unite abnormalities in such apparently disparate structures as the heart, face, and thymus through primary (or secondary) involvement of the neural crest highlights the utility of searching for the developmental basis of human abnormalities as far back into development as the primary embryonic layers from which the affected cells arise. One hopes that research into such associations will not stop at their classification, but will lead to a greater understanding of the mechanisms that cause the defects and ultimately to improved treatment or even prevention. Analysis of animal models has revealed a role for *Hox* genes.

A Role for *Hox* Genes

Disruption of *Hoxa-5* by gene-targeting in mice results in homozygous embryos that lack thymus and parathyroid glands and have throat, heart, and craniofacial defects (Chisaka and Capecchi 1991). These symptoms in mice parallel both the DiGeorge syndrome in humans and defects seen after ablation of the cardiac neural crest.

Hoxa-3 also plays a major role in the development of the thymus and thyroid, regulating the differentiation of mesenchymal neural crest cells (Manley and Capecchi 1995, 1998). *Hoxa-3* mutant mice are athymic and have severely underdeveloped thyroid glands, defects that may be mediated through *Hoxa-3* control of *Pax-1*. *Hoxa-3*, *Hoxb-3*, and *Hoxd-3* have partly overlapping functions in regulating migration of the ultimobranchial bodies that contribute the C-cells to the thyroid.

Polycomb (*Pc-G*) response elements within the regulatory regions of homeotic genes are important regulators of *Hox* genes in vertebrates as indeed they are of homeotic genes in *Drosophila*. *Polycomb* response elements maintain heritable states of transcriptional activity through modification of chromatin structure. At least nine such elements are known in mice. Targeted disruption of *Rae28*, the mouse homologue of *Drosophila* polyhomeotic, results in a rostral shift of the rostral expression border of *Hox* genes and subsequent posterior skeletal transformations. Defective neural crest derivatives include the eye, cleft palate, ectopic occipital bones, abnormal parathyroids, and thymus and heart defects. *Trithorax* (*TrxG*) response elements also regulate transcription of neural crest cells.[29]

12

Birth Defects

Seventy percent of all known birth defects involve the head and neck. Of these, 10% result from chromosomal abnormalities, 20% are single gene defects producing complex syndromes, while the remaining 70% are polygenic or multifactorial. Clinical geneticists recognize four classes of birth defects, summarized by Spranger et al. (1982) as:

- *malformations*—inherited defective developmental processes;
- *deformations*—mechanical disruptions to embryos;
- *disruptions*—external interference with developmental processes, as when a teratogen is administered; and
- *dysplasias*—neoplastic changes such as those discussed in the previous chapter.

A fifth class, *syndromes*, is often added, although some workers prefer to regard all four classes, along with neurocristopathies, as syndromes.

In an insightful analysis, Dufresne and Richtsmeier (1995) propose a parallel classification based on the responses of individuals to surgery—a poor response reflects a growth disorder, enhanced response a defect in which growth processes are not affected (Fig. 12.1). Although their scheme is based on clinical response,

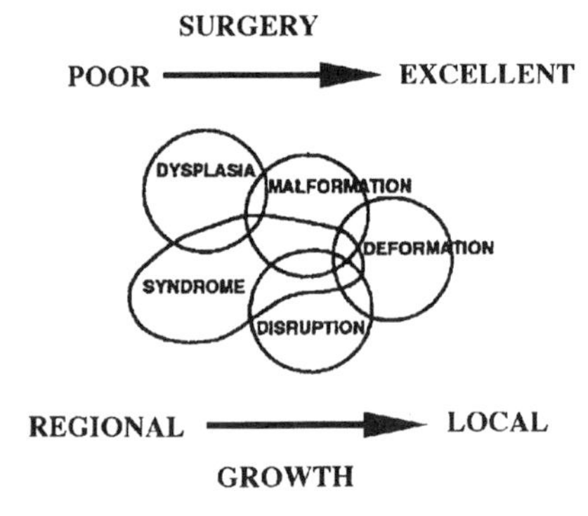

FIGURE 12.1. This classification of defects proposed by Dufresne and Richtsmeier (1995) is based on the response of individuals to surgery (from poor to excellent). Dysplasias and syndromes reflect regional influences on growth and poor response to surgery; deformations reflect local influences on growth and excellent response to surgery. See text for details.

they convincingly argue that differential responses to surgery are a true reflection of the causes of craniofacial defects. Dysplasias and syndromes—where the response to surgery is often poor—reflect regional influences on growth. Deformations, where response to surgery is usually very good, reflect local influences, while malformations and disruptions fall somewhere between local and regional influences on growth and poor to excellent aesthetic or functional response to surgery (Fig. 12.1).

In this chapter I discuss birth defects, especially craniofacial malformations and disruptions, by:

- providing a general outline of the various stages of neural crest cell development and how disruption of processes operating at these stages could result in craniofacial defects;
- considering the induction of defects following administration of excess vitamin A to pregnant mammals or to embryos in vitro;
- discussing mutations that affect neural crest cells (some of which were introduced in the previous chapter);
- pointing out some elements of species-specificity in mechanisms of abnormal craniofacial development; and, finally,
- considering regulation, the ability of embryos to compensate for effects of agents that would otherwise disrupt normal development. Regulation brings us full circle to controls of normal neural crest cell development and differentiation.

Susceptible Stages of Neural Crest Development

As with all developmental defects, the earlier in embryonic life that a neural crest defect occurs the more likely it is to be life threatening, major, and/or to produce a structural abnormality. Craniofacial defects involving neural crest cells could arise from:

- Total failure of development of the neural crest because of failure of the inductive interactions that specify neural crest cells during neurulation. Such a failure would almost certainly be lethal or result in major structural abnormalities to the entire embryo.
- Abnormalities in neural crest cell migration because of defects in the neural crest cells themselves, abnormal extracellular matrices, or abnormal cells with which neural crest cells normally interact. Such defects result in major structural abnormalities to entire units such as the face, cranium, jaws, ears, heart, or peripheral nervous system.
- Abnormal differentiation of neural crest cells because of intrinsic problems such as altered cell surface adhesion, defective structural genes, or defective inductive interactions. Individual neural crest derivatives such as bones, cartilages, teeth, or glands would either fail to form or form abnormally.

- Abnormalities in tissues such as muscles, nerves, or blood vessels with which neural crest cell derivatives normally interact. Minor structural defects or defective function would likely result.

Defective Migration

Neural crest cells migrate away from the neural tube by decreasing their adhesion to neural and epidermal epithelia, degrading the basal lamina, becoming mesenchymal, selectively adhering to specific components of the extracellular matrix, and moving along extracellular pathways. Migration of neural crest cells could be abnormal if any of these steps was defective. For example, a delay in onset of migration could result in neural crest cells failing to attain their final site at all, arriving in reduced numbers, or arriving too late to undergo the tissue interactions necessary for differentiation.[1]

Defective migration may occur not because of any abnormality in the neural crest cells themselves, but because of an abnormality in the extracellular environment through which the cells migrate or with which they interact. The extracellular spaces through which neural crest cells normally migrate fail to form in a number of mutation-based craniofacial defects. Consequently, migration is prevented and development is defective. One example, albinism in the axolotl, was discussed in Chapter 9.[2]

Human craniofacial defects with a mutational basis, such as agnathia (absence of jaws), first arch malformations, or Treacher Collins syndrome, are interpreted as resulting from varying degrees of defective migration of cranial neural crest cells. As one example, abnormal migration of mesencephalic neural crest at stages 10 and 11 leads to incomplete development of the chondrocranium and defective midline fusions in human embryos with holoprosencephaly and cyclopia. Defective craniofacial development can be induced experimentally by delaying the migration of neural crest cells. High in vitro concentrations of glucose inhibit migration of cranial neural crest cells from both normal and diabetic rats, suggesting that glucose concentration may be a contributing factor to defects arising in embryos born to diabetic mothers. Exposure of murine embryos to ethanol disrupts migration to such an extent that neural crest cells migrate inwards toward the lumen of the neural tube rather than away from the tube.[3]

Defective Proliferation

Embryos are particularly sensitive to perturbation away from normal development when growth is rapid or when growth rate is changing. Such times are *critical* or *sensitive periods*. Consequently, disruption of the normal rate of division of neural crest cells can result in craniofacial defects. Inhibition of cell division and/or migration, which can led to extensive cell death, follows exposure of embryos to X-rays, steroid hormones, or alcohol. Although neural crest cells divide as they migrate, cell division increases once they reach their final site.

Such enhanced proliferation reflects both increasing interactions among neural crest cells and interactions with epithelia and is critical for the formation of cellular condensations and initiation of differentiation.[4]

Enhanced Cell Death

Apoptosis (programmed cell death) has achieved much prominence in recent years as a mechanism that removes excess cells during differentiation and that molds and sculpts embryonic tissues during morphogenesis. As indicated in Chapter 6, Béard described cell death as a normal part of neuronal development in Rohon-Béard cells in skate embryos more than 100 years ago.

BMP-4, which promotes migration and differentiation of murine trunk neural crest, induces apoptosis in hindbrain neural crest cells. Apoptosis is very specific—cells from rhombomeres 3 and 5 undergo apoptosis as a result of interactions with even-numbered rhombomeres, activating *Msx-2* and BMP-4 in the odd-numbered rhombomeres. *Msx-2*$^{-}$ transgenic mice die before birth and have multiple craniofacial malformations. The balance between cell survival and apoptosis, which is regulated by *Msx-2*, is disrupted in these transgenic embryos. Injection of BMP-2 or BMP-4 into the neural tube increases expression of *Msx* genes, while expression of *Msx-2* in rhombomeres from which neural crest cells normally emigrate induces cell death and eliminates crest cell migration. Even-numbered rhombomeres may use BMP-4 as the signaling molecule; they contain BMP-4, while BMP added to rhombomeres 3 and 5 up-regulates both BMP-4 and *Msx-2*, as do even-numbered rhombomeres.[5]

Later in development, BMP-2 and BMP-4 selectively regulate apoptosis in such derivatives of cranial and trunk neural crest as the first and second arches (especially the mandibular arch) and progenitor cells of sympathoadrenal neurons. Neuronal survival requires FGF and NGF, a growth factor dependence that is preconditioned by exposure to BMP-2 and BMP-4. The role of these and other growth factors in differentiation of neural crest derivatives is discussed in Chapter 10. Early activation of these normal patterns of cell death or their expansion to cells that would not normally undergo apoptosis, can contribute to defective development. As discussed, below vitamin A–induced craniofacial defects are mediated, in part, through enhanced cell death.[6]

Trithorax-group response elements maintain stable states of transcription in *Drosophila* and in mammals. *Mll*, the mammalian equivalent of *Drosophila trx* prevents excess cell death in murine branchial arch mesenchyme. *Mll* is also a posterior regular of *Hox* genes. Indeed, *Hox* expression is abolished in *Mll* homozygous null (–/–) embryos. Embryos in which *Mll* has been knocked out have severely reduced mandibular and defective maxillary arches because of enhanced apoptosis in arch mesenchyme but not in arch ectoderm. Gene expression in mouse embryos become dependent on *Mll* between 9.5 and 9 days of gestation, an interval that marks an important phase of chromatin reorganization (Yu et al. 1998).

Defective Induction

As discussed in Chapter 10, epithelial–mesenchymal interactions play key roles in initiating the differentiation of cells that arise from the neural crest. Abnormal interactions can lead to defects. The neurocristopathies discussed in the previous chapter are prime examples, as are many craniofacial defects involving the skeleton. Specific craniofacial defects interpreted as resulting from defective secondary tissue interactions are von Recklinghausen neurofibromatosis, tumors and neurocristopathies, holoprosencephaly, and facial dysmorphogenesis.

In the following section, defects that follow administration of vitamin A will be used to illustrate susceptibility of neural crest cells at the various stages of neural crest cell development.[7]

Vitamin A, Craniofacial Defects, and the Neural Crest

Direct Action In Vivo

At low concentrations, vitamin A is involved in determination of cell number, differentiation and dedifferentiation status during the regeneration of neural crest cells (Baranowitz 1989). When administered to pregnant women at concentrations of 50,000 to 100,000 IU/day, however, vitamin A is teratogenic, producing abnormalities that simulate inherited craniofacial defects (first-arch syndromes) that are also known to arise through mutations. Administration of vitamin A to other pregnant mammals or cultivation of mammalian embryos in the presence of vitamin A produces craniofacial defects similar to those seen in humans. Retinoic-acid and cellular retinoic-acid-binding protein (CRABP-I and -II) localize from 9.5 days onward in murine embryos to neural crest, neural crest derivatives (craniofacial mesenchyme, the visceral arches, dorsal root ganglia, enteric ganglia of the gut), and to rhombomeres 2 and 4 to 6 of the hindbrain, somites, major blood vessels, limb bud mesenchyme, and flank and facial muscles (Fig. 12.2). Although retinoic acid affects derivatives of both cranial and trunk neural crest, the effect on cranial derivatives is much greater and leads to more major anomalies.[8]

Respecification by retinoic acid of the identity of neural crest cells is stage-dependent. This was demonstrated by Lee et al. (1995) in rat embryos to which retinoic acid was administered at 9 or 9.5 days of gestation and in which both cell migration and cell fate were assayed following injection of DiI into premigratory neural crest cells. Administration of retinoic acid at 9 days decreased the size of the first visceral arch and initiated ectopic migration of crest from rhombomeres 1 and 2 into the second arch. (Normally these crest cells would migrate into the first arch.) Administration at 9.5 days was followed by fusion of the first and second arches that was not associated with ectopic migration of neural crest cells.

Administration of retinoic acid (60 mg/kg body weight) to female golden Syrian hamsters on day 8 of pregnancy produced defects of the facial skeleton in all

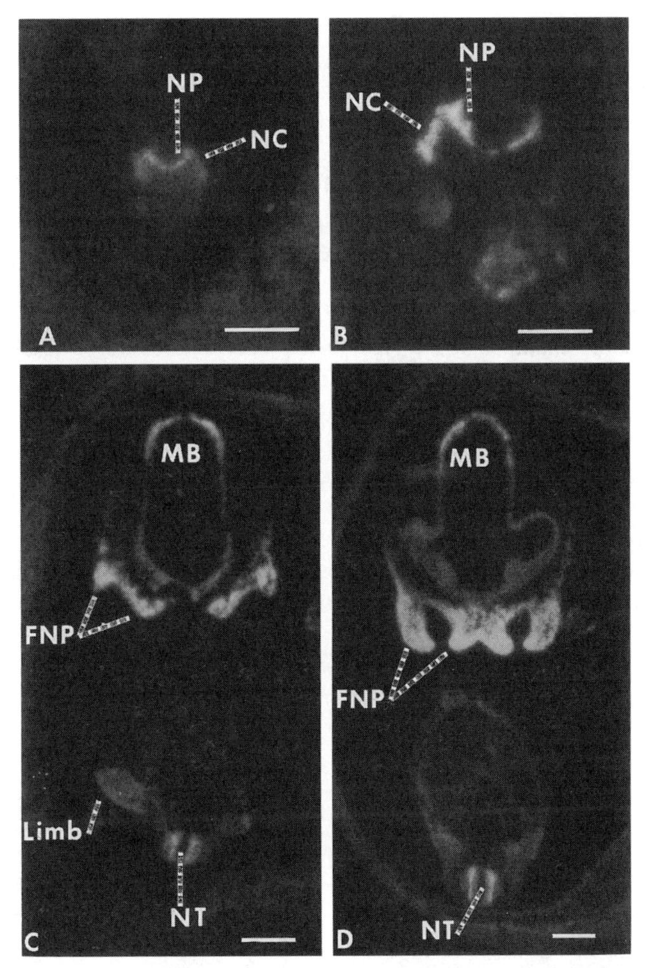

FIGURE 12.2. Injection of ^{14}C-labeled retinoid to pregnant mice allows retinoid to be visualized during embryonic development, as can be seen in these sections through embryos between 8.5 and 11.5 days of gestation. At 8.5 and 9.5 days (A, B), radioactivity (shown in white) is seen in the neural plate (NP) and in early migrating neural crest cells (NC). At 10.5 and 11.5 days (C, D), radioactivity is seen in the frontonasal processes (FNP), the roof of the midbrain (MB) and in the neural tube (NT). Some radioactivity is also seen in the limb buds. Bar = 500 μm. Reproduced from Dencker et al. (1990), *Development* 110:346, with the permission of The Company of Biologists Ltd., from figures kindly supplied by L. Dencker.

surviving embryos. As early as 4 hours after administration, neural crest cells displayed such ultrastructural signs of cytological damage as swollen mitochondria and dilated nuclear and endoplasmic reticular membranes. Dead neural crest cells were first seen 8 hours after retinoic acid administration with maximal numbers

24 hours after administration. At later times, surviving neural crest cells had not migrated as far as those in control embryos, producing local deficiencies in cell density and subsequent skeletal defects. This study demonstrates the dramatic, early, and extensive effect of retinoic acid on cranial neural crest cell survival and migration and consequent defects in the facial skeleton.[9]

A model developed by Dickman and colleagues (1997) combines depletion of vitamin A and then genetic crossing to unaffected individuals, who are given oral supplements of vitamin A at times coinciding with predicted stages of embryonic development. Effects on the embryos are assayed. Defects in the neural crest, eye, and nervous system resulted from the direct action of vitamin A on neural crest and is an example of the type of approach that could be used to analyze individuals who carry mutated retinoic acid receptors or genes.

Nonmammalian vertebrates have also been used in teratogenic studies to understand mechanisms of action of vitamin A. Administration of vitamin A to embryonic chicks or quail disrupts segmentation of rhombomeres 4 to 8 and caudal hindbrain development (but does not disrupt dorso-ventral patterning of the neural tube), prevents migration of neural crest cells, and leads to craniofacial defects such as severe underdevelopment of the branchial arches (Figs. 12.3 and 12.4). In another avian model, 13-cis-retinoic acid is injected into embryos

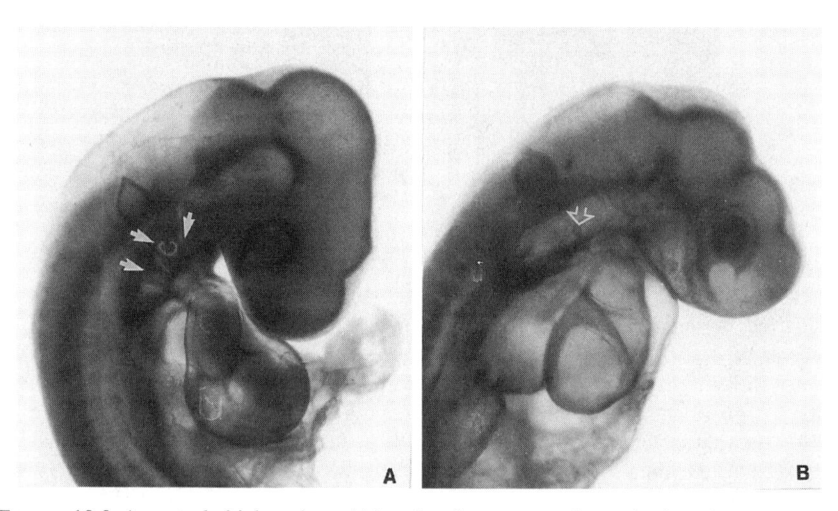

FIGURE 12.3. A control chick embryo (A) and embryo exposed to retinoic acid (B) to show the effect of retinoic acid on the branchial arches, which are well-developed in the control embryo (arrows in A) but severely underdeveloped in the embryo treated with retinoic acid (arrow in B). Also note the underdeveloped forebrain and displaced otic vesicle in the treated embryo. Reproduced from Moro-Balbás et al., Retinoic acid induces changes in the rhombencephalic neural crest cells migration, and extracellular matrix composition in chick embryos, *Teratology* 48:197-206, Copyright © (1993), from a figure kindly supplied by Jose Moro-Balbás. Reprinted by permission of Wiley-Liss Inc., a subsidiary of John Wiley & Sons, Inc.

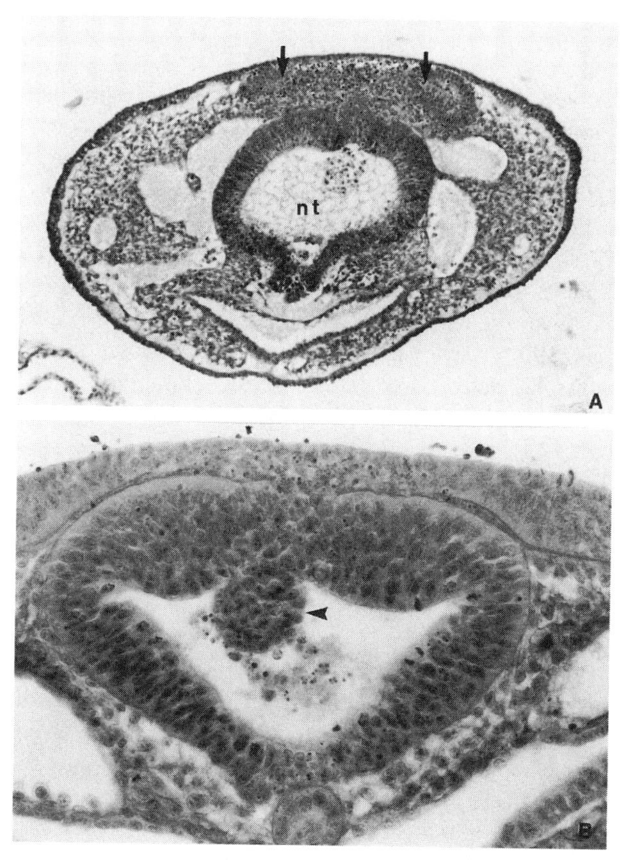

FIGURE 12.4. Cross sections through the neural tube of a chick embryo to illustrate the effect of retinoic acid on neural crest cell migration. (A) is a transverse section through the anterior rhombencephalon. Neural crest cells (arrows) have accumulated above the neural tube (nt), their normal migration having been inhibited. (B) is a transverse section through the posterior rhombencephalon. Neural crest cells (arrowhead) have migrated ectopically into the lumen of the neural tube. Reproduced from Moro-Balbás et al., Retinoic acid induces changes in the rhombencephalic neural crest cells migration, and extracellular matrix composition in chick embryos, *Teratology* 48:197-206, Copyright © (1993), from a figure kindly supplied by Jose Moro-Balbás. Reprinted by permission of Wiley-Liss Inc., a subsidiary of John Wiley & Sons, Inc.

between 2 and 6 days of incubation and embryos examined at 14 days. Craniofacial and cardiovascular defects were both induced. Interestingly, defects affecting craniofacial mesenchyme are greatest if retinoic acid is injected *after* neural crest cell migration, implying a greater action on the localization and differentiation of neural crest cells than on migration in this species. Problems associated with species specificity are taken up later in this chapter.[10]

Craniofacial Defects

Vitamin A has specific actions on the facial processes, affecting outgrowth of the facial skeleton by acting specifically on the neural-crest-derived mesenchyme of those processes. By varying the dose of vitamin A administered (within the ranges of 50,000 to 100,000 IU) varying levels of severity are obtained—from defects that affect the whole facial complex to those that only affect one side of the face. The former defects are analogous to Treacher Collins syndrome in humans, the latter to hemifacial microsomia or mandibulofacial dysostosis, in which the mandible, zygomatic arch, and middle ear ossicles are abnormal. Grant and colleagues (1997) showed that mouse embryos exposed to retinoic acid on day 9 do not display defects associated with the neural crest of the frontonasal processes, indicating temporal specificity of retinoid action, in addition to the dosage effects.

In typical experiments with mammalian embryos, administration of vitamin A to pregnant mice results in the failure of development of elements of the ear and craniofacial skeleton or development of skeletal tissues in abnormal or ectopic positions. For example, cartilage does not normally develop in mammalian maxillary processes, but ectopic maxillary cartilages are found in rodent embryos treated with vitamin A. In fact, an entire craniofacial structure can develop in an abnormal position in rat embryos treated with vitamin A; for example, the development of the otic capsule in the first rather than the second visceral arch. In the ear defects, the tympanic ring (a membrane bone derived from neural crest cells) is lost while the external acoustic meatus is altered. Both structures are developmentally dependent, the tympanic ring inducing the external acoustic meatus. This developmental coupling is reflected in mutant phenotypes: *Hoxa-2* mutant mice have duplicated tympanic rings and an altered external acoustic meatus; retinoic acid and the *Hox* gene play roles in the developmental interactions between these two ear elements; see below.[11]

In an innovative test of how structures become integrated as a result of shared developmental or functional attributes, Miriam Zelditch (1988) examined ontogenetic variation in skull and limb development in laboratory rats and the cotton rat, *Sigmodon fulviventer*. In part, her interest is in the switch from developmental to functional integration during ontogeny. A growth-based model provided the major developmental explanation for covariance of limb and skull development in the laboratory rat, but various models, including those that incorporate patterns of embryonic induction, explained some of the ontogenetic variation. For the cotton rat, Zelditch and Carmichael (1989) compared models based on embryological origin (neural crest vs. mesoderm), visceral arch of origin (mandibular vs. maxillary), and functional integration (muscle function). Characters derived from the same visceral arch co-varied early in ontogeny; embryological origin from neural crest is a major determinate establishing developmental fields of action. Later in ontogeny, repatterning is explained by functional rather than developmental integration. As might be expected, features of the neurocranium varied less than craniofacial or limb skeletal characters. Structures therefore develop as independent, integrated units because of selection operating both on early developmental and later functional features.[12]

Mechanisms of Action

Abnormal migration of subpopulations of neural crest cells is the most ready explanation for several defects involving ectopic cartilages in the visceral arches in human embryos, although alternate explanations for such defects as hemifacial microsomia include local hemorrhaging. It seems more likely, however, that hemorrhaging is a secondary consequence of primary defects in chondrogenic neural crest cells, which in hemifacial microsomia illustrate that mandible, zygomatic arch, and middle ear ossicles form a developmental field of action.[13]

Primate embryos, such as fetal macaques (*Macaca nemestrina*), exhibit defective mandibular development if vitamin A is administered maternally during neural crest cell migration and craniofacial morphogenesis (20 to 44 days of gestation), establishing a further link between nonhuman animal models and human syndromes.[14]

The conclusion that vitamin A alters the migration of neural crest cells is based on the following:

- Administration of the vitamin when neural crest cell are migrating evokes the syndromes, administration earlier or later does not.
- Neural crest cells accumulate in abnormal embryonic locations after administration of vitamin A.
- Migration of neural crest cells is inhibited when they are maintained in vitro in the presence of vitamin A.[15]

As might be expected, given its multitude of roles in cell metabolism, vitamin A affects other important events in neural crest cell development. Vitamin A:

- slows the division of neural crest cells by increasing the length of the mitotic cycle;
- enhances naturally occurring programmed cell death in the neuroepithelium;
- causes blebbing of neural crest cells;
- slows deposition of such extracellular matrix products as hyaluronan;
- inhibits interaction between neural crest cells and extracellular matrix products (an interaction that is required for normal migration to take place);
- acts directly on neural crest mesenchyme to inhibit differentiation of membrane bones in the embryonic chick; and
- is taken up by and accumulates in neural crest cells themselves, in a manner that is both time- and site-specific.

In part, retinoic acid may exert its actions through growth factors such as TGFß-1 and -2, both of which are expressed in the neuroepithelium of 8.5- to 10.5-day-old mouse embryos and both of which are down-regulated by retinoic acid at the neural plate stage of embryogenesis. Down-regulation is greatest in neuroepithelium and cranial neural crest. FGF-3 may also be a downstream target of retinoic acid. In avian embryos, FGF-3 is expressed in the epiblast, in rhombomeres and then at rhombomere boundaries, in pharyngeal endoderm, and in the ectoderm of the second and third visceral arches. Similarly, in murine embryos, FGF-3 is expressed in the brain, cranial surface ectoderm, second arch ectoderm, and otic placodes, and can be used as a marker of changes after administration of retinoids. Correlations between FGF-3 expression and craniofacial defects were used to demonstrate the dual identity of the fused visceral arches and that pro-rhombomere 2, the second visceral arch, and the otic ectoderm represent a unit with respect to response to retinoids. That unity has an early developmental history. The gene for a bZIP transcription factor (*kreisler* in mice, *valentino* in the zebrafish) regulates *Hox* gene expression, and so plays a role in subdividing the pro-rhombomere into rhombomeres 5 and 6. *Valentino*-mutant zebrafish and *kreisler*-mutant mice display similar phenotypic defects in visceral arch derivatives and in the inner ear, indicating that regional behavior can be traced back to very early patterning events in the neural tube.[16]

Indirect Effects

Vitamin A also affects neural crest cells indirectly. Localized hematomas are associated with vitamin-A-induced craniofacial defects in laboratory animals. These hematomas, which are often quite large, prevent craniofacial mesenchyme from forming in sites such as the mandibular arch or developing ear, with the consequence that mandibular or ear structures fail to form.

Vitamin A may act on the epithelia that surround developing craniofacial structures. Epithelia restrain mesenchymal growth so that a defective epithelium could lead to abnormal growth or differentiation of the subjacent mesenchyme and a craniofacial defect. Defects that follow administration of vitamin A are likely to be based on one or more mechanisms, depending, in part, on time of exposure and dose.[17]

Evolutionary Origins of Sensitivity to Retinoic Acid

Reproducible defects and deletions in neurocranial and pharyngeal skeletal elements occur in fish (*Fundulus heteroclitus*, *Danio rerio*) exposed to exogenous retinoic acid. The correlations between the defects and the dose of retinoid administered, and the fact that retinoic acid shifts the expression boundary of *Hoxd-4* rostrally in a dose-dependent manner in both the central nervous system and in the pharyngeal arches of the flounder (*Paralichthys olivaceus*), suggests an ancient origin for these actions of retinoic acid.[18]

Patterned effects of retinoic acid on neural and facial structures did not arise with the craniates but were present in such chordates as amphioxus (*Branchiostoma*), in which, as in vertebrates and as discussed in Chapter 3, retinoids activate *Hox* genes. If neural tube sensitivity to vitamin A is a chordate feature, we would expect it to be present in jawless vertebrates and indeed, exposure of embryos of the lamprey *Lampetra japonica* to all-*trans* retinoic acid leads to rostral truncation of the nervous system and loss of the mouth, pharynx, esophagus, heart, and endostyle, i.e., loss of many of the structures that characterize the vertebrate head (Kuratani et al. 1998). Response was both dose- and stage-dependent, but did not extend to the myomeres, which were normal.

Synthetic Retinoids

A synthetic retinoid, 13-cis-retinoic acid, under the trade names Isotretinoin or Accutane®, initially commercially available in the United States, is now used worldwide by millions of individuals as one of the most effective treatments for severe acne. Side effects of the treatment for adults are said to be minimal; 10 of 96 patients treated for 4 months with low doses of Accutane developed hyperostoses on the anterior margins of their cervical, thoracic, or lumbar vertebrae. There are, however, side effects associated with the eyes including inflammation, dry eye syndrome, intolerance to contact lenses, alterations in refraction, photosensitivity, and reduced night vision.

Accutane was licensed for distribution with the warning that it not be taken by pregnant women. In the mid-1980s, Webster and colleagues documented birth defects in 24 children born to 200 women exposed to the vitamin during the first weeks of pregnancy. One of these children, who would now be 14 years old, was blind, deaf, and demonstrated no neurological advance after birth. Others had major auricular malformations (including anotia and severe microtia) and central nervous system malformations. Three and a half days exposure at 1 month of gestation is sufficient to produce a defect(s). Although topical application appears not to be associated with birth defects, systemic administration is associated with birth defects, fetal resorption, and stillbirths. Defects embrace craniofacial and cardiovascular anomalies—cleft palate, depressed midface, jaw anomalies, hypoplastic aortic arches, and septation defects of both atria and ventricles. The external ear is very commonly affected.[19]

Sensitivity of different animals to Accutane varies more than 100-fold. Humans are the most sensitive, followed by monkeys, rabbits, hamsters, mice, and rats; humans are 16 times more sensitive than hamsters. Clearly, one has to extrapolate results from other animal models to humans with care.

In a series of laboratory animal studies undertaken by Webster and colleagues, Accutane was administered to pregnant mice during early embryogenesis and craniofacial development. Malformations induced included abnormal development of the external ear, underdevelopment of the mandibles, maxillae, and cleft palate, and heart and thymus defects. These are all structures known to arise from the neural crest, and indeed SEM and TEM studies by this group documented degeneration or death of migrating neural crest cells in treated embryos. Defective neural crest, mesenchyme, and placodally derived ganglia (such as the trigeminal ganglion) produced syndromes resembling the human Treacher Collins and hemifacial microsomia syndromes.

In a second series, embryos cultured in media containing Accutane at concentrations of 500 ng/ml exhibited decreased cell-substrate adhesion, providing a clear indication that the vitamin targets neural crest cells and inhibits their migration. Other migratory cell populations such as myoblasts and cells of epidermal placodes are also inhibited.[20]

Accutane and a second synthetic retinoid, tretinoin (Retin-A), affect the expression of several key genes in visceral arch mesenchyme of murine embryos, including retinoic acid receptor γ, CRABP-1, the cAMP response element protein, TGFß-1, and nicotine acetylcholine receptor sub unit-α. The retinoid X receptor-γ nuclear receptor gene is expressed very early in migrating chick neural crest cells such as the cells entering the somites (and at H.H. stages 24 to 27 in the peripheral nervous system, dorsal root, and cranial ganglia), and may be an earlier marker than HNK-1 for migrating neural crest cells. Expression in murine embryos is even more ubiquitous, initially in presomitic mesoderm at 8 days postconception, then in the frontonasal processes, visceral arches, limb, and sclerotomal mesenchyme at 9.5 to 11.5 days postconception; in all precartilaginous mesenchyme at 12.5 days, and in all cartilages, keratinized epithelia, teeth, and vibrissae from 13.5 days onward. Such ubiquitous distribution speaks to why exogenous retinoids have such widespread actions on so many tissues.

A further vitamin A derivative, the ester etretinate (Tigason®) was released in December 1986 for the treatment of psoriasis. Administration of etretinate (100 mg/kg body weight) to mice at 8.5 days of gestation produces first and second arch defects, including reduction of Meckel's cartilage of the lower jaws and ear abnormalities.[21]

Serum drawn from human subjects within 5 hours of ingestion of 30,000 µg of vitamin A retards craniofacial growth and produces craniofacial defects in rat embryos cultured in the serum. Serum not supplemented with vitamin A has no adverse effect on embryonic development. Unlike Accutane, which has a half-life of 16 to 24 hours and is rapidly metabolized, etretinate is stored in tissues, its metabolites having been detected up to 3 years after initial ingestion. It is unclear whether such long-lived metabolites could cause abnormalities in embryos that develop from eggs fertilized several years after administration of the vitamin. Because of the prolonged half-life of etretinate and a potential for birth defects that persists, women are advised not to conceive for at least 2 years following discontinuation of treatment.[22]

In a comparative study of the effect of various synthetic retinoids on mouse embryos, Kochhar and his colleagues (1996) determined that three compounds that acted exclusively on retinoic acid receptors and two that affected neither retinoic acid nor its receptors were nonteratogenic. While these results highlight multiple modes of action of synthetic retinoids, such studies do raise the hope that nonteratogenic compounds effective in the treatment of acne well be developed.

Defects Following Disruption of the *Hox* Code

Confirmation that *Hox* genes pattern the branchial region (see Fig. 2.15 [color plate] and Chapter 2) comes from studies in which *Hox* genes are disrupted following homologous recombination in embryonic stem cells and introduction into the germ line of mouse embryos. Elegant as such experiments are, functional overlap between *Hox* genes and the ability of one *Hox* gene to compensate for the inactivation of another means that loss of function need not necessarily result in an altered phenotype, even though the eliminated gene is involved in the production of that phenotype. A sufficient connection between *Hox* genes and congenital malformations had been established by 1992 that Redline and associates were able to produce an insightful overview of the potential role of *Hox* genes in such malformations.

Hox genes that are regulated by vitamin A have also been uncovered. *Msx-1*, which is actively involved in craniofacial development, is expressed more strongly and ectopically in avian embryos that are deficient in vitamin. Expression in premigratory neural crest cells is especially affected, suggesting one molecular pathway activated by vitamin A. One or two examples of studies disrupting *Hox* genes are discussed below; *Hoxa-1* and *Hoxa-3* play especially important roles in the neural crest and its mesenchymal derivatives and/or in the tissue interactions that induce them.

Using zebrafish as the model organism, the retinoic acid phenotype was phenocopied by overexpression of *Hoxa-1* by Alexandre et al. (1996). Overexpression after injection of RNA into zebrafish eggs alters rostral hindbrain growth and changes the fate of neural and neural crest cells arising in rhombomere 2. Resulting abnormalities early in development include abnormal growth of the hindbrain and defective development of neural crest cells that arise from rhombomere 2. Older embryos show diminished growth or complete absence of the mandibles and an increase in second arch structures as mandibular-destined cells are rerouted. Meckel's cartilage and the palatoquadrate of the mandible (first visceral arch) fail to form, while second arch hyoid cartilages are enlarged and partially duplicated; the mandibular arch skeleton in the chick has a composite origin from neural crest, arising from midbrain and rhombomeres 1, 2, and 4 of the hindbrain.[23]

Only five rhombomeres develop in the hindbrain of mice mutant for *Hoxa-1*. Rhombomere 5 is absent and r4 is reduced. Disruption of *Hoxa-1* results in ear, hindbrain, and visceral arch defects consistent with a role for the gene in specific neural crest populations. Mice with compound mutations of *Hoxa-1* and *Hoxb-1* display patterning defects of r4-6 and major disruptions of middle and external ear development, and lack all second arch structures. These results of interaction between first and second arch, coupled with the demonstration from my laboratory of the development of the ear from first and second arch elements, provide evidence of genes that regulate developmental interactions between the first and second arches.[24]

Disruption of *Hoxa-3* in mice produces a different suite of defects centered on the neural crest, with defects in the visceral arches, thymus, thyroid, hyoid bone, and hyoid arch skeleton. While such changes are not equivalent to homeotic mutations, homeotic transformations can be elicited if boundaries of *Hox* genes are expressed more rostrally; disruption of *Hoxa-2* induces defects in the branchial region corresponding to the altered rostral expression. Neural crest mesenchyme derived from rhombomeres that normally populate first and second visceral arches, fails to produce normal second arch derivatives but does produce first arch structures. The result is a homeotic transformation of the entire second arch skeleton into a first arch skeleton.

The normal rostral level of expression of *Hoxd-4* is at the first cervical somite. When *Hoxd-4* is expressed more rostrally, the occipital bones (which lie rostral to the cervical vertebrae) are transformed into bones that mimic cervical vertebrae.[25]

Gale and her colleagues used ectopic expression of retinoic acid in rhombomere 4 of stage 10 rat embryos to demonstrate that retinoic acid alters expression of zinc-finger genes such as *Krox-20*, eliminates expression of *Hoxb-1*, but leaves expression of *Hoxa-2* unaffected, resulting in development of facial ganglia more rostrally than is normal. (*Krox-20* is also the upstream regulator of *Hoxb-2* in the hindbrain.) Retinoic acid modification of the *Hox* code of the hindbrain leads to transformation of rhombomeres 2 to 3 into rhombomeres 4 to 5.[26]

Mutations and Birth Defects

Craniofacial defects similar to those produced by exposure of laboratory animals to excess vitamin A also arise as the result of mutations. Examples of such mutant genes are *Amputated, Looptail*, and *Patch* in the mouse, and the mutations associated with neurocristopathies in humans discussed in Chapter 11.[27]

Amputated, a mutation that results in development of a foreshortened face, results from defective interactions between and among neural crest cells, with resulting inhibition of craniofacial growth and morphogenesis.

Looptail (*Lp*) and *Patch* led to premature differentiation; neural crest cells migrate along abnormal pathways because of precocious alterations in the extracellular matrices they encounter. Consequently, neural crest cells begin to differentiate before sufficient cells have accumulated for normal craniofacial development and growth. The hindbrain and spinal cord fail to close in *Lp*, while neural crest cells remain attached to the neural folds, failing to migrate, indicating the importance of neural fold closure for initiation of migration.[28]

The *Splotch* mutation, an intragenic deletion of the transcription factor *Pax-3*, is associated with missing or small spinal ganglia and spina bifida, although ganglionic reduction is independent of the neural tube defects. *Splotch* prevents neural crest cells from migrating away from the neural tube; neural tubes from *Sp* or *SP*^{4} mutant mice cultured in vitro exhibit a 24-hour delay in the initiation of neural crest cell migration, in part because of alterations in N-CAM, an effect that is also seen after retinoic acid treatment of avian embryos, when neural crest cells retain N-CAM and so cannot migrate normally. Migration of vagal neural crest is especially affected. Transplantation of neural tubes between *Splotch* mice and chick embryos reveals that the migration defect is not intrinsic to the neural crest but rather arises from abnormal cell interactions. *Splotch* in humans is manifested as Waardenburg syndrome (see Table 11.1).[29]

Mice homozygous for *Splotch* have defective aortic arches, thymus, and thyroid and parathyroid glands. Normally, cells from the cardiac neural crest migrate through visceral arches 3, 4, and 6 to colonize the outflow tract of the developing heart. Sp^{2H}, one of five mutant alleles of the *Splotch* locus, results in persistent truncus arteriosus and related conotruncal heart defects. The transient expression of *Pax-3* normally seen in the developing hearts of mice at 10.5 days of gestation is not seen in embryos carrying the *Splotch* mutation. From their studies with Sp^{2H} mutant mice, Conway and his colleagues argue that *Pax-3* is required for cardiac neural crest cell migration. Indeed, because of important actions at the condensation stage of organogenesis (see Fig. 10.6), the *Pax* family of genes are key players in the development of many organs—kidneys, eyes, ears, nose, limbs, vertebrae, and brain.[30]

An interesting addendum to the action of the *Splotch* mutant, one that may apply to many mutants, is reduction in the frequency of neural tube defects in *Splotch* mice exposed to retinoic acid, provided that exposure is at a specific stage of neural tube development. Thus exposure to 5 mg/kg maternal body weight at day 9 of gestation stimulates growth of the neural tube with concomitant reduction in neural tube defects, while exposure at day 8 is ineffective (Kapron-Bras and Trasler 1985). Environmental agents can reduce defects that we normally consider and describe as genetically determined.

Other inherited defects have specific effects on particular populations of neural crest cells:

- Familial facial osteodysplasia in humans is an inherited syndrome in which bone with a neural crest origin is entirely absent, but in which neural-crest-derived cartilages and bones and cartilages of mesodermal origin are present and normal.
- *Lethal-Spotting* and *Piebald-Lethal*, mutations in the mitogenic growth factor endothelin-3 and the endothelin-B receptor, respectively, result in megacolon, an absence of the enteric ganglia in the terminal 2 to 3 mm of the bowel. (Fourteen of some 50 syndromes involving neural crest cells demonstrate aganglionic megacolon. Hirschsprung disease [see Table 11.1] is perhaps the most common.) Migration of neural crest cells is normal in these mutants, the defects arising because an abnormal non-neural-crest cell environment in the bowel will not permit neural crest cells to colonize the bowel wall, resulting in the development of an aganglionic segment.[31]

Continued study of these mutations affecting laboratory animals will shed further light on how defective neural crest cells or a defective neural crest cell environment produces the mutant phenotype and, *en passant*, will increase our understanding of normal neural crest cell development.

Species Specificity

Having presented data on the validity of other animal models for human syndromes, but having noted differential sensitivity of animals to Accutane, I sound a word of caution about species specificity.

Chemicals that produce phenotypically identical craniofacial defects in species from various vertebrates classes—reptiles, birds, mammals—may exert their teratogenic action at quite different stages during the migration, differentiation, and growth of neural crest cells and neural crest derivatives. This point is illustrated by a study that utilized 5-fluoro-2'-deoxyuridine (FUDR), an antimetabolite that blocks the synthesis of DNA. Administration of FUDR results in micromelia (reduced or shortened lower jaws) in representative reptiles, birds, and mammals, but affects quite different developmental processes:

- selectively destroying migrating neural crest cells in alligator embryos;
- inhibiting mandibular growth in embryonic chicks; and
- selectively inhibiting the differentiation of chondroblasts and chondrocytes of Meckel's cartilage in embryonic rats.[32]

Even when administered coincident with neural crest cell migration, FUDR fails to elicit micromelia in the embryonic chick (Hall 1987b). In contrast, micromelia is induced in alligator embryo exposed to FUDR in vitro only when cells destined for the mandibular arches are migrating. Palate development is also disrupted by exposing alligator embryos to FUDR when palate-destined neural crest cells are migrating, but not earlier or later. Such species, and indeed class specificity, argues for caution when extrapolating mechanisms of teratogen action from one vertebrate group to another.

Regulation

Even when a mutant gene or teratogen has a deleterious effect on an embryonic population of cells, embryos may still develop perfectly normally. This is because of the developmental phenomenon of *regulation*, the ability of embryos to compensate for loss that would otherwise result in abnormal development.

It has long been known that amphibian and avian embryos can compensate for removal of neural crest cells. Avian and mammalian embryos can also regulate for deficiencies that occur during neural crest cell migration. Exposure of rodent embryos of 9 days of gestation to Mitomycin C temporarily blocks cell division, which is reinitiated when the drug is metabolized. Snow (1981) demonstrated that the number of cells in an embryo is reduced by as much as 30% during the 24 hours following Mitomycin administration. Such a dramatic reduction in embryo size at such a sensitive stage of development would be expected, if not to kill the embryos, then certainly to lead to the development of either very small or very abnormal embryos. However, in an amazing demonstration of regulative ability, by 12 or 13 days of gestation these embryos have undergone compensatory cell division and growth, restoring them to the size of normal, untreated litter mates. Neural crest derivatives develop normally in these embryos, indicating that regulation must also have occurred among the neural crest cells. More detailed examination of the regulative capability of the mammalian neural crest would enormously aid interpretation of the etiology of embryonic defects, provide one of the best means to compensate for abnormalities that would otherwise follow exposure to teratogens, and provide a possible means of preventing such defects in future generations.

A study by Bronner-Fraser (1986) beautifully documented how regulation can compensate for an initial defect in neural crest cell migration. She utilized the CSAT antibody to a cell surface receptor that recognizes fibronectin and laminin, two of the chief extracellular matrix components involved in neural crest cell migration. CSAT antibody was injected lateral to the midbrain of embryonic chicks immediately before the onset of neural crest cell migration. Twenty four hours after antibody injection Bronner-Fraser observed:

- a reduction in the number of neural crest cells—indicative of reduced cell division;
- aggregation of neural crest cells within the lumen of the neural tube—indicative of defective initiation of cell migration; and
- neural crest cells in ectopic locations—indicative either of defective pathways of migration or of abnormal trapping of neural crest cells.

The only adverse effects seen 36 to 48 hours after antibody injection, however, were ectopic condensations of neural crest cells and abnormally developed neural tubes. The neural crest cells that had migrated ectopically remained in those ectopic sites, but otherwise the neural crest derivatives developed perfectly normally. Defective proliferation and delayed onset of migration of the majority of the neural crest cells was compensated for.

Sources of Cells

The source of the cells that replace missing neural crest cells depends, in part, on the level along the neural axis from which neural crest cells are deleted, and in part on how much neural crest is deleted. Compensation can take the form of:

- migration of neural crest cells across the midline, as occurs if neural crest cells are deleted from one side of the neural tube only;
- migration of neural crest cells from a more cranial or caudal region of the neural crest, as occurs if neural crest cells are deleted only from one cranio-caudal region of the neural folds;
- increased proliferation of the remaining neural crest cells, as occurs if neural crest cells are removed only from one region of the neural tube;
- regulation from neural ectodermal cells that would not otherwise have formed neural crest cells; or

- regulation from placodal or epidermal ectoderm.[33]

An approach that may help to increase understanding of neural crest regulation is to use mutations that affect neural tube development. As demonstrated by Chapman and Papaioannou (1998), mutation of the murine T-box gene *Tbx-6*, a gene that produces a DNA-binding protein, converts axial mesoderm into neural tissue, resulting in embryos with three neural tubes—one in the normal axial position and one on each side of the area where paraxial mesoderm would normally develop. A mutation that switches mesodermal tissues from somitic to neuronal to produce ectopic neural tubes with normal dorso-ventral patterning is of interest, in part because of the potential for the analysis of neural crest cell development in these ectopic neural tubes. Would such mesodermal turned neural tissue be capable of forming neural crest, given that neural crest normally arises at the boundary between neural and epidermal ectoderm? If capable, could such ectopic neural tubes regulate for removal of neural crest?

Completeness of Regulation

Regulation, or the degree of regulation, is not only influenced by availability of a source of competent cells, but also by time of removal, the site from which neural crest is removed, and the amount of crest removed, as we have demonstrated in zebrafish embryos (Vaglia and Hall 1997). For any regulation to occur after ablation of chick neural crest, the neuroepithelium and epidermis must come into apposition. For cranial neural crest, regulation is maximal at the 3 to 7 somite stage. Regulation is quite complete if caudal midbrain or hindbrain neural crest is removed but less complete after ablation of caudal forebrain (which does not produce neural crest) or rostral midbrain (which does). Regulation after hindbrain ablation in chick embryos requires that closure of the dorsal midline be reestablished, a cellular process that is associated with modulation of expression of both *Pax-3* and *Slug*.[34]

These studies on mechanisms underlying the completeness of regulation are not unique. Ablation of the brachial neural crest from chick embryos of H.H. stages 13 to 14 is followed by some regulation and formation of dorsal root ganglia from the remaining neural crest, although dorsal root ganglia 15 to 17 are small or missing. The proportion of sensory neurons in dorsal root ganglia is regulated by interactions between ganglionic neurons and peripheral target tissues. An unexpected finding in the ablation study was that in addition to this local regulation, ganglia remote from the crest removed underwent hypertrophy, some by as much as 220%. Ganglia as far removed as cervical ganglia showed compensatory hypertrophy; indeed they demonstrated the greatest hypertrophy. Regulation of the number of cells contributing to the dorsal root ganglia in avian embryos was demonstrated further by transplanting quail neural crest cells into the equivalent position in chick embryos; a variation of the classic transplantation experiments undertaken with amphibian embryos by Harrison and Wagner earlier in the century and discussed in Chapter 5. Quail embryos are smaller than chick embryos and so have smaller dorsal root ganglia. However, the size of the ganglia in the chimera was appropriate for the host embryo—local compensation mechanisms regulated ganglion size.

Although somitic mesoderm is not required for precursors of dorsal root ganglia to begin to migrate, it is required to regulate their growth. Differences in the sizes of dorsal root ganglia along the neural axis have been demonstrated in avian embryos: ganglia adjacent to brachial segments 14 and 15 are 80% larger than are those adjacent to segments 5 and 6—differences that are evidently secondary consequences of the local sclerotomal environment rather than intrinsic properties of the neural crest cells forming the ganglia; quail ganglia transplanted into chick hosts adapt to the size of chick (host) ganglia. Part of that host environment is distribution of the ligand ephrin-B2 and the Eph family of receptor tyrosine kinases and their transmembrane ligands, which mediate interactions between sclerotome and migrating neural crest cells and confine the latter to specific rostro-caudal territories. Complementary expression of EphA4/EphB1 receptors and the ligand prevents cells from the second and third arches from intermingling and so targets third arch neural crest cells to their destination (Fig. 12.5 [color plate]). The tyrosine kinase receptor, which was cloned from the chick, plays a role in regionalization of the vagal neural tube and somites (notably the dermamyotome) and is expressed in placodal ectoderm, where it may play a role in segregation or regionalization. Subpopulations of neural crest cells have specific receptor tyrosine kinases and so respond differentially to different growth factors.[35]

The visceral arch skeleton is also replaced by regulation following bilateral extirpation of neural crest from avian embryos. Regulation from the edges of the neural tube forms a visceral arch skeleton of normal size and morphology and has been shown in two studies to be preceded by restoration of the *Hox* gene code appropriate to the region of neural crest removed. *Hoxa-2* has the most rostral expression of any *Hox* gene, extending to the boundary between r1 and r2 in neural ectoderm but not in neural crest (Fig. 2.15). In r4, expression is in both neural tube and neural crest, even when rhombomeres are transposed, indicating both intrinsic patterns of expression and independent regulation in neural tube and neural crest. Patterns of migration of crest cells from rhombomeres 3 and 5 are also segmentally restricted. Both Paul Hunt and Gérald Couly and their colleagues found that surgical removal of the rhombencephalic neural crest from chick embryos was followed by normal visceral arch development and normal expression patterns of *Hoxa-2*, *a-3*, and *b-4*. Regulation from the neural epithelium on either side of the surgical site compensated for neural crest cells removed through enhanced proliferation and migration; i.e., there is regulation of *Hox* genes as well as of cells and tissues.

Removing the quail mesencephalon and metencephalon from chick embryos and replacing them in inverted orientation, reverses the normal rostro-caudal polarity of *Engrailed*, but within 20 hours of the operations a normal pattern of *Engrailed* is reestablished. Interestingly, while ablation of the dorsal hindbrain alters pathways of neural crest migration and patterns of gene expression, regulation results in normal development of the hyoid arch skeleton.[36]

Regulation of Cardiac Neural Crest

Although cranial (especially caudal midbrain/rostral hindbrain) neural crest can be replaced completely by regulation, replacement of ablated trunk neural crest cells is not complete. The potential to replace cardiac neural crest by regulation is even more limited.

As discussed in Chapter 7, removal of cardiac neural crest results in cardiovascular anomalies, a result that speaks to limited regulation or lack of regulation. Regulation can occur, provided that some cardiac neural crest cells remain. Thus, back-transplantation of cardiac neural crest cells to replace extirpated cardiac crest produces normal cardiac crest and rescues embryos that would otherwise exhibit heart anomalies (Kirby et al. 1993).

Suzuki and Kirby (1997) explicitly tested for the ability of the cardiac neural crest of chick embryos to regulate following bilateral ablation (using either laser surgery or microdissection) and DiI-labeling of the ventral neural tube adjacent to the ablated crest. Ablation before the 5-somite stage evoked little regulation. Crest ablated after the 6-somite stage was not replaced at all (Fig. 12.6 [color plate]). Trunk crest cells that would normally have followed a ventral pathway of migration were also not replaced following ablation, although melanocytes from adjacent regions migrated to fill the void so that pigment patterns were not disturbed despite loss of the neural crest precursors for pigment cells in a local region.

The nodose placode (see Fig. 2.13) can regulate to provide autonomic neurons and mesenchyme to the chick heart (normally this placode produces neurons but not mesenchyme) but cannot prevent all the cardiac defects that follow removal of the cardiac neural crest. Neurons that form from the nodose placode are indistinguishable physiologically from neural-crest-derived neurons that would have developed without the extirpation. Removal of both cardiac neural crest and the nodose placode leads to severe cardiovascular abnormalities. However, neither mesencephalic nor trunk neural crest transplanted in place of ablated cardiac neural crest can eliminate the truncal septation that follows cardiac crest ablation. Indeed, mesenchyme derived from mesencephalic neural crest and which has invaded the heart primordium, forms cartilage ectopically within the heart and interferes with cardiac mesenchyme, indicating a predetermination of mesenchyme that parallels the predetermination of mandibular arch skeletal structures reported by Noden in 1983 and discussed in Chapter 7.[37]

Placodal Regulation

Placodes can also be replaced by regulation, as can be illustrated by studies on the nodose ganglia.

Nodose ganglia arise from the placodal ectoderm under the influence of the sympathetic neural crest adjacent to somites 10 to 20; *interleukin-2* (*Int-2*), which is produced in rhombomeres of the embryonic chick hindbrain, diffuses to the ectoderm where it induces formation of both nodose ganglia and otocyst. Early in development, before placodal ectoderm is determined (1.5 days of incubation in avian embryos), trunk non-neurogenic ectoderm transplanted in place of placodal ectoderm can replace placodal ectoderm and form nodose sensory neurons. Ablation of the placodal precursor of the nodose ganglion from chick embryos at H.H. stage 9 is followed by partial replacement by 12 days of incubation. Use of quail/chick chimeras demonstrates that the cells that regulate for the nodose ganglion arise from cardiac neural crest adjacent to the third pair of somites. Thus, at least in the embryonic chick, nodose placodes can compensate for extirpation of cardiac neural crest and vice versa. Such regulation occurs because of shared properties among neural crest cells destined to form different cell types; a monoclonal antibody generated against nodose ganglia recognizes peripheral ganglia, Schwann cells, some sensory and some autonomic neurons, as well as 25% of migrating neural crest cells. Mybp75, a transcriptionally regulated oncoprotein expressed in otic and epibranchial placodes of avian embryos and correlated with proliferation and neurogenesis from the placode, may be a useful marker to monitor changes associated with regulation of ablated placodes.[38]

Notes

1. Discovery

1. His was professor of anatomy at Leipzig from 1872 until his death in 1904. See His (1870,1895a, b) for the microtome, *Nomina Anatomica*, and the discovery of the remains of J. S. Bach. Between 1866, when he invented his microtome, and 1870, when he described it in print, His made over 5000 sections of embryos. Based on a modified microscope substage supported on the microscope stand, his microtome is illustrated in his 1870 paper and in Bracegirdle (1986).

2. Marshall was educated at London University and Trinity College, Cambridge, where he began his academic career teaching the first combined lecture and practical class in animal morphology with Francis Balfour in 1874. Marshall subsequently obtained a medical degree and was elected FRS in 1885. He wrote a beautifully illustrated and produced manual on the frog (published in 1882), one third of which is devoted to what was then the most thorough description of frog embryology. He also wrote the text *Vertebrate Embryology*, published in 1893.

3. Three terms—*visceral, branchial,* and *pharyngeal*—are often used interchangeably for the multiple arches of vertebrates. The following is the convention I follow throughout the text: visceral arches, when *all* arches (mandibular, maxillary, hyoid, and gill) are implied; branchial (pharyngeal) arches for *gill* arches that are retained in the adult, as in fish.

4. See Box 5.1 for further details of Julia Platt's scientific achievements.

5. For major reviews of the neural crest, see Neumayer (1906), Landacre (1921), Stone (1922), Holmdahl (1928, 1934), Harrison (1938), Hörstadius (1950), Weston (1970), Le Douarin (1982), Maderson (1987), Hall and Hörstadius (1988), and Le Douarin et al. (1992). See Platt (1893, 1894, 1897), Hörstadius and Sellman (1941, 1946), and de Beer (1947) for seminal studies on the contribution of the neural crest to the skeleton.

6. The generic name *Ambystoma*, often misspelled *Amblystoma*, has had a checkered and fascinating history. The name *Ambystoma* was assigned to the genus by Tschudi in 1838, but *Ambystoma* has had 24 different names (some generic synonyms, some preoccupied senior synonyms) starting with *Gyrinus* in 1789 and running through *Axolotus, Philhydrus, Siredon, Axolotl, Sirenodon, Stegoporus, Xiphonura, Salamandroidis, Axoloteles, Amblystoma, Heterotriton, Limnarches, Xiphoctonus, Plagiodon, Camarataxis, Pectoglossa, Linguaelapsus, Plioambystoma, Bathysiredon, Lanebatrachus,* and *Ogallalabatrachus*, the latter two in 1941. Louis Agassiz caused even greater confusion when he renamed the genus *Amblystoma*. Agassiz appears to have coined the new name only because he believed Tschudi had erred in his knowledge of Latin, really intending to name the genus *Ambylstoma* from *amblys* (wide) and *stoma* (mouth). Be that as it may, the final

authority in such matters, the International Commission on Zoological Nomenclature, ruled in 1963 that the correct generic name is *Ambystoma*, not *Amblystoma* (see *Bull. Zool. Nomencl.* 1963, 20, 102–104, Opinion 649). See Duellman and Trueb (1986) for details of the various names assigned to *Ambystoma*, and see Melville and Smith (1987) for lists and indexes of official names approved by the Commission on Zoological Nomenclature. The basic source book for names and ranges of amphibian species is Frost (1985). See Stejneger (1907) for an evaluation of Tschudi, and see H. M. Smith (1969) and D. B. Wake (1976) for details and discussions.

7. Active in scientific administration at the international level, Hörstadius was secretary-general and organizer of the Tenth International Ornithological Congress in Uppsala in 1950, a founding member of the Council of the World Wildlife Fund, chairman of the European Section of the International Council for Bird Preservation, president of the Swedish Ornithological Society, president of the International Union of Biological Sciences (1953–1958), and president of the International Council of Scientific Unions (1962–1963).

8. An assessment of the experimental work by Hörstadius on the neural crest was that: "This is the first time that the phenomenon of complex and additive inductive action by different structures has been demonstrated under experimental conditions" (*Nature* 1950, 169, p. 821). Hörstadius was elected a Fellow of the Linnean Society on 24 May 1950 (*Nature* 1950, 165, p. 919) and a Fellow of the Royal Society on 1 May 1952 (*Nature* 1952 169, p. 821). See Ebendal (1995) for biographical details. A symposium and publication ("Regulatory Processes in Development: The Legacy of Sven Hörstadius") to mark his contributions to the developmental biology of echinoderms and the neural crest and to chart future directions for developmental biology was organized by C.-O. Jacobson, L. Olsson, and T. Laurent for the Wenner-Gren Foundation and the Royal Swedish Academy of Sciences in Stockholm in 1998, the centenary of his birth. The symposium will be published for the Wenner-Gren Foundation by Portland Press, Ltd., London.

9. See Johnston and Listgarten (1972), Bolande (1974), Vincent, Duband, and Thiery (1983), and Morriss-Kay and Tan (1987) for early studies in these areas.

10. We have known for over 110 years that the hindbrain is organized into segments, which Orr (1887) termed *rhombomeres* on the basis of his studies on lizard brain development.

11. See Lankester (1873, 1877) for the tripartite subdivision of the animal kingdom on the basis of germ layers, and see Hall (1998a, b) for neural crest as the fourth germ layer and for evaluations of the work of Lankester and Balfour. See Russell (1916), Oppenheimer (1940), Churchill (1986), and Hall (1998a, b) for discussions of various aspects of germ-layer theory.

12. See Sedgwick (1894a, b, 1895a, b) for his views on both the cell and the germ-layer theories. Sedgwick (1910) wrote what stands as one of the most insightful summaries of embryology—the entry for the eleventh edition of the Encyclopaedia Britannica.

13. See Hyman (1940, 1951), Salvini-Plawen (1980), Starck and Siewing (1980), and Hall (1998b) for discussions of the use of the term *mesenchyme*.

14. For examples of experimental induction of organs from different germ layers, see Oppenheimer (1940) and Hall (1998b).

15. For discussion on development of homologous structures from different embryonic regions or by different developmental processes, see E. B. Wilson (1894), Sedgwick (1910), de Beer (1958, 1971), and Hall (1994b, c, 1995, 1998a).

16. James P. Hill (1873–1954) was an eminent student of comparative marsupial embryology, Jodrell Professor of Zoology at University College, London, from 1906 to

1921, and, with the reorganization of the department in 1921, professor of histology and embryology until 1938. His extensive collection of marsupial and primate embryos is housed in the Hubrecht Laboratory in the Netherlands. Katherine Watson was a student of Hill's. Some unpublished drawings by Hill of his specimens were published by Klima and Bangma (1987) in the context of a discussion of marsupial embryos as larvae, a conclusion they draw because initial deciduous claws are replaced by permanent claws, and because the breast-shoulder arch is temporary. See Hall and Wake (1999) for up-to-date discussions of definitions of larvae.

17. See Holmdahl (1925a–c) and von Kölliker (1879, 1884, 1889) for secondary development of the caudal region; Schoenwolf and Nichols (1984), Schoenwolf, Chandler, and Smith (1985 and references therein), Copp (1993) and Le Douarin, Grapin-Botton, and Catala (1996) for secondary neurulation in avian embryos; Wilson and Wyatt (1988) for murine embryos, Müller and O'Rahilly (1987) for human embryos, and Hall (1998a, b) and Kanki and Ho (1997) for further discussions of phases of development and secondary neurulation. For current understanding of epithelial–mesenchymal transitions, see Hay (1989) and the papers in the special issue of *Acta Anatomica* edited by Newgreen (1995). See Pasteels (1943), Gont et al. (1993) and De Robertis et al. (1994) for tail development as a continuation of primary neurulation, and see Hall (1998a, b) for discussion of this issue.

18. See Hall (1978) and Hinchliffe and Johnson (1980) for early studies on mechanisms of limb bud development; Grüneberg (1956), Kostovic-Knezevic, Gajovic, and Svajger (1991), and Gajovic and Kostovic-Knezevic (1995) for the ventral ectodermal ridge; and Searls and Zwilling (1964) and Amprino, Amprino Bonetti, and Ambrosi (1969) for common signaling mechanisms between limb and tailbuds.

19. See Grüneberg (1956), Grüneberg and Wickramaratne (1974), Cogliatti (1986), and Johnson (1986) for *vestigial tail* and *Brachyury* (*T*), Salzgeber and Guénet (1984) for *repeated epilation*, and Peeters et al. (1998) for *curly tail*. See Wilson and Wyatt (1988) for closure of the posterior neuropore in *vL* mutant mice and the suggestion that defective secondary neural crest may contribute to spina bifida in these mutants. See Schoenwolf and Nichols (1984) and Schoenwolf, Chandler, and Smith (1985) for dual origins of the neural crest; and see Holmdahl (1925a–c) and Griffith, Wiley, and Sanders (1992) for primary and secondary body development.

20. For recent overviews of mesoderm induction, see Chen and Grunz (1997), Sasai and de Robertis (1997), and Stennard, Ryan, and Gurdon. (1997).

2. Embryological Origins

1. Hörstadius (1950, pp. 10–23, 75–78), Weston (1970), Le Douarin (1982), Langille and Hall (1988a, b), and Vaglia and Hall (1999) discuss studies in which the neural crest was extirpated.

2. See Hörstadius (1950, pp. 4–6) for early studies (mostly amphibian) on the origin of the neural crest. Rosenquist (1981) and Garcia-Martinez, Alvarez, and Schoenwolf (1993) mapped the neural crest in the chick. Schoenwolf and Sheard (1990) and Hatada and Stern (1994) also mapped the avian epiblast but without identifying future neural crest cells. Schoenwolf and Alvarez (1991) used quail/chick chimeras to establish the timing of determination of neural-epidermal cell fate. See Lawson, Meneses, and Pedersen (1991) for a comparison of avian, mouse, and urodele fate maps. Although murine neural crest was not labeled in these studies, the site of neural crest between epidermal and neural ectoderm can be inferred.

3. See Thiery, Duband, and Delouvée (1982), Tucker et al. (1984), and Jungalwala et al. (1992) for characterization of the HNK-1 antigen; Kuratani (1991) for HNK-1 expression in alternate rhombomeres; and Bronner-Fraser (1987) for perturbation studies with HNK-1 antibody. Several HNK-1 antigens are available (Luider et al. 1992) and may recognize different antigenic sites in different species. Information on HNK-1 labeling in different groups may be found in the relevant sections in chapters 5 to 8.

4. See Edelman (1983), Levi, Crossin, and Edelman (1987), Prieto et al. (1989), and Akitaya and Bronner-Fraser (1992) for the distribution of N-CAM and N-cadherin, and see Holtfreter (1933), Raven (1935), and Holtfreter and Hamburger (1955) for experimental studies on relationships between neural tissue and neural crest. Bronner-Fraser, Wolf, and Murray (1992) used antibodies against N-CAM and N-cadherin to explore the role of these molecules in cranial neural crest cell migration.

5. See Duband et al. (1995) for studies with *Slug* using avian embryos. Nieto et al. (1994) and Duband et al. (1995) demonstrated involvement of *Slug* at the onset of neural crest cell migration; Sechrist et al. (1995) used *Slug* to monitor regulation of the neural crest, while Mancilla and Mayor (1996) and Mayor, Guerrero, and Martínez (1997) analyzed *Slug* (*XSlug*) in *Xenopus*.

6. See McMahon and Bradley (1990), Parr et al. (1993), and Ikeya et al. (1997) for studies on *Wnt*; and see Krauss et al. (1992), Joyner (1996), and Urbánek et al. (1997) for gene families that establish the midbrain/hindbrain boundary.

7. For induction of the neural crest in *Ambystoma* and the action of FGF on neural crest, see Moury and Jacobson (1990), Murphy, Drago, and Bartlett (1990), Kengaku and Okamoto (1993), Tiedermann et al. (1994), and Jacobson and Moury (1995). For transdifferentiation in response to antisense bFGF, see Sherman et al. (1993).

8. See Mikhailov and Gorgolyuk (1988) for the studies with concanavalin A. For induction of neural crest in birds and combined use of quail/chick chimeras and HNK-1 or *Slug* expression, see Dickinson et al. (1995), Bronner-Fraser (1995), and Selleck and Bronner-Fraser (1995). For a recent overview of neural crest induction, see Baker and Bronner-Fraser (1997a).

9. Throughout the text, I use rostral and caudal rather than anterior and posterior when referring to the organization of axial embryonic structures such as the neural tube, neural crest, or somites.

10. See Ruiz i Altaba and Jessell (1991), Doniach (1995), McGrew, Lai, and Moon (1995), Hemmati-Brivanlou and Melton (1997), Sasai and de Robertis (1997), Weinstein and Hemmati-Brivanlou (1997), and Marchant et al. (1998) for neural induction, and Clement et al. (1995) for the studies with BMP-2.

11. For cascades of signals in embryonic induction, see Chandebois and Faber (1983), Hall (1983a–c, 1987a, 1988a, 1990a, 1994a, d), Nieuwkoop (1985, 1992), Nieuwkoop, Johnen, and Albers (1985), Gurdon (1987, 1992), Nieuwkoop and Koster (1995), and Grunz, Schüren, and Richter (1995). See Holtfreter and Hamburger (1955) and Saxén and Toivonen (1962) for thresholds of neural and neural crest induction.

12. For development of neural crest at the boundary between neural and epidermal ectoderm, see Moury and Jacobson (1989, 1990), Sharpe (1990), Yamada (1990), Scherson et al. (1993), Selleck, Scherson, and Bronner-Fraser (1993), Bronner-Fraser (1995), Selleck and Bronner-Fraser (1995), and Graveson, Smith, and Hall (1997). See Albers (1987) and Nieuwkoop and Albers (1990) for altered ectodermal responsiveness, Xu et al. (1995) for the BMP-4 receptor, Wilson and Hemmati-Brivanlou (1995) for activin, and Weinstein, Honoré, and Hemmati-Brivanlou (1997) for eIF-4AIII.

13. See Bonstein, Elias, and Frank (1998) for these studies with paraxial mesoderm.

14. See Dickinson et al. (1995) for the chick studies, Chang and Hemmati-Brivanlou (1998), and La Bonne and Bronner-Fraser (1998) for those with *Xenopus* and Jiang et al. (1998) for mouse. *Xenopus* twist (*Xtwist*) in normal development is shown in Figure 2.11 [color plate].

15. For involvement of BMP-4 and BMP-7 in neural or neural crest induction in *Xenopus*, chick, mouse, and zebrafish, see Hawley et al. (1995), Sasai et al. (1995), Selleck and Bronner-Fraser (1995), Dickinson et al. (1995), Liem et al. (1995), Monsoro-Burq et al. (1996), Arkell and Beddington (1997), Liem, Tremml, and Jessell (1997), Shimamura and Rubenstein (1997), and Nguyen et al. (1998). For an up-to-date overview of molecular control of neural induction, see Sasai and de Robertis (1997). See Watanabe and Le Douarin (1996) for BMP-4 and chondrification of paraxial mesoderm. See Korade and Frank (1996) for neural crest from ventral neural tube cells.

16. See Basler et al. (1993) and Kingsley (1994) for the role of *Dorsalin* in specification of neural crest; see Bronner-Fraser and Fraser (1997) for a review.

17. See Nakata et al. (1997) and Suzuki, Ueno, and Hemmati-Brivanlou (1997) for *Zic-3* and *Msx-1* and their relationship to BMPs and neural crest induction. *Zic-2*, which encodes a zinc-finger trancription factor, promotes neural crest over neural differentiation (Brewster, Lee, and Ruiz i Altaba, 1998).

18. See Jones, Lyons, and Hogan (1991) and Bennett, Hunt, and Thorogood (1995) for distribution of BMP in murine orofacial tissues. See Le Douarin (1988b), Hall and Ekanayake (1991), and Chapter 10 for other growth factors that regulate differentiation of neural crest cells.

19. See Ekker et al. (1997) and Shimeld, McKay, and Sharpe (1996) for *Msx* genes in zebrafish and mice; see Wittbrodt, Meyer, and Schartl (1998) for an overview of gene numbers in fish.

20. See Albers (1987) and Servetnick and Grainger (1991) for transplantation between *Ambystoma mexicanum* and *Triturus alpestris*.

21. See Mitani and Okamoto (1991) and Mayor, Morgan, and Sargent (1995) for separate induction of neural tube and neural crest, and see Lamb et al. (1993) for *Noggin* as inducer of anterior neural structures. See Holtfreter (1968), Epperlein and Lehmann (1975), Minuth and Grunz (1980), and Nieuwkoop, Johnen, and Albers (1985) for homoiogenetic induction affecting the neural crest.

22. See Albers (1987) for placode formation at boundaries, and see Le Douarin, Fontaine-Perus, and Couly (1986), Northcutt (1992, 1996), Noden (1993), Webb and Noden (1993), Northcutt, Catania, and Criley (1994), Osumi-Yamashita et al. (1994), Northcutt and Brändle (1995), Northcutt, Brändle, and Fritzsch (1995), and Northcutt and Barlow (1998) for overviews of placode development. See Meier (1978a, b) for otic placodes in chick embryos and Vogel and Davies (1993) for nodose neurons from non-neurogenic ectoderm. See M. M. Smith and Hall (1990), S. C. Smith, Graveson, and Hall (1994), Parichy (1996c), and S. C. Smith (1996) for discussions of the developmental and evolutionary links between placodal and neural crest ectoderm.

23. See Croucher and Tickle (1989) for the nasal placodes; Tongiorgi et al. (1995) for tenascin-C in zebrafish placodes; Heath and Thorogood (1989) for keratan sulfate in placodal ectoderm; Robertson and Mason (1995) and León et al. (1992) for the tyrosine kinase receptor and Mybp75 in chick placodal ectoderm; and Mahmood et al. (1995) and Mahmood, Mason, and Morriss-Kay (1996) for FGF-3 in murine otic placodes. Woods and Couchman (1998) review the role in cell adhesion of such syndecans as heparan sulfate proteoglycan.

24. See Selleck and Stern (1991) for the fate map and lineage analysis of Hensen's

node and Shah et al. (1997) for the studies with chick Vg1.

25. For induction and/or regionalization of neural tube by notochord or Hensen's node in avian embryos, see Schoenwolf et al. (1989), Dias and Schoenwolf (1990), Storey et al. (1992), and Yuan and Schoenwolf (1998). The studies with immortalized Hensen's node cells were undertaken by Darland and Leblanc (1996); those on respecification of trunk as cranial crest by Rogers et al. (1992) and Leblanc and Holbert (1994, 1996); and those on TGF-ß, cell-substrate adhesion, and specification of cranial and trunk crest by Delannet and Duband (1992) and Leblanc, Holbert, and Darland (1995). See Grainger (1992) for lens induction and Dias and Schoenwolf (1990) for studies on relationships between endoderm, notochord, neural tube, and neural crest induction.

26. See Dunn, Mercola, and Moore (1995) for the studies with cyclopamine, Milos et al. (1990, 1998) and Evanson and Milos (1996) for the lectin studies, and Milos and Wilson (1986) and Milos et al. (1987) for cell surface carbohydrates.

27. See Bronner-Fraser and Fraser (1988), Fraser and Bronner-Fraser (1991), Selleck and Bronner-Fraser (1995), and Mujtaba, Mayer-Proschel, and Rao (1998) for evidence for the common lineage of central neurons and neural crest cells.

28. See Khare and Choudhury (1985) and Couly and Le Douarin (1990) for these studies on patterning of neural fold ectoderm. See Alvarado-Mallart (1993) for timing of commitment of the mesencephalic-metencephalic neuroepithelium, a commitment that requires at least two steps.

29. See Kuratani (1991) for HNK-1 expression in alternate rhombomeres. See Hunt et al. (1991a–c), Kessel and Gruss (1990, 1991), Krumlauf et al. (1991), Hunt and Krumlauf (1992), Kessel (1992), and Krumlauf (1993) for discovery of the *Hox* code.

30. See Saldivar et al. (1996) for cell-autonomous rather than environmentally regulated expression of *Hox-3a*; Guthrie et al. (1992), Kuratani and Eichele (1993), and Wilkinson (1993, 1995) for intrinsic patterning of cranial nerves and rhombomere *Hox* gene expression patterns.

31. See Oxtoby and Jowett (1993) and Nieto et al. (1995) for *Krox-20* in the zebrafish and chick hindbrain.

32. See Hemmati-Brivanlou et al. (1991), Gardner and Barald (1991) and Hemmati-Brivanlou, Stewart, and Harland (1990) for expression of *Engrailed.* See Simeone et al. (1992) for nested expression domains of *Otx-1* and *Otx-2* in the murine rostral hindbrain, and Rhinn et al. (1998) for *Otx-2* in murine visceral endoderm and its role in specification of fore- and midbrain territories. For gene families that establish the midbrain/hindbrain boundary, see Krauss et al. (1992), Joyner (1996), and Urbánek et al. (1997).

33. See Dirksen, Mathers, and Jamrich (1993) for *Dll* expression in *Xenopus*, and Qiu et al. (1997) for the proximo-distal *Dll* code.

34. See Grapìn-Botton et al. (1995) and Grapìn-Botton, Bonnin, and Le Douarin (1997) for the transplantation studies. For cranial nerve patterning, see Kuratani and Aizawa (1995).

3. Evolutionary Origins

1. For *Sog* and *chordin*, see Nübler-Jung and Arendt (1994), Holley et al. (1995), Lacalli (1995, 1996), Arendt and Nübler-Jung (1996), Salzberg and Bellen (1996), and Hall (1998a). For a discussion of shared genes between arthropods (*Drosophila*) and vertebrates (*Danio*, *Xenopus*), including *decapentaplegic* and its vertebrate homologue BMP,

see Holley and Ferguson (1997).

2. I do not provide an extensive discussion of the various theories for the origin of the craniates; they have recently been the topic of several major treatments, especially Bowler (1996) and Gee (1996). For a discussion in the context of the origin of the neural crest, see Baker and Bronner-Fraser (1997b).

3. Amphioxus, like lancelet, is a common, not a generic name. Hence amphioxus should neither be capitalized nor italicized. The generic name is *Branchiostoma*. See Stokes and Holland (1995) and Presley, Horder, and Slipka (1996) for analyses of lancelet embryonic and larval development.

4. See Corbo et al. (1997) and a pers. comm. from J. Langeland in Baker and Bronner-Fraser (1997b) for *Snail* in amphioxus and ascidians.

5. See P. W. H. Holland (1992, 1996), Holland and Garcia-Fernàndez (1996), Sharman and Holland (1996, 1998), and Hall (1998a) for the significance of *Hox* gene clusters for vertebrate evolution, and see P. W. H. Holland (1996) for *Hox* and other genes cloned from *Branchiostoma*. See Panopoulou et al. (1998) for AmphiBMP2/4.

6. See P. W. H. Holland et al. (1992) and Holland and Garcia-Fernàndez (1996) for these studies and interpretations of amphioxus *Hox* (*AmphiHox*) gene expression. See Kuratani (1997) and Kuratani et al. (1998) for comprehensive analyses of the importance of the postotic neural crest as the boundary between cranial and trunk crest and so between head and trunk in vertebrates. For further studies on homology of the portions of the nervous system of amphioxus with the vertebrate nervous system (based either on shared patterns of gene expression or on shared morphology), see Berrill (1987a, b), Lacalli, Holland, and West (1994), N. D. Holland (1996), P. W. H. Holland and Garcia-Fernàndez (1996), Williams and Holland (1996), and L. Z. Holland et al. (1997).

7. See Simeone et al. (1992), Matsuo et al. (1995), and Rhinn et al. (1998) for *Otx-2* in murine embryos; and Williams and Holland (1996, 1998) for expression and organization of *AmphiOtx*. The *cis*-acting elements of *Otx-2*, which regulate gene expression in mesencephalic neural crest but which differ between premandibular and mandibular segments, are conserved between puffer fish and mice (Kimura et al. 1997).

8. See Dirksen et al. (1994) and N. D. Holland et al. (1996) for expression of *AmphiDll* and Hall (1994b, c, 1995, 1998a) for the homology problem. For discussions of the significance of common patterns of gene expression for our understanding of the origins of the vertebrates and of the neural crest, see P. W. H. Holland (1992, 1996), P. W. H. Holland and Graham (1995), N. D. Holland (1996), P. W. H. Holland and Garcia-Fernàndez (1996), L. Z. Holland et al. (1997), and Hall (1998a).

9. See L. Z. Holland and N. D. Holland (1996) for the studies with retinoic acid on amphioxus. For evolution of tissue interactions at the outset of craniate evolution, see Hall (1975, 1999b), Maderson (1975, 1983), Maisey (1986), and M. M. Smith and Hall (1990, 1993).

10. See Mackie (1995) for the nervous system of *Ciona* and Thorndyke and Probert (1979) for calcitonin reactivity.

11. See Maclean and Hall (1987) and Whittaker (1987) for determination of cell fate in ascidians. For recent discussion of ascidian development, see Berrill (1987b), Swalla (1992), Okamura, Okado, and Takahashi (1993), Satoh (1994), Swalla and Jeffery (1995), Jeffery (1997), and Chapter 22 in Hall (1998a).

12. For suppression of BMP dorsally and for BMP-2, -4, and -7 in ascidians, see Hammerschmidt, Serbedzija, and McMahon (1996), and Miya et al. (1996. 1997). For AmphiBMP2/4, see Panopoulou et al., 1998.

13. See Mansouri et al. (1996) for *Pax-7*$^{-}$ mutant mice. For the homologues of verte-

brate *Pax* genes in the nervous system of *Ciona* and *Halocynthia*, see Wada, Holland, and Satoh (1996), Corbo et al. (1997), and Wada et al. (1998).

14. See Chen et al. (1995) and Shu, Zhang, and Chen (1996), respectively, for the original description and reinterpretation of *Yunnanozoon.*

15. See Gans and Northcutt (1983, 1985), Northcutt and Gans (1983), and Gans (1987, 1989, 1993) for their pivotal scenario of the evolution of the neural crest. See Goodrich (1930), Berrill (1955, 1987a, b), Romer (1972), and Schaeffer (1977) for earlier studies. The symposium proceedings edited by Maderson (1987), and the three-volume series on the vertebrate skull edited by Hanken and Hall (1993) review and synthesize much of this evidence. Hall (1990b) provides a review in relation to evolutionary issues in craniofacial biology.

16. See Maisey (1986) and Thomson (1993) for studies on segmentation of the vertebrate head, and see Hall (1988b) and Stern (1990) for the different bases for segmentation of the hindbrain (which is based in the neural ectoderm) and the trunk (which is based in mesoderm).

17. See Stefansson et al. (1982), Cocchia et al. (1983), and Landry, Youson, and Brown (1990) for S-100 acidic protein; Cheah et al. (1991) for type II collagen; Asher et al. (1995) for aggregan; and Gallo, Bertolotto, and Levi (1987) for cartilage-specific chondroitin sulfate. In 14-day rat embryos, ß-S100 protein is expressed in Meckel's cartilage, in cartilages of the nasal region, maxillary processes, and skull base, and in the ventricle of the myelencephalon.

18. See Bowler (1996), Gee (1996), Northcutt (1996), Baker and Bronner-Fraser (1997b), and the papers in Hall and Wake (1999) for evaluations of the various theories for chordate origins.

19. For overviews of the skeletogenic neural crest, see Le Douarin (1982), Gans and Northcutt (1983), Northcutt and Gans (1983), Gans (1987, 1989, 1993), Halstead (1987), Maderson (1987), M. M. Smith and Hall (1990, 1993), M. M. Smith (1991), and the chapters in the three volumes on the skull edited by Hanken and Hall (1993).

20. For the evolutionary origins of vertebrate skeletal tissues, see Hall (1975, 1978, 1987a, 1999b), Person (1983), Hanken and Hall (1983), Thomson (1987), M. M. Smith and Hall (1990, 1993), Hanken and Thorogood (1993), Langille and Hall (1993a, b), Presley, Horder, and Slipka (1996), and Sansom et al. (1997).

21. For the earliest chordate skeletal remains, see M. M. Smith and Hall (1990, 1993), Janvier (1996a, b), Sansom, Smith, and Smith (1994, 1996), and Sansom et al. (1997). See Sansom et al. (1992), Sansom (1996), and Donoghue (1998) for histology and growth of conodont elements; and see Briggs, Clarkson, and Aldridge (1983), Aldridge and Theron (1993), Aldridge et al. (1993), and Gabbott, Aldridge, and Theron (1995) for reconstruction of the conodont animal as a chordate.

22. See Aldridge et al. (1993), Janvier (1996a, b), and Aldridge and Purnell (1996) for chordate phylogenies incorporating conodonts.

23. See Koole, Bosker, and van der Dussen (1989), Koole (1994a, b), and Borstlap et al. (1990) for healing bony defects, Hardesty and Marsh (1990) for resorption, and Gorski (1998) for woven versus lamellar bone.

4. Agnathans

1. For discussions of lamprey and hagfish relationships, see Schaeffer and Thomson (1980), Hall and Hanken (1985), Maisey (1986), Thomson (1987), Schaeffer (1987),

Langille and Hall (1989), Hanken and Hall (1993), Forey and Janvier (1993, 1994), Peterson (1994), Forey (1995), Janvier (1996a, b), and Sower (1998).

2. See Hardisty (1979) for an overview of hagfish diversity and life history; see Wicht and Northcutt (1995) and Braun and Northcutt (1997) for studies on the lateral line system in hagfishes.

3. For a sample of analyses of embryonic development as an indication of phylogenetic relationships, see Gould (1977), Rachootin and Thomson (1981), Alberch (1985, 1987), Schaeffer (1987), Thomson (1987), Gilbert (1997), and Hall (1998a, b).

4. Franz Doflein of Munich University, who was also in California in 1896, obtained embryos from the same fisherman who were collecting for Dean. These were the specimens used by von Kupffer in his studies of head development in *Eptatretus* (von Kupffer 1906 and references therein). Dean may also have collected embryos of *Eptatretus burgeri* in Japan in 1900 and 1901 (Conel 1931, pp. 97–98). If he did, they remain unstudied and their location unknown.

5. See Strahan (1958) and Thompson (1942) for the transformational approach to animal form and development; see Gorbman (1983) and Gorbman and Tamarin (1985) for pituitary development in the Californian hagfish.

6. See Wright et al. (1984) and Moss and Moss-Salentijn (1983) for hagfish skeletal structure, and Hall (1978) and Person (1983) for invertebrate cartilages.

7. See Bytinski-Salz (1937) for the grafting studies, and Damas (1944, 1951) and Newth (1956) for reviews of the early literature.

8. See Wright et al. (1988) for elastin in lamprey cartilages, and see Robson et al. (1997) for the various types of cartilage in the sea lamprey. See Wright and Youson (1982, 1983), Wright, Keeley, and Youson (1983), and Wright et al. (1988) for studies on the nature of the skeleton in agnathans. Lamprey and hagfish cartilages, unlike the cartilages of jawed vertebrates, are unmineralized. So too is invertebrate cartilage. Interestingly, as previously found for invertebrate cartilage, lamprey cartilage can mineralize in vitro under appropriate conditions of temperature and medium; see Langille and Hall (1993b) for mineralization of lamprey cartilage in vitro and Hall (1978) and Person (1983) for invertebrate cartilages.

9. The four papers on neural crest transplantation are Newth (1950, 1951, 1955, 1956). For recent experimental studies see Langille and Hall (1986, 1988b, 1989).

10. See *Biol. Abst.* (1955, 31, no. 11178) for the English abstract of Newth's Russian paper.

11. For further discussions of regulation in the context of neural crest cell extirpation see Weston (1970), Chibon (1970), and Vaglia and Hall (1999) and Chapter 12.

12. See Langille and Hall (1988b, 1989) for these studies on *Petromyzon marinus*.

13. For mesodermal origin of the parachordals and otic capsules in lampreys, see Damas (1944), Johnels (1948), de Beer (1937), Weston (1970), and Noden (1978a, b). For the chimeric (mesodermal and neural crest) origin of avian otic capsules, see Le Lièvre (1978), Noden (1975, 1983a), Couly, Coltey, and Le Douarin (1993), and Le Douarin, Ziller, and Couly (1993).

14. See Jungalwala et al. (1992) for the HNK-1 antigen, Hirata, Ito, and Tsuneki (1997) for the studies with HNK-1 in lampreys; Percy and Potter (1991) for heart development in lampreys, and Nakagawa et al. (1993) for HNK-1 in the rat heart.

15. See N. D. Holland et al. (1993) for *Engrailed* expression in lamprey and gnathostome muscles, and see P. W. H. Holland (1992, 1996), Miyake, McEachran, and Hall (1992), and Hall (1998a) for more general discussions of the issues raised by such shared specific patterns of gene expression. For discussions of the possibility that the velum of

lampreys may be a homologue of the mandibular arches of gnathostomes, see Johansen and Strahan (1963), Hardisty (1979), Forey and Janvier (1993, 1994), and Mallatt (1984, 1996, 1997).

5. Amphibians

1. For pigment cells as markers for transplanted neural crest, see figures 37–41 in Hörstadius (1950), Epperlein and Claviex (1982), Graveson, Hall, and Armstrong (1995), and Northcutt (1996). See figures 3–5 in Graveson, Hall, and Armstrong (1995) and Figure 6 in Northcutt (1996) for illustrations of the transplantation of neural crest from pigmented into albino axolotls.

2. See Harrison (1910), Weidenreich (1912), Holtfreter (1929), Mangold (1929) and DuShane (1943, 1944 and references therein) for the early studies; Huszar, Sharpe, and Jaenisch (1991) for sources of pigment, and Terentiev (1941), Twitty and Bodenstein (1941), and Woerdeman (1945) for the neural crest and the dorsal fin.

3. Also see Stearner (1946) for changes in pigmentation at metamorphosis.

4. Readers interested in further details of the origin of neuronal and pigmented cells from the amphibian neural crest will find that chapters 2 and 4 of Hörstadius (1950), comprising some 40% of his monograph, provide an excellent evaluation and synthesis of the earlier work. See Le Douarin (1986) for early studies on the neuronal neural crest derivatives in avian embryos. More recent reviews include Bronner-Fraser (1993a ,b, 1995) and Bronner-Fraser and Fraser (1997).

5. See Detwiler (1937) and Detwiler and Kehoe (1939) for studies with vital dyes, and see Harrison (1904, 1924), Raven (1931a–c, 1933a, b, 1936, 1937), and DuShane (1938) for experimental confirmation of the neural crest origin of spinal ganglia.

6. Van Campenhout (1930a, b, 1946), Hollinshead (1940), and Yntema and Hammond (1947) provide comprehensive reviews of the early studies on the sympathetic nervous system.

7. See Harrison (1904, 1907a, b, 1910), Müller and Ingvar (1921, 1923), and Detwiler and Kehoe (1939) for studies on the neural crest origin of Schwann cells, and see Hörstadius (1950) for a discussion of early studies that gave contrary results.

8. See Balfour (1878a), Rohon (1884), and Béard (1892, 1896) for discovery and descriptions of Rohon-Béard cells; DuShane (1938) and Chibon (1967) for their neural crest origin; Hughes (1957) for their disappearance in *Xenopus*; and Laudel and Lim (1993) for their development in *Oreochromis mossambicus*.

9. See M. Jacobson and Moody (1984) for lineage analysis of Rohon-Béard cells; Lamborghini (1980) for their origin during gastrulation; and Jacobson (1991) for a general discussion.

10. See Kastschenko (1888), Goronowitsch (1892, 1893a, b), and Platt (1893, 1894, 1896, 1897) for the pioneering studies, and de Beer (1947) for a discussion of them.

11. Zottoli and Seyfarth (1994) is the only in-depth analysis of the life and work of Julia Platt.

12. See Landacre (1921), Stone (1922, 1926, 1929), Raven (1931b, 1936), Holtfreter (1933), and de Beer (1947) for confirmation of the neural crest origin of head mesenchyme and craniofacial cartilages.

13. The spotted salamander of North America, *Ambystoma maculatum* is often referred

to in the older literature as *A. punctatum* but *A. maculatum* is the correct name (Wake 1976).

14. For recent application of time-lapse analysis to neural tube closure and neural crest cells, see Jaskoll, Greenberg, and Melnick (1991), Shankar, Jaskoll, and Melnick (1992), and Krull et al. (1995).

15. See Raven (1931b, 1936) for transplantation and xenoplastic grafts in urodeles and for studies establishing differences in potential between cranial and trunk neural crest; Newth (1950, 1951) and Le Lièvre (1971a, b) for similar studies in lampreys and Hall (1987a) for a discussion.

16. See Holtfreter (1935a, b, 1936) for these classic studies, and see Sellman (1946) for further confirmation. See Ströer (1933), Balinsky (1925, 1940), and Sellman (1940, 1946) for the role of oral endoderm in amphibian tooth induction.

17. See Harrison (1929), Twitty and Schwind (1931), Stone (1932), Twitty (1932), and Twitty and Elliott (1934) for studies on determination of the size of organ primordia.

18. See Harrison (1935a, b, 1936, 1938) for exchange of neural crest between different species of urodeles.

19. For discussions of intrinsic patterning of the neural crest, see Lumsden (1987, 1988), Hall (1988b), and Hall and Hörstadius (1988). See Thorogood (1993a) for an insightful analysis of the (not necessarily congruent) studies on regionalization of the skeletogenic neural crest in amphibians and birds. See Shimamura et al. (1995) for a recent overview of various theories seeking to explain the rostro-caudal organization of the neural tube, including those of Wilhelm His.

20. See Steen (1968, 1970) and Foret (1970) for the regeneration studies

21. See Chibon (1964, 1966, 1967, 1974) for the transplantation studies and Hall (1987a) and Thorogood (1993a) for discussions of them.

22. See Krotoski, Fraser, and Bronner-Fraser (1988) for crest cell migration in *Xenopus* spp. For patterns of neural crest cell migration in coqui and comparisons with other amphibians, see Moury and Hanken (1995), Olsson and Hanken (1996), Hanken, Jennings, and Olsson (1997), and Hall (1999a). For skeletal development in coqui, see Hanken et al. (1992) and the summaries in Hall (1999a, b).

23. For patterning of the caudal brain by *Dlx* see Dollé, Price, and Duboule (1992), Dirksen, Mathers, and Jamrich (1993), and Akimenko et al. (1994). See Fang and Elinson (1996) for patterns of *Dlx* gene expression in *E. coqui*, and see Elinson (1990) for the value of study of direct-developing frogs for understanding the evolution of ontogeny. For a discussion of whether the vertebrate forebrain is segmented, see Northcutt (1995) and Smith Fernandez et al. (1998).

24. See Lumsden (1987, 1988), M. M. Smith and Hall (1990, 1993), and Graveson, Smith, and Hall (1997) for tooth-forming trunk neural crest.

25. See LaFlamme and Dawid (1990) for keratins associated with cement gland induction, and see Jamrich and Sato (1989) and Drysdale and Elinson (1991, 1992, 1993) for induction of cement glands in *Xenopus*.

26. See Blitz and Cho (1995) for the studies with *XOtx-2*. For a discussion of positive and negative controls regulating the development of cement glands in *Xenopus*, see Bradley, Wainstock, and Sive (1996). See Fang and Elinson (1996) for induction of cement glands in coqui. For examples of loss of ectodermal competence, see Maclean and Hall (1987) and Hall (1987a, 1998a).

6. Bony and Cartilaginous Fishes

1. See Schmitz, Papan, and Campos-Ortega (1993) and Papan and Campos-Ortega (1994) for neurulation in *Danio rerio*. See Eisen and Weston (1993) and Schilling and Kimmel (1994) for the studies on the zebrafish, and see Ekker and Akimenko (1991) for a fate map of the zebrafish gastrula. For neurulation by cavitation of a neural keel in *Amia*, *Serranus*, and *Petromyzon*, see Dean (1896), H. V. Wilson (1899), and de Selys-Longchamps (1910).

2. See Reichenbach, Schaaf, and Schneider (1990) for *Cichlasoma nigrofasciatum*.

3. See Sadaghiani and Vielkind (1989, 1990a, b) and Sadaghiani, Crawford, and Vielkind (1994) for *Xiphophorus* and *Oryzias*, and Hirata, Ito, and Tsuneki (1997) for the studies on the swordtail.

4. For the origin of the dorsal roots in the spinal cord, see Balfour (1876, 1878a, b), Sagemehl (1882), Kastschenko (1888), and Sobotta (1935). For other early studies on neural crest contribution to dorsal root ganglia, see Cook and Neal (1921) and Edwards (1929).

5. See Borcea (1909) for the studies on *Belone acus*, Lopashov (1944) for the extirpations and grafts, and Kemp (1990) for ablation of lungfish neural crest.

6. See Raible et al. (1992) for migration, Raible and Eisen (1994, 1996) for lineage restriction in zebrafish trunk neural crest, and Gimlich and Braun (1985) for an assessment of fluorescent dextran tracers.

7. See Langille and Hall (1987, 1988a) for the fate map of skeletogenic neural crest in the Japanese medaka. For the neural crest origin of the branchial arch skeleton in *Xiphophorus*, see Sadaghiani and Vielkind (1989, 1990a, b).

8. See de Villers (1947) for aggregation of osteogenic cells under neuromasts. See Holmgren (1940), Coombs, Gorner, and Munz (1989), Vischer (1989a, b), Webb (1990), and Wonsettler and Webb (1997) for lateral line patterns in sharks and rays, *Eigenmannia* (a gymnotiform), cichlid, and hexagrammid fishes; and Webb and Northcutt (1997) for neuromasts in non-teleost bony fishes in which multiple canal neuromasts between pore positions is the evolutionarily primitive condition. See Merrilees and Crossman (1973), Reif (1982), and Meinke (1986) for scale and tooth induction in fishes, and see Puzdrowski (1989) for canal neuromasts. For possible induction of bone or cartilage by organs of the lateral line, see Allis (1898), Merrilees and Crossman (1973), Merrilees (1975), Graham-Smith (1978), Hanken and Hall (1983), Hall and Hanken (1985), and Webb and Noden (1993).

9. See Collazo, Fraser, and Mabee (1994) for the DiI-labeling study and see S. C. Smith, Lannoo, and Armstrong (1988, 1990), Lannoo and Smith (1989), and Northcutt (1996) for associations between placodal and neural crest ectoderm.

10. See M. M. Smith and Hall (1993) and M. M. Smith et al. (1994) for the neural crest origin of fin mesenchyme and the dermal skeleton. For discussions of the cellular and molecular transitions required to go from fin to limb, see Hall (1991b, 1998a), Thorogood (1991), Coates (1994), Hinchliffe (1994), Shubin (1991, 1995), Sordino, van der Hoeven, and Duboule (1995), and Sordino and Duboule (1996). See Hall (1975, 1978, 1999b), M. M. Smith and Hall (1990, 1993), and M. M. Smith (1991) for evolutionary relationships among the skeletal tissues.

11. A number of papers, in addition to Matsumoto et al. (1983) and Matsumoto, Wada, and Akiyama (1989), have now been published developing culture systems for fish cells. Among these are Bols and Lee (1991) for a variety of cell types including pigment and fin

mesenchyme but not skeleton, and Miyake and Hall (1994) and Koumans and Sire (1996) for skeletal tissues and scales.

12. Whiteley and Armstrong (1991) showed that ectopic expression in the axolotl of a genomic fragment containing a homeobox caused defects in the anterior neural plate and the failure of eye development.

13. For entries into the rapidly growing literature on targeted mutagenesis of the zebrafish genome, see Currie (1996), Neuhass et al. (1996), Piotrowski et al. (1996), Schilling et al. (1996), and Schilling (1997). For *Chinless* (*Chn*) mutation and *Dlx* in the zebrafish, see Eisen and Weston (1993), Schilling, Walker, and Kimmel (1996), and Ellies et al. (1997). See Qiu et al. (1995) for similar studies with *Dlx-1* and *Dlx-2* in mice. See Schilling et al. (1996) for jaw and branchial arch mutants, Malicki et al. (1996) for mutations affecting ear development, and Kelsh et al. (1996) for mutations affecting pigmentation.

14. See Thisse, Thisse, and Postlethwait (1995) and Jesuthasan (1996) for studies with the *Spadetail* mutant.

7. Reptiles and Birds

1. See Ferguson (1984, 1985) for neural crest cell migration in alligator embryos, and Ferguson (1981) for the studies with FUDR in alligators.

2. For the neural crest origin of sympathetic ganglia, see Kuntz and Batson (1920), Kuntz (1922), Müller and Ingvar (1921, 1923), Jones (1937, 1939, 1941, 1942), Yntema and Hammond (1945), Hammond and Yntema (1947), and the literature summarized in Le Douarin (1982) and Anderson (1989a, 1993b). For enteric and visceral ganglia, see Van Campenhout (1931, 1932, 1941), Yntema and Hammond (1945), Peters-van der Sanden et al. (1993), and Peters-van der Sanden (1994). For chromaffin cells, see Hammond and Yntema (1947).

3. See Dorris (1936, 1938, 1939) for the demonstration of neural crest formation of pigment cells. See Hörstadius (1950, pp. 81–88) for a discussion of pioneering studies grafting between and among breeds and species of avian embryos. For an insightful comparison of the primitive streak and the neural crest with respect to migration and regionalization along the A–P axis, see Bellairs (1987).

4. Nieto et al. (1994) and Duband et al. (1995) demonstrated the involvement of *Slug* at the onset of neural crest cell migration, while Sechrist et al. (1995) used *Slug* to monitor regulation of the neural crest. See Shankar, Jaskoll, and Melnick (1992) and Shankar et al. (1994) for retention of N-CAM by neural crest cells in response to retinoic acid; and see Nakagawa and Takeichi (1998) for the studies with N-cadherin and cadherin-7.

5. See Reissmann et al. (1996) for BMP-4 and BMP-7 within the dorsal aorta and differentiation of sympathetic neurons. BMP-2 is also present in the murine dorsal aorta (Shah, Groves, and Anderson 1996), where it may also promote neuronal differentiation.

6. Although called chimeras, the term is not strictly correctly applied if a chimera is thought to be an organism with large contributions from each species, usually half from one species, half from another. These grafted embryos are neural crest chimeras, not chimerical animals. Nevertheless, the term quail/chick chimera is usually used as a shorthand for an embryo containing cells of another species.

7. See Kinutani et al. (1989) for avian spinal cord chimeras after hatching and Le Douarin (1988a) for chimeras and the immune system.

8. For these pioneering studies on quail/chick chimeras, see Le Douarin (1969, 1971a, b, 1974, 1988a) and the summary in Le Douarin (1982). For mapping of the fate map of avian neural crest, see Le Lièvre (1971a, b, 1974, 1976, 1978), Le Lièvre and Le Douarin (1974, 1975), and Le Douarin, Ziller, and Couly (1993). For pathways of neural crest migration in avian embryos, see Noden (1973, 1975, 1978a, 1980, 1982a, b, 1983a).

9. Kameda (1995) argues that the C-cells of the avian ultimobranchial gland arise from neuronal cells of the distal vagal ganglion that subsequently enter the primordium of the ultimobranchial gland where they differentiate into C-cells. Accordingly, C-cells initially differentiate along a neuronal pathway.

10. For fate mapping of the ganglia, see Noden (1978b, c, 1984a) and Johnston et al. (1979). See Sohal et al. (1996) and Stark et al. (1997) for contribution of ventral neural tube cells to formation and development of the trigeminal ganglion.

11. For segregation of individual muscles and the patterning of muscles by neural-crest-derived connective tissue, see Noden (1983b, 1984b, 1985, 1986a, b, 1987) and McClearn and Noden (1988). For detailed fate-mapping of avian cephalic mesoderm, see Noden (1983b, 1986b), McClearn and Noden (1988), Couly, Coltey, and Le Douarin (1992) and Huang et al. (1997).

12. See Baroffio, Dupin, and Le Douarin (1988), Ito and Sieber-Blum (1993), and Ekanayake and Hall (1994) for clonal culture on feeder layers.

13. See Pomeranz and Gershon (1990) and Serbedzija et al. (1991) for the sacral neural crest. These cells are multipotential; they form neuronal and glial cells if cultured on laminin but are normally inhibited from forming catecholaminergic neurons (Pomeranz et al. 1993).

14. See Maxwell, Forbes, and Christie (1988), Sanders and Cheung (1988), and Maxwell and Forbes (1991) for isolation of HNK-1-positive cells using cell sorting.

15. See Charlebois et al. (1990) for the chick cytokeratin cDNA, Bevan et al. (1996) for the cDNA libraries, and Martinsen and Bronner-Fraser (1997) for differential display of cDNA fragments.

16. See Bockman et al. (1987) and Bockman, Redmond and Kirby (1989) for indirect effects of cranial neural crest on heart and vascular development.

17. See Le Lièvre and Le Douarin (1975), Kirby and Stewart (1983), and Kirby et al. (1983) for early studies on neural crest contributions to the aortic arches and the heart. For later studies establishing the cardiac neural crest, see Nishibatake, Kirby, and van Mierop (1987), Phillips, Waldo, and Kirby (1989), Sumida, Akimoto, and Nakamura (1989), Kirby and Waldo (1990, 1995), Filogamo, Corvetti, and Daneo (1990), Rosenquist et al. (1990), Takamura et al. (1990), Kappetein et al. (1991), Kuratani and Kirby (1991, 1992), Kuratani, Miyagawa-Tomita, and Kirby (1991), Miyagawa-Tomita et al. (1991), Noden (1991), Hood and Rosenquist (1992), Kirby (1993), Kirby et al. (1993), Olson and Srivastava (1996), and Waldo et al. (1998). See Fishman and Chien (1997) for an excellent and beautifully illustrated review of heart development. See Beall and Rosenquist (1990) for the studies with smooth muscle α-actin.

18. See Le Douarin (1982) and Bockman and Kirby (1984, 1985) for neural crest contributions to the avian thymus; T. J. Wilson et al. (1992) for the microenvironment in which the avian thymus develops; and Kuratani and Bockman (1991) for these studies on axial levels.

8. Mammals

1. For neural crest and craniofacial defects, see the papers in the symposium volumes edited by Bergsma (1975), Opitz and Gorlin (1988), Vig and Burdi (1988), and Copp (1993); also see Pierce (1985, 1987), Copp et al. (1990), the *Festschrift* for M. Michael Cohen Sr. edited by Cohen and Baum (1997), and Chapter 12 herein.

2. Hall (1987a) and Morriss-Kay and Tan (1987) summarized the state of knowledge concerning the mammalian neural crest as it was known at the beginning of 1986; Osumi-Yamashita and Eto (1990) as it was known at the end of the decade.

3. See Quinlan et al. (1995), Tam and Quinlan (1996), and Tam and Selwood (1996) for these studies.

4. See Beddington (1981) for the ^{3}H-thymidine-labeling study, Schoenwolf and Nichols (1984), and Schoenwolf, Chandler, and Smith (1985) for caudal neural crest and tailbud formation in mice and avian embryos. See Chapter 1 and Hall (1998a, b) and references therein for secondary neurulation. In a recent analysis, Thomas and Beddington (1996) determined that anterior primitive endoderm may be responsible for the initial patterning of the rostral neural plate, especially the prosencephalon in murine embryos. See Beddington and Robertson (1998) for mechanisms of anterior patterning of mice and other vertebrate embryos.

5. See Nozue (1974), Nozue and Tsuzaki (1974a, b), and Nozue and Kayana (1977a, 1978) for the studies with Mitomycin C and see Pearse (1969) for uptake of decarboxylate amine precursors.

6. See Ito and Takeuch (1984) and Jaenish (1985) for neural crest origin of melanocytes and adrenergic neurons in mice. In quail embryos, retinoic acid promotes the differentiation of both melanocytes and adrenergic neurons with adrenergic neurons evoked preferentially from early migrating cells.

7. Whether the forebrain is segmented remains controversial; see Northcutt (1995) for a review. See Tan and Morriss-Kay (1986), Smits-van Prooije et al. (1987, 1988), Tam (1989), and Chan and Tam (1998) for pioneering studies using labeled rodent neural crest.

8. See Wake (1993) and Kuratani (1997) for comprehensive analyses of different aspects of neural tube boundaries.

9. See Tam (1989), Trainor, Tan, and Tam (1994), and Trainor and Tam (1995) for mesoderm and neural crest patterning. See Campbell (1989) for periocular mesenchyme and melanogenesis.

10. See Serbedzija et al. (1991), Serbedzija, Fraser, and Bronner-Fraser (1990), and Serbedzija, Bronner-Fraser, and Fraser (1994) for DiI labeling; Kubota, Morita, and Ito (1996) for the monoclonal antibody; Smith-Thomas and Fawcett (1989) for studies with the m217c antibody; Serbedzija and McMahon (1997) for the LacZ reporter; Tremblay, Kessel, and Gruss (1995) for the transgenic mouse line, and Harvey et al. (1992) for Y-chromosome-specific probes.

11. For the origin of odontoblast precursors in the neural crest, see Lumsden (1987) and Imai et al. (1996). The molecular basis of tooth induction, which is now quite well understood, is outside the scope of this review; for summaries, see Lumsden (1987, 1988), Slavkin et al. (1990), Thesleff (1995), Thesleff, Vaahtokari, and Partanen (1995), Thesleff et al. (1995), Thesleff and Sahlberg (1996), Maas and Bei (1997), Thesleff and Sharpe (1997), Tucker, Al Khamis, and Sharpe (1998), and the papers in Ruch (1995).

12. See de Beer (1947) and Hörstadius (1950, pp. 66–70) for neural crest origin of teeth in amphibians; and see Kollar and Baird (1969), Lumsden (1987, 1988), and Hall

(1988b) for patterning of teeth by oral ectoderm. See Kollar (1983) and Lumsden (1987) for overviews of murine tooth induction via epithelial–mesenchymal interactions.

13. See Levak-Svajger and Svajger (1974), Skreb, Svajger, and Levak-Svajger (1976), and Svajger et al. (1981) for teratomas derived from grafts of embryonic ectoderm. For discussions of teratomas and teratocarcinomas, see Hall (1983b, 1994d) and Maclean and Hall (1987).

9. Mechanisms of Migration

1. Erickson (1986, 1993b), Newgreen and Erickson (1986), Rovasio and Thiery (1987), and Bronner-Fraser (1993c) provide excellent and extensive reviews of early studies on crest migration.

2. See Adelmann (1925), Bartelmez (1922, 1960, 1962), Holmdahl (1928), Bartelmez and Blount (1954), and Kallen (1953) for pioneering studies on the mammalian neural crest. See Holmdahl (1928), O'Rahilly and Gardner (1979), and Müller and O'Rahilly (1994) for studies on human embryos. See von Schulte and Tilney (1915) for cat, Chiarugi (1894) and Celestino da Costa (1920) for guinea pig, and Adelmann (1925) for rat.

3. For migration of trunk neural crest cells through the somites, see Lim et al. (1987), Loring and Erickson (1987), and Stern and Keynes (1987). For migration of determined melanocytes along dorsolateral pathways associated with ectoderm, see Erickson, Duong, and Tosney (1992), Erickson and Goins (1995), and Reedy, Faraco, and Erickson (1998). For the role of N-cadherin and cadherin-7 in controlling melanocyte migration, see Nakagawa and Takeichi (1998).

4. For scanning electron microscopic analysis of mammalian neural crest, see Waterman (1975, 1976, 1979), Morriss and Thorogood (1978), Morriss and New (1979), Erickson and Weston (1983), Morriss-Kay and Tuckett (1985), and Schoenwolf (1986). For timing of onset of neural crest cell migration, see Vermeij-Keers and Poelman (1980), Nichols (1981, 1986a, b), Verwoerd, van Oostrōm, and Verwoerd-Verhoef (1981), Tan and Morriss-Kay (1984, 1985), and Smits-van Prooije et al. (1985).

5. The term *basement membrane* refers to the extracellular matrix deposited on the basal surface of epithelial cells and visible in the light microscope. It consists of several subcomponents visible only with transmission electron microscopy, including the basal lamina and lamina lucida; see Hay (1989) and Leblond and Inoue (1989).

6. See Nichols (1981, 1987) for pioneering studies on migration of mammalian neural crest cells. For breakdown of the basal lamina in association with initial neural crest cell migration, see Erickson and Weston (1983), Innes (1985), Nichols (1985), Martins-Green and Erickson (1986, 1987), Sternberg and Kimber (1986), Erickson, Duong, and Tosney (1992), and Erickson (1993a). For migration on basal laminae in vitro, see Halfter, Chiquet-Ehrismann, and Tucker (1989). For the studies on the macaque, see Blankenship, Peterson, and Hendrickx (1996) and Peterson et al. (1996). See Poelman et al. (1990) for attachment of fragments of basal lamina to migrating neural crest cells.

7. See Valinsky and Le Douarin (1985), Erickson (1986), and Menoud, Debrot, and Schowing (1989) for plasminogen activator, and Newgreen and Minichiello (1995) for protein kinase inhibitors and epithelial to mesenchymal transformation.

8. See Richardson, Hornbruch, and Wolpert (1989) and Richardson and Hornbruch (1991) for prepatterns in feather development.

9. See Weston (1970), Noden (1978a), Le Douarin (1982), Graveson, Hall, and Arm-

strong (1995), and Northcutt (1996) for control of directionality of migration.

10. See Yip (1986) and Teillet, Kalcheim, and Le Douarin (1987) for the initial studies on regionalization of ganglionic primordia.

11. See Lahav et al. (1996) and Thomas et al. (1998) for endothelin-3 and melanocyte proliferation and differentiation, and Richardson and Sieber-Blum (1993) and Sieber-Blum and Zhang (1997) for determination of fate late in migration. Knocking out the endothelin-A receptor in mice produces severe anomalies of neural-crest-derived craniofacial and cardiac structures (Clouthier et al., 1998). Endothelin-1 regulates *Msx-1* expression through dHAND, a helix-loop-helix transcription factor (Thomas et al., 1998).

12. See Löfberg et al. (1985) and Brauer and Markwald (1987) for the adsorption studies; Roth (1973) and Shur (1982) for enzymes in the glycocalyx; and Runyan, Maxwell, and Shur (1986) for effects on cell adhesion. See French et al. (1988) for an overview of somitogenesis, and Williamson, Parrish, and Edelman (1991a, b) for the studies with cytotactin.

13. For data on extracellular matrix control over amphibian neural crest cell migration and region-specific differentiation in wild-type and albino, see Löfberg et al. (1985), Brauer and Markwald (1987), Perris, von Boxberg, and Löfberg (1988), Löfberg, Perris, and Epperlein (1989), Perris et al. (1990), Stigson and Kjellen (1991), Thibaudeau and Frost-Mason (1992), Epperlein and Löfberg (1990, 1993), Erickson and Perris (1993), Epperlein, Löfberg, and Olsson (1996), and Parichy (1996a, b). For the older literature on pigment patterning in amphibians, see Twitty (1944, 1945, 1949), Twitty and Bodenstein (1939, 1944), Twitty and Niu (1948), and also Hörstadius (1950, pp. 75–78). See Armstrong (1985) for an analysis of 30 mutants in the axolotl, *Ambystoma mexicanum*.

14. See Jacobson and Meier (1984) for the studies on *Taricha*. For specification of pigment patterns by neural crest cells themselves, see Twitty (1936, 1945, 1949), Twitty and Bodenstein (1939), and Hörstadius (1950, pp. 73–78, especially figures 38 and 39). For altered migration of neural crest cells when grafted to different strains, see Hallet and Ferrand (1984).

15. See Huszar, Sharpe, and Jaenisch (1991) and Huszar et al. (1991) for these studies on murine coat patterns. For mechanisms that may account for pigmentation patterns in vertebrates, see Murray, Deeming, and Ferguson (1990), Olsson and Löfberg (1992), Olsson (1993, 1994), Epperlein, Löfberg, and Olsson (1996), and Parichy (1996a–c).

16. For the composition of the ECM through which neural crest cells migrate, see Frederickson and Low (1971), von der Mark, von der Mark and Gat (1976), Corsin (1977), Bolender et al. (1980), Thiery, Duband, and Delouvée (1982), Newgreen and Erickson (1986), Epperlein, Halfter, and Tucker (1988), Riou et al. (1988), Halfter, Chiquet-Ehrismann, and Tucker (1989), Morriss-Kay and Tuckett (1989), McCarthy and Hay (1991), Perris et al. (1991), Erickson and Perris (1993), and Henderson and Copp (1997). For modification of that ECM by migrating neural crest cells, see Bolande, Seliger, and Markwald (1980), Bolender et al. (1980), Ruiz, Mujwid, and Steffek (1982), Tosney (1982), and Brauer, Bolande, and Markwald (1985).

17. See Lallier et al. (1992) for differences between cranial and trunk neural crest cell attachment to extracellular matrices.

18. See Zagris and Chung (1990) for laminin and primitive streak formation; Hall and Ekanayake (1991) for possible roles for laminin in neuronal differentiation; Tuckett and Morriss-Kay (1986) and Smits-van Prooije et al. (1986) respectively for the antibody and lectin studies of neural crest cell migration; and see Martha et al. (1990) for the axolotl studies.

19. See Loring and Erickson (1987) for the role of fibronectin, and Guillory and Bron-

ner-Fraser (1986), Sanders, Prasad, and Cheung (1988), Bronner-Fraser and Stern (1991), and Bronner-Fraser, Stern, and Fraser (1991) for migration through sclerotome. See Gooday and Thorogood (1985) for contact-mediated interactions between neural crest and somitic cells.

20. For binding of neural crest cells to fibronectin, see Rovasio et al. (1983), Duband et al. (1986), and Krotoski, Domingo, and Bronner-Fraser (1986).

21. See Rovasio et al. (1983) for the antibody-binding studies and Boucaut et al. (1984) for the synthetic peptide. See Bronner-Fraser (1985) for studies using the cell-binding domain of fibronectin, and Pomeranz et al. (1993) for immunoselection.

22. See Krotoski and Bronner-Fraser (1990), Kil, Lallier, and Bronner-Fraser (1996), and Goh, Yang, and Hynes (1997) for the distribution of integrins and integrin ligands and their role in *Xenopus* trunk neural crest cells and apoptosis of neural crest cells.

23. See H. P. Erickson and Bourdon (1989), Turley, Roth, and Weston (1989), Perris and Johansson (1990), and Kerr and Newgreen (1997) for binding properties of fibronectin, tenascin and chondroitin sulfate proteoglycan; Sanders, Prasad, and Cheung (1988), Perris, Krotoski, and Bronner-Fraser (1991), and Perris et al. (1993) for effects of collagens on neural crest and sclerotomal migration.

24. See Manasek and Cohen (1977), Kalcheim and Leviel (1988), and Tucker and McKay (1991) for synthesis of extracellular matrix (ECM) components by neural crest cells.

25. See Erickson (1986) and Newgreen and Erickson (1986) for discussions of possible physical mechanisms guiding migration.

26. See Newgreen and Erickson (1986), Erickson (1987), Erickson and Perris (1993), and Tosney and Oakley (1990) for barriers to migration; Luckenbill-Edds and Carrington (1988) and Morris-Wiman and Brinkley (1990) for hyaluronan, early crest migration, and elevation of the neural folds; Newman and Comper (1990) for physical mechanisms associated with morphogenesis and pattern formation; and Epperlein, Halfter, and Tucker (1988) and Spence and Poole (1994) for fibronectin in the migration pathway and as the basis for vessel-mediated migration.

27. See Newgreen and Erickson (1986) and Tucker and Erickson (1984) for ECM components and migration; Oakley and Tosney (1991), and Oakley et al. (1994) for glycoconjugates as barriers; and Moro-Balbás et al. (1993, 1998) for injection of chondroitin sulfate or retinoic acid into rhombomeres.

28. See Teillet, Kalcheim, and Le Douarin (1987) for migration of ganglionic precursors through the anterior halves of somites; Erickson, Loring, and Lester (1989) for matrix products in anterior sclerotome; and Newgreen, Scheel, and Kastner (1986), Pettway, Guillory, and Bronner-Fraser (1990), Tosney and Oakley (1990), and Stern, Artinger, and Bronner-Fraser (1991) for the role of the notochord and peri-notochordal matrix.

29. See Noden (1983a, 1984b) for the development of ectopic skeletal elements after transplanting first arch crest more caudally along the neural axis.

30. See Landolt et al. (1995) and Stigson, Löfberg, and Kjellen (197) for versican; Rothman and Gershon (1984), Jacobs-Cohen et al. (1987), Payette et al. (1988), and Rothman, Goldowitz, and Gershon (1993) for *Lethal-spotting*; Rothman et al. (1987, 1990, 1993) for back-transplantation of bowel neural crest or embryonic gut wall; Brun (1987) for *eyeless*; Lim et al. (1987) and Lunn et al. (1987) for the grafting studies.

31. For the influence of neural crest cells on neuromast patterning and of lateral line on melanophores, see S. C. Smith, Lannoo, and Armstrong (1988, 1990), Lannoo and Smith (1989), S. C. Smith, Graveson, and Hall (1994), Northcutt (1996), Parichy (1996c), and S. C. Smith (1996). For the *Msx-2* and *Dlx-3* expression study, see Metscher et al. (1997).

See Lannoo and Smith (1989), S. C. Smith (1996), and Northcutt, Catania, and Criley (1994) for an overview of lateral lines in the axolotl. See S. C. Smith, Lannoo, and Armstrong (1988, 1990), Northcutt, Catania, and Criley (1994), Northcutt and Brändle (1995), and Northcutt, Brändle, and Fritzsch (1995) for lateral line and neuromast development in the axolotl and *Rana pipiens*, a frog.

32. For distribution of type II collagen in avian, murine, zebrafish, and *Xenopus* embryos and for the "flypaper model," see Thorogood (1987, 1993a), Thorogood, Bee, and von der Mark (1986), and Wood et al. (1991). See Duband and Thiery (1987), Ishibe, Cremer, and Yoo (1989), Seufert, Hanken, and Klymkowsky (1994), and Yan et al. (1995) for additional studies.

10. Mechanisms of Differentiation

1. For broader discussions of cell commitment, whether determination of cell fate occurs in one or several steps, the temporal separation of commitment and differentiation, and the existence of subpopulations within apparently homogenous cell populations, see Maclean and Hall (1987), Anderson (1989b, 1993a, b, 1997), Sieber-Blum (1989a, b, 1990), Bronner-Fraser and Fraser (1991), Weston (1991), Stemple and Anderson (1992, 1993), Bronner-Fraser (1993a, b, 1995), Groves and Anderson (1996), and Sieber-Blum and Zhang (1997).

2. See Le Douarin (1986), Weston (1986), Anderson (1989b, 1997), Sieber-Blum (1990), and Bronner-Fraser (1993b, c) for heterogeneity of premigratory neural crest cells.

3. See Chapter 3 in Hörstadius (1950), Noden (1983a, 1984b), and Chapter 5 herein for exchange of neural crest in amphibians and avian embryos, and see Thorogood (1993a) for an analysis of patterning in the two groups. For segregation of cell lines, see Ziller et al. (1983), Vincent and Thiery (1984), Ciment and Weston (1985), Le Douarin (1986), Weston (1986), and Atchley and Hall (1991).

4. See MacLean and Hall (1987) and Hall (1997, 1998a) for the fact that lineage tracing does not necessarily equate with determination of cell fate.

5. See Anderson (1997) for a concise discussion of mammalian neural crest cells as multipotent stem cells; Smith-Thomas, Davis, and Epstein (1986) for transformation of chondrogenic to neuronal cells; Maclean and Hall (1987) and Fang and Hall (1997) for de- and redifferentiation; and Ciment et al. (1986) for the studies with phorbol ester. See Stocker et al. (1991) for the studies with growth factors and DRG.

6. See Cohen and Konigsberg (1975) and Sieber-Blum and Sieber (1985) for clonal cell culture; Norr (1973), Fauquet et al. (1981), Howard and Bronner-Fraser (1985), and Kalcheim and Le Douarin (1986) for differentiation of adrenergic neurons; Bronner-Fraser, Sieber-Blum, and Cohen (1980) and Sieber-Blum and Cohen (1980) for adrenergic neuronal differentiation following exposure to somatic extracellular matrix; and Jessell and Melton (1992) for diffusible inductive signals.

7. See Ito and Sieber-Blum (1991, 1993) and Ito, Morita, and Sieber-Blum (1993) for these clonal analyses.

8. See Baroffio et al. (1991) for the lineage studies and Lo, Birren, and Anderson (1991) for the studies with *V-myc*. The mammalian *achaete-scute* homologue is expressed transiently in a subset of neural precursors and may be a useful marker for, or determinate of, the sympathoadrenal lineage (Lo et al. 1991). Its expression in *Xenopus* converts neural crest or epidermal ectoderm cells to a neuronal fate (Turner and Weintraub, 1994).

9. See Krotoski and Bronner-Fraser (1986), Bronner-Fraser and Fraser (1988, 1989, 1991), and Krotoski, Fraser, and Bronner-Fraser (1988) for the labeling studies, and see Le Douarin (1986), Patterson (1990), and Mujtaba, Mayer-Proschel, and Rao (1998) for neuronal lineages.

10. See Bronner-Fraser and Fraser (1991), Artinger and Bronner-Fraser (1992a), and Asamoto, Nojyo, and Aoyama (1995) for the studies on early- and late-migrating avian trunk neural crest cells; Serbedzija, Bronner-Fraser, and Fraser (1989) for labeling of cells in the ventral pathway using DiI; Artinger and Bronner-Fraser (1992b) for the influence of the notochord on the ventrally migrating cells; and Hirata, Ito, and Tsuneki (1997) for the lamprey study.

11. See Ciment and Weston (1985) for the E/C 8 antibody; Sieber-Blum (1989a) for premigratory pluripotential cells; and Duff et al. (1991) for the clonal analysis.

12. See Baker et al. (1997) for these studies with early- and late-migrating avian cranial neural crest cells. Also see Henion and Weston (1997) for lineage restriction early in migration.

13. For differentiation of pigment cells in response to extracellular matricial components, see Loring, Glimelius, and Weston (1982), Löfberg et al. (1985), Perris and Löfberg (1986), Epperlein (1988), Perris, von Boxberg, and Löfberg (1988), Löfberg, Perris, and Epperlein (1989), Epperlein and Löfberg (1990, 1996), Olsson, Svensson, and Perris (1996), and Olsson et al. (1996). For clonal analysis using *Xenopus* neural crest, see Akira and Ide (1987). For the role of melanocyte-stimulating hormone, see Dean and Frost-Mason (1995).

14. See Erickson and Goins (1995) for back-transplantation; Lahav et al. (1996) for endothelin-3 and melanocyte proliferation and differentiation; and Richardson and Sieber-Blum (1993) and Sieber-Blum and Zhang (1997) for determination of fate late in migration.

15. For the role of growth factors and hormones in generating and maintaining diversity among neural crest cells, see Le Douarin (1986), J. Smith and Faquet (1984), Heine et al. (1987), Hall (1988a), Hall and Ekanayake (1991), Sieber-Blum (1991), and Sieber-Blum and Zhang (1997). See Brauer and Yee (1993) for neural crest synthesis of TGF-ß.

16. See Pinco et al. (1993) for the role of NT-3. See Sieber-Blum, Kumar, and Riley (1988) for differentiation of sensory neurons from avian neural crest cells maintained in vitro; Sieber-Blum (1989a, b, 1991) and Sieber-Blum and Zhang (1997) for the studies with growth factors and neurotransmitter-uptake inhibitors; and see Hall and Ekanayake (1991) for a discussion of the evidence for the role of BDNF and NGF in sensory neuronal differentiation.

17. See Landis and Patterson (1981) and End et al. (1983) for NGF and sympathetic neuronal/glial differentiation; Shah, Groves, and Anderson (1996) and Anderson (1997) for studies with the mammalian cells; and Hall (1978, 1986), Tran and Hall (1989), and Fang and Hall (1995, 1996, 1997) for chondrogenesis from periosteal cells, including the role of N-CAM. A monoclonal antibody against the NGF receptor was used to isolate multipotential precursors from mammalian neural crest that form neurons and Schwann cells as they transit from self-renewing stem cells to clonal progeny.

18. See Yntema (1955), Holtfreter (1968), and Hall (1978, 1983b, 1991a) for the ability of the otic vesicle to induce chondrogenesis in head mesoderm but not in mesenchyme from the neural crest. See Lewis (1907), Luther (1924), Kaan (1938), and Holtfreter (1968) for interactions between the otic vesicle and otic capsule.

19. For the requirement for proximity of neural crest and pharyngeal endoderm, see Epperlein (1974) and Epperlein and Lehmann (1975). For diffusible factors in induction, see Jessell and Melton (1992).

20. See Andrew and Gabie (1969) for hanging-drop cultures; Ferguson and Honig (1984) for alligator studies; and Okada (1955), Wilde (1955), Seno and Nieuwkoop (1958), Cusimano-Carollo (1963), Holtfreter (1968), Drews Kocher-Becker and Drews (1972), Epperlein (1974), Corsin (1975), Epperlein and Lehmann (1975), Cassin and Capuron (1979), Minuth and Grunz (1980), Graveson and Armstrong (1987, 1990), Seufert and Hall (1990), and Graveson (1993) for the amphibian studies. For restriction of induction to pharyngeal endoderm, see Wilde (1955) and Graveson and Armstrong (1987). For co-induction by endoderm and dorsal mesoderm, see Wilde (1955) and Corsin (1975).

21. See Hall (1978, 1983a, b, 1987a, 1988a, 1994a,d) for epithelial–mesenchymal interactions, and see Hall (1999c) for how to set up such tissue recombinations. The importance of timing in development is discussed by Hall and Miyake (1995a, 1999).

22. Although basement membranes and basal laminae are epithelial extracellular matrices, not all their products are produced by epithelial cells. Mesenchyme contributes such important structural components of basement membranes as fibronectin. For a detailed analysis in relation to the origin of the basement membrane of the epiblast, which receives contributions from both epi- and hypoblasts, see F. Harrison (1986, 1993). For heterogeneity of basement membranes, see Leu and Damjanov (1988).

23. For binding of growth factors (TGF-ß1, BMP-2) to collagen IV within basement membranes, see Paralkar, Vukicevic, and Reddi (1991), Paralkar et al. (1992), and Vukicevic et al. (1992, 1994). The inductive activity of the endoderm during cartilage induction in *Xenopus laevis* acts across Nuclepore filters of 0.4 μm porosity. The absence of cell processes in the pores of these filters would appear to rule out a cell-contact-mediated interaction and favor an extracellular-matrix-mediated interaction (Seufert and Hall 1990).

24. See Hall (1983a, b, 1988a) and Thorogood and Smith (1984) for matrix-mediated tissue interactions, and Kollar (1983), and Lumsden (1987, 1988) for interactions during tooth development.

25. For overviews of the research on which this conclusion is based, see Hall (1991a, 1994a,d).

26. See Thorogood, Bee, and von der Mark (1986) for trapping of mesenchymal cells, and Hall (1978), Hall and Miyake (1992, 1995b), and Miyake, Cameron, and Hall (1996, 1997a, b) for the importance of condensation.

27. Maxwell and Forbes (1987, 1988, 1990a, b) performed the Matrigel studies, Vukicevic et al. (1994) and Varley et al. (1995) the studies with BMP-7. See Grant et al. (1985) and Vukicevic et al. (1992, 1994) for the constitution of Matrigel. See Paralkar, Vukicevic, and Reddi (1991) for binding of TGF-ß1 to basement membranes; Hall (1994a,d) for matrix-mediated interactions in skeletogenesis and for growth factors bound to Matrigel; and Leu and Damjanov (1988) for heterogeneity of neoplastic basement membranes, especially with respect to laminin and type IV collagen.

28. See Ho et al. (1994) for the localization of PDGFR-α in *Xenopus*, and Orr-Urtreger and Lonai (1992), Schatteman et al. (1992), and Soriano (1997) for expression of PDGF-α and its receptor in mesenchyme and/or in adjacent cell layers.

29. For patterns of expression and the role of *Msx-1* in avian craniofacial development, see Takahashi and Le Douarin (1990), Suzuki et al. (1991), Takahashi, Bontoux, and Le Douarin (1991), and Mina et al. (1995). For BMP-4 and *Msx-1* in tooth development, see Tucker, Al Khamis, and Sharpe (1998).

30. See Hall (1980) for induction of murine mandibular bone and see Hill et al. (1989), MacKenzie, Ferguson, and Sharpe (1991), and MacKenzie et al. (1991) for murine *Msx-1*. See Mahmood et al. (1995) and Mahmood, Mason, and Morriss-Kay (1996) for FGF expression.

31. See Mitchell et al. (1991), Morriss-Kay (1996), Schorle et al. (1996), and Zhang et al. (1996) for the transcription factor AP-2.

32. See Fagone (1959, 1960), Cusimano-Carollo (1962, 1963, 1967, 1969, 1972), and Cusimano, Fagone, and Reverberi (1962) for these studies on development of the mouth and associated structures in *Discoglossus pictus*, and see Kraemer (1974) for the development of the chondrocranium in *Discoglossus*. Thibaudeau and Altig (1988) provide the sequences of ontogenetic changes in the development of the oral apparatus of six species of anurans. Truncated development of oral structures in egg-brooding hylids is discussed by Wassersug and Duellman (1984). Experimental studies on the toad *Bufo bufo* provide further support for interactions between neural crest and endoderm, in this case in regulating the migration of endodermally derived primordial germ cells (Gipouloux and Girard 1986).

33. See Barlow and Northcutt (1994, 1995, 1997), Barlow et al. (1996), and Northcutt and Barlow (1998) for studies on the development of amphibian taste buds.

34. See Grunz and Tacke (1986) and Seufert and Hall (1990) for studies with *Xenopus*. Cassin (1975), Cassin and Capuron (1972, 1977, 1979), and Minuth and Grunz (1980) demonstrated the requirement for prechordal and lateral plate mesoderm. Development of the chondro- and osteocrania of *P. waltl* is documented by Eyal-Giladi and Zinberg (1964) and Corsin (1966a). Trueb and Hanken (1992) argue that the rostral cartilages in *Xenopus* are homologous to the suprarostrals.

35. See Helff (1940) and references therein for studies on tympanic cartilage and membrane differentiation.

36. Epigenetic control of development is discussed by Hall (1983c, 1987a, 1990a, 1998a), Nieuwkoop, Johnen, and Albers (1985), and Maclean and Hall (1987), morphogenesis of bone by Richman (1994). See Reiss (1990) for studies on *R. pipiens* and Teichmann (1955, 1959, 1961, 1962) and Corsin (1966b) for those on *Triturus alpestris*. See Medvedeva (1975, 1986a, b and citations therein) and the discussions in Hall (1983b, 1994d). For literature on intermediate sized nasal capsules, see Yntema (1955), Holtfreter (1968), and Hall (1983b).

37. See Howes (1977, 1978a, b) and Howes and Eakers (1984) for the association between bone of attachment and teeth.

38. Effects of anomalous migration were demonstrated in the re-analysis by Levy, Detwiler, and Copenhaver (1956) of the histological specimens from a study reported by Detwiler and Copenhaver in 1940.

39. For studies on the *premature death* (*p*) mutant in *A. mexicanum*, see Mes-Hartree and Armstrong (1980) and Graveson and Armstrong (1990, 1996).

40. See Nakamura and Ayer-Le Lièvre (1982) for the grafting of trunk neural crest into the cranial region. See M. M. Smith and Hall (1990), Atchley and Hall (1991), and Thesleff (1991) for development of teeth in ectopic sites and for the origin of alveolar bone from the dental follicle.

41. See M. M. Smith and Hall (1990, 1993), Graveson, Smith, and Hall (1997), and Hall (1998a, 1999b) for further elaboration of tooth formation from trunk neural crest.

42. See Jacobson and Duncan (1968) for inhibition of heart development, and Etheridge (1968, 1972) and Green (1970) for inhibition of amphibian and murine kidney development. See Woellwarth (1961), Jacobson (1966, 1987), Etheridge (1972), Jacobson and Sater (1988), and Hall (1997, 1998a) for further discussions of inhibition by neural crest.

43. See Gipouloux, Girard, and Delbos (1992) for the positive influence of neural crest on primordial germ cell migration.

11. Neurocristopathies

1. See Persaud, Chudley, and Skalko (1985) and Ortiz-Monasterio (1978) for depiction of malformations from antiquity; Gould and Pyle (1896), Bergsma (1975, 1979), Gorlin (1978), Kalter (1980), Melnick et al. (1980), and Gorlin, Cohen, and Levin (1988) for catalogues of birth defects; and Toriello (1988) for the incidence of description of new syndromes. See Cohen (1989, 1990) for a ten-part update on syndromology; Copp et al. (1990) for an overview of neural tube defects; Pierce (1985, 1987) and the papers in Hodges and Rowlatt (1994) for developmental aspects of tumors and syndromes; Corcoran (1998) for those neural tube defects that are resistant to folic acid; Jablonski (1992) for ultraviolet light/melanin/folic acid and neural tube defects; and Fleming and Copp (1998) for the metablic action of folate metabolism on pyrimidine synthesis. See Wilson and Wyatt (1988) for the contribution of defective neural crest to failure of closure of the posterior neuropore and to spina bifida in *vacuolated lens* (*vL*) mutant mice.

2. See Grüneberg (1947), Willis (1962), Cohen (1982), Opitz and Gorlin (1988), Cohen and Cole (1989), and Hall (1998b) for germ layer of origin and the etiology of organismal defects. See Cohen (1989) and Cohen and Cole (1989) for types of classifications of syndromes, including the use of embryological criteria; and Kjaer (1998) for embryological origins and interactions between the skeletal and nervous systems.

3. See Carpenter and Hollyday (1992a, b) for Schwann-cell-coated nerves in limb buds, and Saxén and Saxén (1960) for the case study.

4. For the CHARGE association, see Pagon et al. (1981), Siebert, Graham, and MacDonald (1985), and Cohen (1989). For neurocristopathies, see Bolande (1981, 1984), Couly (1981), Johnston, Vig, and Ambrose (1981), Opitz and Gorlin (1988), M. C. Jones (1990), and Johnston and Bronsky (1995).

5. For the isoallelic status of the *far* locus in mice, see Juriloff, Harris and Froster-Iskenius (1987) and Harris and Juriloff (1989).

6. See Rogers et al. (1994) for the studies with TGF-ß; Wahlström and Saxén (1976) for the survey of 90,000 biopsies; Mathew (1982), Kindblom, Lodding and Angervall (1983) and Mii et al. (1989) for the sarcomas; Redekop, Elisevic, and Gilbert (1990) for the ventricular Schwannoma; and Pascualcastroviejo (1990) for tumors of the neural crest.

7. See Tsokos et al. (1987) and Hill and Robertson (1997) for the in vitro studies with neuroblastoma cell lines.

8. See Bolande (1984) for tumor regression to nonmalignant cell type and see Maclean and Hall (1987) and Hodges and Rowlatt (1994) for the distinctive features of neoplastic cells.

9. See Hoefnagel et al. (1987) for treatment and diagnosis and Bertani-Dziedzic et al. (1979) for urinary VMA levels.

10. See Riopelle, Haliotis, and Roper (1983) for 7S NGF receptor complex; Stemple and Anderson (1992) for the monoclonal antibody against the NGF receptor to isolate stem cells; Oppedal, Brandtzaeg, and Kemshead (1987a) for the monoclonal antibody against neuroblastomas; and Kemshead et al. (1983) and Oppedal, Brandtzaeg, and Kemshead (1987b) for the use of such antibodies to screen for neuroblastomas.

11. See Pons, O'Dea, and Mikkin (1982) for the C-1300 neuroblastoma, Liwnicz (1982) for the lectin studies, and Sieber-Blum and Sieber (1981) and Nishihira et al. (1981) for response of neural crest cells to tumor promoters.

12. For basic studies on neurofibromatosis, see Crowe, Schulle, and Neel (1956), Riccardi and Mulvihill (1981), Riccardi and Eichner (1986), and Rubenstein, Bunge, and Housman (1986). See Huson (1987) for the first European symposium.

13. Friedrich Daniel von Recklinghausen (1833–1910), who studied medicine in Bonn, Würzburg, and Berlin, began his career at the Pathological Institute in Berlin as assistant to the great pathologist Rudolf Virchow and ended it as an eminent German pathologist who attracted postgraduate students from all over the world to the University of Strasbourg, to which institution he moved in 1871. However, von Recklinghausen was aloof and distant, rarely deigning to speak to his students (Beighton and Beighton 1986, p. 165).

14. For the diagnosis of von Recklinghausen neurofibromatosis, see Mautner et al. (1988). See Ponder (1990) for an overview of cloning of the neurofibromatosis gene, and see Brannan et al. (1994) for chromosome mapping and targeted gene disruption. See Riccardi and Eichner (1986) for categories of neurofibromatoses; Cohen and Hayden (1979) for pooling of syndromes; Seizinger, Martuza, and Gusella (1986), Barker et al. (1987), Rouleau et al. (1987), Di Simone and Berman (1989), and the proceedings of the 9th Human Gene Mapping Workshop held in Paris (*J. Med. Genetics*, 1987, **24**, 513–544) for chromosome mapping.

15. For non-neural-crest involvement in neurofibromastosis, see Crawford (1986) and Hope and Mulvihill (1981). For orthopedic manifestations, see Di Simone, Berman, and Schwentker (1988), while for bone in tumors, see Helson (1984) and Jaffe et al. (1984). See Lizard-Nacol et al. (1989) for Ewing sarcoma as a mix of mesenchymal and neural tissues.

16. See Treves (1923), Montagu (1971), Pomerance (1979), and Howell and Ford (1980) for popular treatments of the "Elephant Man." See Anastassiades, Puzic, and Puzic (1978) for the stimulating factor(s). Merrick's skeleton is on display in the Museum of the London Hospital Medical College. Such is the interest in the Elephant Man that the singer Michael Jackson was reported to have offered the college $500,000 for the skeleton to add to his collection of "curious objects" (*The Toronto Globe and Mail*, 2 June 1987).

17. For the evidence that Joseph Merrick had Proteus syndrome and not von Recklinghausen neurofibromatosis, see Tibbles and Cohen (1986) and Cohen (1987, 1988a, b). I thank Michael Cohen for the information on unreported cases of individuals with Proteus syndrome.

18. For transgenic mouse models, see Jensen et al. (1993). See Green (1966) and Deol (1970) for pleiotropy in mouse mutants; Weston (1980), Morrison-Graham and Weston (1989), Morrison-Graham et al. (1992), and Soriano (1997) for *Patch*.

19. See Morrison-Graham, Bork, and Weston (1990), Morrison-Graham, West-Johnsrud, and Weston (1990), Jessell and Melton (1992), Murphy et al. (1992), and Morrison-Graham and Weston (1993) for *Steel*, steel factor, and *White spotting*, which also affects extracellular matrix encountered by neural crest cells. See Wehrle-Haller and Weston (1995) for soluble and membrane-bound steel factor. See Helwig et al. (1995) for double *Patch-Undulated* mutants.

20. See Hayashi and Arima (1987), Zelkowitz (1981), Riccardi and Margos (1981), and Riccardi (1988) for culture of cells from patients with neurofibromatosis.

21. See Schlumberger (1951), Schmale, Hensley, and Udey (1983), Riccardi and Eichner (1986), and Schmale and Hensley (1988) for neurofibromas in fish.

22. See Pearse (1977a) for APUD cell types and Nozue (1988c) and Nozue and Ono (1989) for serotonin-induced APUdomas and adenosine-induced neural crest tumors.

23. For the neural crest origin of calcitonin-synthesizing and other APUD cells, see Pictet and Rutter (1972), Pictet et al. (1976), Pearse (1977b), Andrew and Kramer (1979), Le Douarin (1982), and Ayer-Le Lièvre and Perus (1982).

24. See Lauder and Zimmermann (1988), Nozue (1988c), Sieber-Blum (1989b), Shuey, Sadler, and Lauder (1992), Yavarone et al. (1993), and Moiseiwitsch and Lauder (1995) for supporting studies on serotonin.

25. For colonization of the pancreas by neural precursors, see Kirchgessner, Adlersberg, and Gershon (1992). For production of neurites by endodermal cells of the pancreas, see Teitelman (1990).

26. See Nozue (1974) and Nozue and Kayano (1977b,c) for induction of neoplastic changes in APUD cells, and see Atsumi et al. (1990) for studies with the AT805 teratocarcinoma cells.

27. For deletion of neural crest and compensation for that deletion, see Kirby et al. (1993), Kirby, Aronstam, and Buccafusco (1985a), Kirby, Turnage, and Hays (1985), Kirby and Stewart (1984), Besson et al. (1986), and Stewart, Kirby, and Sulik (1986). For other aspects of cardiac neural crest, see Kirby and Waldo (1990, 1995), Waldo and Kirby (1993), and Waldo, Kumiski, and Kirby (1994, 1996). For regulation of the heart field in zebrafish, see Serbedzija, Chen, and Fishman (1998). For apoptotic cells in the chick outflow tract, see Poelmann, Mikawa, and Gittenberger-de Groot (1998).

28. See Kirby and Bockman (1984) and Lodewyk, Van Mierop, and Kutsche (1986) for developmental changes in DiGeorge syndrome; Halford et al. (1993) and Budarf et al. (1995) for the chromosome mapping; and Gailil et al. (1998) for discussion of genes including *Goosecoid-like* (*Gscl*) from the equivalent region of the mouse genome. The parathyroid arises in the third visceral arch but can be traced back to an epipharyngeal placode (Manley and Capecchi 1995, Mérida-Velasco et al. 1996). See Thomas et al. (1998) for an animal model for CATCH-22.

29. See Takihara et al. (1997) for targeted disruption of the polycomb element in mice. See Orlando and Paro (1995), A. Gould (1997), and Schumacher and Magnuson (1997) for *Polycomb* response elements and transcriptional regulation in mammals, including nine genes identified in mice. See Yu et al. (1998) for knockout of the trithorax-group gene *Mll* and subsequent dysplasia or hypoplasia of the branchial arches.

12. Birth Defects

1. See Poswillo (1974, 1975a, b, 1978) and McLeod et al. (1980) for defective cell migration as a major factor in craniofacial defects, and see Newgreen and Gibbins (1982) and Tosney (1982) for mechanisms underlying early stages of migration.

2. For induction of birth defects by altering cell migration, see Keith (1977), Hassell, Greenberg, and Johnston (1977), Morriss and Thorogood (1978), and Morriss-Kay (1993). For defective ECM as a factor in altered migration, see Frost and Malacinski (1980) and Weston (1980) and the discussion in Chapter 9.

3. See Suzuki, Svensson, and Eriksson (1996) for the studies on high glucose concentrations. See Rovasio and Battiato (1995) for the effect of ethanol of migration of neural crest cells, and Müller and O'Rahilly (1989) for holoprosencephaly.

4. For proliferation of migrating neural crest cells, see Vermeij-Keers (1972), Osman and Ruch (1975), Maxwell (1976), Minkoff and Kuntz (1977), Noden (1980, 1982b), and Hall (1983a, b). See Hall (1985) and Scott (1986) for sensitive periods, and see Johnston (1975), Been and Lieuw Kie Song (1978), Johnston and Sulik (1979), Sulik and Johnston (1983), Sulik (1984), and Johnston and Bronsky (1995) for craniofacial defects associated with defective proliferation.

5. See Winograd et al. (1997) for transgenic *Msx-2* mice. For BMP-4 and apoptosis of hindbrain neural crest cells, see Lumsden, Sprawson, and Graham (1991), Jeffs, Jaques, and Osmond (1992), Graham, Heyman, and Lumsden (1993), Thorogood (1993b), Gra-

ham et al. (1994), Lumsden and Graham (1996), and Lumsden and Krumlauf (1996). For similar roles in neural crest derivatives in visceral arches and neuron precursors, see Barlow and Francis-West (1997), Ekanayake and Hall (1997), and Song, Mehler, and Kessler (1998). See Monsoro-Burq et al. (1996) for up-regulation of *Msx* by BMP and see Takahaski et al. (1998) for *Msx-2*-induced apoptosis in even-numbered rhombomeres.

6. See Jones, Lyons, and Hogan (1991) and Bennett, Hunt, and Thorogood (1995) for distribution of BMP in murine orofacial tissues. See Le Douarin (1988b), Hall and Ekanayake (1991), and Chapter 10 for other growth factors that regulate differentiation of neural crest cells.

7. For defective tissue interactions and craniofacial defects, see Cohen et al. (1971), Riccardi (1979), Hall (1982, 1991a, 1994d), and Trosko, Chang, and Netzloff (1982).

8. See Dencker et al. (1990), Vaessen et al. (1990), Tan (1991), Maden et al. (1992), and Ruberte et al. (1992) for localization of CRABP to neural crest and other cell types; Eichele (1997) for an overview of retinoids and neural tube patterning, dysmorphogenesis and disease; and Ito and Morita (1995) for differential effects on cranial and trunk neural crest.

9. See Wiley, Cauwenbergs, and Taylor (1983) for the studies with Golden Syrian hamsters.

10. See Hart et al. (1990) and Moro-Balbás et al. (1993) for the avian model for retinoic acid embryopathy. For effects of vitamin A on neural crest cells and craniofacial development, see Keith (1977), Hassell, Greenberg, and Johnston (1977), Wedden and Tickle (1986), Wedden (1987), and Maden et al. (1996).

11. See Mallo (1997, 1998) and Mallo and Gridley (1996) for ear defects that arise from lack of migration, failure of formation or respecification of neural crest cells, or defective developmental interactions. See Johnston and Bronsky (1991, 1995) for animal models for retinoic-acid-based human craniofacial malformations.

12. See Atchley and Hall (1991) and Hall (1998a) for selection operating throughout development and see Hall (1998a) for a general discussion of transitions between classes of control in ontogeny.

13. For homeotic transformation of craniofacial structures after administration of vitamin A, see Poswillo (1974), Morriss (1972, 1975, 1976, 1980), Tassin and Weill (1981), and Kay (1986, 1987). For defective migration as a major factor in human visceral arch malformations, see Clark (1976), Belensky and Medina (1980), Olsen, Maragos, and Weiland (1980), and McLeod et al. (1980). For pros and cons of hemorrhaging as a mechanism for hemifacial microsomia, see Poswillo (1975a–c, 1978) and Cousley and Wilson (1992). For the field concept in hemifacial microsomia, mandibulofacial dysostosis, and embryological interpretations of carcinomas, see Pierce (1985, 1987), Kay and Kay (1989), Hall (1994d), and Hodges and Rowlatt (1994). For a possible genetic model for hemifacial microsomia, the *far* mutation in the mouse, see Juriloff, Harris, and Froster-Iskenius (1987) and Harris and Juriloff (1989).

14. See Yip, Kokich, and Shepard (1980) and Steele, Plenefisch, and Klein (1982) for links between vitamin A and primate craniofacial defects.

15. See Morriss and Thorogood (1978), Thorogood et al. (1982), and Morriss-Kay (1992, 1993) for direct affects of vitamin A on neural crest cell migration, and Dencker (1986) and Dencker et al. (1987) for mechanistic studies.

16. See Mahmood, Flanders, and Morriss-Kay (1992) for retinoic acid and TGFß-1 and -2; Mahmood et al. (1995) and Mahmood, Mason, and Morriss-Kay (1996) for retinoids and FGF-3; and Moens et al. (1998) and Prince et al. (1998) for *valentino* and *kreisler*. Ruberte, Wood, and Morriss-Kay (1997), who claim that pro-rhombomeres are not func-

tional precursors of rhombomeres, query whether identification of pro-rhombomere is functionally meaningful.

17. See Kutsky (1973) for multiple effects of vitamin A, Tyler and Dewitt-Stott (1986) for direct action, and Poswillo (1974, 1975a–c, 1978) and Johnston et al. (1977) for indirect action of vitamin A on neural crest cells.

18. See Vandersea et al. (1998) and Suzuki, Oohara, and Kurokawa (1998) for these fish studies.

19. See Kilcoyne et al. (1986), Webster et al. (1986), Willhite, Hill, and Irving (1986), Jahn and Ganti (1987), and Caffery and Josephson (1988) for the human studies, and Monga (1997) for an overview. Under appropriate conditions retinoic acid can promote survival and proliferation of neurogenic precursor cells of neural crest origin; see Henion and Weston (1994).

20. See Smith-Thomas, Lott, and Bronner-Fraser (1987) for exposure of neural crest cells to Accutane in vitro. See Johnston, Sulik, and Webster (1985, 1986), Lammer et al. (1985), Sulik, Johnston, and Webster (1986), Willhite, Hill, and Irving (1986), Sulik, Cook, and Webster (1988), Jarvis, Johnston, and Sulik (1990), Johnston and Bronsky (1991, 1995), Kochhar and Christian (1997), and Eckhoff and Willhite (1997) for studies with laboratory animals. Sohal et al. (1996) used DiI-labeling to show that ventral neural tube cells (after neural crest cell migration) also contribute to the avian trigeminal ganglion.

21. See Adams (1993) for a comparison of the effects of Accutane and Tigason; Taylor, Bennet, and Finnell (1995) for specific genes whose expression is affected by isotretinoin or tretinoin, and Ruberte et al. (1990), Rowe, Eager, and Brickell (1991), and Rowe and Brickell (1995) for the retinoid X receptor-γ gene. See Granström, Jacobsson, and Magnusson (1991) for the mouse studies with etretinate.

22. See Steele, Plenefisch, and Klein (1982) for the effects of vitamin A in serum.

23. See Chen et al. (1995) for the studies on vitamin A and *Msx-1* and Davidson (1995) for a review of the role of *Msx* genes. See Köntges and Lumsden (1996) and Miyake, Cameron, and Hall (1996) for skeletal structures with a composite origin within the mandibular arch (Meckel's cartilage, mandibular membrane bones) or from two visceral arches (middle ear ossicles).

24. See Mark et al. (1993) for the *Hoxa-1* mutant mice; Gavalas et al. (1998) for *Hoxa-1/Hoxb-1* double mutants; and Miyake, Cameron, and Hall (1996) for contributions of first and second arch elements to mammalian ear development. For an insightful analysis of whether knocking out patterning genes from murine embryos produces evolutionary reversals of the first arch, see K. K. Smith and Schneider (1998).,

25. For studies on disruption of *Hoxa-1*, *Hoxa-2*, *Hoxa-3* see Lufkin et al. (1991), Chisaka and Capecchi (1991), Mark et al. (1995), Alexandre et al. (1996), and Kanzler et al. (1998). For homologous recombination of *Hoxd-4* and *Hoxa-2*, see Lufkin et al. (1992), Gendron-Maguire et al. (1993), and Rijli et al. (1993). For the recent demonstration that *Hoxd-4* specifies regional identity between visceral arches 1 and 2 in the flounder, see Suzuki, Oohara, and Kurokawa (1998).

26. See Gale et al. (1996) for the studies on rhombomere 4; Marshall et al. (1992) for transformation of rhombomere identity; and Sham et al. (1993) and Wilkinson (1993, 1995) for *Krox-20* regulation of *Hoxb-2*.

27. See Bergsma (1979), Flint (1980), Copp and Wilson (1981), Weston (1980), Erickson and Weston (1983), and Morrison-Graham and Weston (1989) for mutation-induced craniofacial defects.

28. See Wilson and Wyatt (1995) for cranial morphogenesis in *Lp* mutant mice. See

also Wilson and Wyatt (1988) for closure of the posterior neuropore in *vL* mutant mice and the contribution of defective neural crest cell migration to spina bifida.

29. See Shankar, Jaskoll, and Melnick (1992) and Shankar et al. (1994) for changes in N-CAM and cell migration in embryos exposed to retinoic acid. See Tassabehji et al. (1992) for *Pax-3* and Waardenburg syndrome.

30. See Auerbach (1954), Moase and Trasler (1989, 1990, 1991), Franz and Kothary (1993), Franz et al. (1993) and Serbedzija and McMahon (1997) for the basic defects in the *Splotch* mutant; Conway, Henderson, and Copp (1997) for the *Pax-3* studies; Dahl, Koseki, and Balling (1997) for an overview of *Pax* genes and vertebrate organogenesis, including their involvement in such neural-crest-based defects as Waardenburg syndrome; and Epstein (1996) for an overview of *Splotch*, *Pax-3*, the neural crest, and cardiovascular development. Interestingly, if the *Splotch* allele is induced with radiation, neural tube and tail defects can be induced without affecting the neural crest (Franz 1992).

31. See Anderson et al. (1972) for familial facial osteodysplasia, and see Webster (1973), Rothman and Gershon (1984), Jacobs-Cohen et al. (1987), Hosoda et al. (1994), and Puffenberger et al. (1994) for *Piebald-Lethal* and *Lethal-Spotting*.

32. See Ferguson (1978, 1981, 1984) and Hall (1987b) for these comparative studies among the vertebrates,

33. See Hörstadius (1950, pp. 41-43), Chibon (1970), Weston (1970), Scherson et al. (1993), Hunt et al. (1995), Sechrist et al. (1995), Raible and Eisen (1996), Buxton et al. (1997), and Vaglia and Hall (1999) for mechanisms of regulation of the neural crest.

34. See Vaglia and Hall (1997, 1999) for regulation of zebrafish neural crest, and see Sechrist et al. (1995) and Buxton et al. (1997) for regulation of chick neural crest.

35. See Carr (1984) for the studies on compensatory hypertrophy; Marusich, Pourmehr, and Weston (1986a, b), Asamoto, Nojyo, and Aoyama (1992), and Gvirtzmnn, Goldstein, and Kalcheim (1992) for identification of regulation of the size of dorsal root ganglia; Goldstein, Avivi, and Geffen (1995) for the intrinsic size differences; A. Smith et al. (1997) and Krull et al. (1997) for the Eph-family of receptors; and Robertson and Mason (1995) and Wehrle-Haller and Weston (1997) for tyrosine kinase receptors in the chick.

36. See Scherson et al. (1993), Hunt et al. (1995), Couly et al. (1996), Buxton et al. (1997), and Saldivar et al. (1997) for regulation of the avian neural crest; Snow and Tam (1979) for regulation of the mouse neural crest; Prince and Lumsden (1994) for intrinsic patterning of *Hoxa-2*; Sechrist et al. (1993) and Sechrist, Scherson, and Bronner-Fraser (1994) for the segmental migration of neural crest from rhombomeres 3 and 5; Martinez and Alvarado-Mallart (1990) for regulation of *Engrailed* expression in chick/quail hybrids and Maclean and Hall (1987), Gilbert (1997), and Vaglia and Hall (1999) for general discussions of regulation.

37. See Kirby (1988a, b), Kirby, Creazzo, and Christiansen (1989), and Rosenquist, Kirby, and Van Mierop (1989) for regulation from nodose placodes, and Kirby (1989) for the ability of neural crest from other regions to replace cardiac neural crest.

38. See Qin and Kirby (1995) for the studies with *Int-2*; Vogel and Davies (1993) for replacement of placodal ectoderm by trunk non-neurogenic ectoderm; T. A. Harrison et al. (1995) for regulation of nodose ganglia; Barbu et al. (1986) for the monoclonal antibody generated against nodose ganglia; and León et al. (1992) for the oncoprotein Mybp75.

References

Abdulla, R. I., Slott, E. F., and Kirby, M. L. 1993. Proteins associated with cardiac neural crest in the pharyngeal region of early chick embryos. *Pediatr Res* 33:43–47.

Adams, J. 1993. Structure-activity and dose-response relationships in the neural and behavioral teratogenesis of retinoids. *Neurotoxicol Teratol* 15:193–202.

Adelmann, H. B. 1925. The development of the neural folds and cranial ganglia of the rat. *J Comp Neurol* 39:19–171.

Akimenko, M.-A., Ekker, M., Wegner, J., Lin, W., and Westerfield, M. 1994. Combinatorial expression of three zebrafish genes related to *Distalless*: Part of a homeobox gene code for the head. *J Neurosci* 14:3475–3486.

Akira, E., and Ide, H. 1987. Differentiation of neural crest cells of *Xenopus laevis* in clonal culture. *Pigment Cell Res* 1:28–36.

Akitaya, T., and Bronner-Fraser, M. 1992. Expression of cell adhesion molecules during initiation and cessation of neural crest cell migration. *Dev Dyn* 194:12–20.

Alberch, P. 1985. Problems with the interpretation of developmental sequences. *Syst Zool* 34:46–58.

Alberch, P. 1987. Evolution of a developmental process: Irreversibility and redundancy in amphibian metamorphosis. In: Raff, R. A., and Raff, E. C., eds. *Development as an Evolutionary Process*. New York: Alan R. Liss, Inc. pp. 23–46.

Albers, B. 1987. Competence as the main factor determining the size of the neural plate. *Dev Growth Differ* 29:535–545.

Aldridge, R. J., Briggs, D. E. G., Smith, M. P., Clarkson, E. N. K., and Clark, N. D. L. 1993. The anatomy of conodonts. *Philos Trans R Soc Lond B Biol Sci* 340:405–421.

Aldridge, R. J., and Purnell, M. A. 1996. The conodont controversies. *Trends Ecol Evol* 11:463–468.

Aldridge, R. J., and Theron, J. N. 1993. Conodonts with preserved soft tissue from a new Ordovician *Konservat-Lagerstätte*. *J Micropalaeont* 12:113–117.

Alexandre, D., Clark, J. D. W., Oxtoby, E., Yan, Y.-L., Jowlett, T., and Holder, N. 1996. Ectopic expression of *Hoxa-1* in the zebrafish alters the fate of the mandibular arch neural crest and phenocopies a retinoic acid-induced phenotype. *Development* 122:735–746.

Alfandari, D., Wolfsberg, T. G., White, J. M., and DeSimone, D. W. 1997. ADAM 13: A novel ADAM expressed in somitic mesoderm and neural crest cells during *Xenopus laevis* development. *Dev Biol* 182:314–330.

Allis, A. P. 1898. On the morphology of certain bones of the cheek and snout of *Amia calva*. *J Morphol* 14:425–466.

Allman, G. J. 1853. On the anatomy and physiology of *Cordylophora*: A contribution to our knowledge of Tubularian zoophytes. *Philos Trans R Soc Lond B Biol Sci* 143:367–384.

Alvarado-Mallart, R.-M. 1993. Fate and potentialities of the avian mesencephalic/metencephalic neuroepithelium. *J Neurobiol* 24:1341–1355.

Amprino, R., Amprino Bonetti, D., and Ambrosi, G. 1969. Observations on the developmental relations between ectoderm and mesoderm of the chick embryo tail. *Acta Anat* 56(suppl):1–26.

Anastassiades, T. P., Puzic, O., and Puzic, R. 1978. Effect of solubilized bone matrix components on cultured fibroblasts derived from neonatal rat tissues. *Calcif Tissue Res* 26:173–179.

Anderson, D. J. 1989a. Development and plasticity of a neural crest-derived neuroendocrine sub lineage. In: Landmesser , L. T., ed. *Assembly of the Nervous System.* Chichester: John Wiley & Sons. pp. 17–36.

Anderson, D. J. 1989b. The neural crest lineage problem: Neuropoiesis? *Neuron* 3:1–12.

Anderson, D. J. 1993a. Cell and molecular biology of neural crest cell lineage diversification. *Curr Opin Neurobiol* 3:8–13.

Anderson, D. J. 1993b. Molecular control of cell fate in the neural crest: The sympathoadrenal lineage. *Annu Rev Neurosci* 16:129–158.

Anderson, D. J. 1997. Cellular and molecular biology of neural crest cell lineage determination. *Trends Genet* 13:276–280.

Anderson, L. G., Cook, A. J., Coccard, P. J., Coro, C. J., and Bosma, J. F. 1972. Familial facial osteodysplasia. *JAMA* 220:1687–1693.

Andrew, A., and Gabie, V. 1969. Hanging drop culture of *Xenopus laevis* neural crest. *Acta Embryol Exp* 2–3:123–136.

Andrew, A., and Kramer, B. 1979. An experimental investigation into the possible origin of pancreatic islet cells from rhombencephalic neurectoderm. *J Embryol Exp Morphol* 52:23–38.

Andriokopoulos, K., Suzuki, H. R., Solursh, M., and Ramirez, F. 1992. Localization of pro-a-2 (V) collagen transcripts in the tissues of the developing mouse embryo. *Dev Dyn 195:113–120.*

Arendt, D., and Nübler-Jung, K. 1996. Common ground plans in early brain development in mice and flies. *BioEssays* 18:255–259.

Arkell, R., and Beddington, R. S. P. 1997. BMP-7 influences pattern and growth of the developing hindbrain of mouse embryos. *Development* 124:1–12.

Armstrong, J. B. 1985. The axolotl mutants. *Dev Genet* 6:1–26.

Artinger, K. B., and Bronner-Fraser, M. 1992a. Partial restriction in the developmental potential of late emigrating avian neural crest cells. *Dev Biol* 149:149–157.

Artinger, K. B., and Bronner-Fraser, M. 1992b. Notochord grafts do not suppress formation of neuronal crest cells or commisural neurons. *Development* 116:877–886.

Asamoto, K., Nojyo, Y., and Aoyama, H. 1992. Regulation of cell number in formation of the dorsal-root ganglion revealed by transplantation of quail neural crest cells into chick embryos. *Dev Growth Differ* 34:553–560.

Asamoto, K., Nojyo, Y., and Aoyama, H. 1995. Restriction of the fate of early migrating trunk neural crest in gangliogenesis of avian embryos. *Int J Dev Biol* 39:975–984.

Asher, R. A., Scheibe, R. J., Keiser, H. D., and Bignami, A. 1995. On the existence of a cartilage-like proteoglycan and link proteins in the central nervous system. *Glia* 13:294–308.

Atchley, W. R., and Hall, B. K. 1991. A model for development and evolution of complex morphological structures and its application to the mammalian mandible. *Biol Rev Camb Philos Soc* 66:101–157.

Atsumi, T., Miwa, Y., Kimata, K., and Ikawa, Y. 1990. A chondrogenic cell line derived from a differentiating culture of AT805 teratocarcinoma cells. *Cell Differ Dev* 30:109–116.

Auerbach, R. 1954. Analysis of the developmental effects of a lethal mutation in the house mouse. *J Exp Zool* 127:305–327.

Ayer-Le Liévre, C., and Perus, J.-F. 1982. The neural crest: Its relation with APUD and paraneuron concepts. *Arch Histol Japan* 45:409–427.

Baker, C. V. H., and Bronner-Fraser, M. 1997a. The origins of the neural crest. Part I: Embryonic induction. *Mech Dev* 69:3–11.

Baker, C. V. H., and Bronner-Fraser, M. 1997b. The origins of the neural crest. Part II: An evolutionary perspective. *Mech Dev* 69:13–29.

Baker, C. V. H., Bronner-Fraser, M., Le Douarin, N. M., and Teillet, M.-A. 1997. Early- and late-migrating cranial neural crest cell populations have equivalent developmental potential in vivo. *Development* 124:3077–3087.

Baldwin, C. T., Hoth, C. F., Amos, J. A., da-Silva, E. O., and Milunsky, A. 1992. An exonic mutation in the HuP2 paired domain causes Waardenburg's syndrome. *Nature* 355:637–638.

Balfour, F. M. 1876. On the development of the spinal nerves in elasmobranch fishes. *Philos Trans R Soc Lond B Biol Sci* 166:175–195.

Balfour, F. M. 1878a. On the development of elasmobranch fishes. *J Anat Physiol* 11:128–172.

Balfour, F. M. 1878b. *A Monograph on the Development of Elasmobranch Fishes.* London: Macmillan & Co.

Balfour, F. M. 1880. *A Treatise on Comparative Embryology.* Vol. 1. London: Macmillan & Co. (Translated into German in 1881 by B. Vetter as *Handbuch der Vergleichenden Embryologie.* Jena: Verlag von Gustav Fischer.) Original edition as 13 micro cards, Readex Microprint, New York, 1969.

Balfour, F. M. 1881. *A Treatise on Comparative Embryology.* Vol. 2. London: Macmillan & Co. (Translated into German by B. Vetter in 1882 as *Handbuch der Vergleichenden Embryologie.* Jena :Verlag von Gustav Fischer.) Original edition as 13 micro cards, Readex Microprint, New York, 1969.

Balinsky, B. I. 1925. Transplantation des Ohrbläschens bei *Triton. Wilhelm Roux Arch EntwMech Org* 105:718–731.

Balinsky, B. I. 1940. Experiments on total extirpation of the whole endoderm in *Triton* embryos. *C R Acad Sci URSS* 28:196–198.

Baltzer, F. 1941. Untersuchungen an Chimären von Urodelen und *Hyla.* 1. Die Pigmentierung chimärischen Molch-und Axolotl larven mit *Hyla* (Laubfrosch). Ganglienleiste. *Rev Suisse Zool* 48:413–482.

Baranowitz, S. A. 1989. Regeneration, neural crest derivatives and retinoids: A new synthesis. *J Theor Biol* 140:231–242.

Barbu, M., Ziller, C., Rong, P. M., and Le Douarin, N. M. 1986. Heterogeneity in migrating neural crest cells revealed by a monoclonal antibody. *J Neurosci* 6:2215–2225.

Barker, D., Wright, E., Nguyen, K., Cannon, I., Fain, P., Goldgar, D., Bishop, D. T., Carey, J., Baty, B., Kivlin, J., Willard, H., Waye, J. S., Greig, G., Leinwand, L., Nakamura, Y., O'Connell, P., Leppert, M., Lalouel, J.-M., White, R., and Skolnick, M. 1987. Gene for von Recklinghausen neurofibromatosis is in the pericentromeric region of chromosome 17. *Science* 236:1100–1102.

Barlow, A. J., and Francis-West, P. H. 1997. Ectopic application of recombinant BMP-2 and BMP-4 can change patterning of developing chick facial primordia. *Development* 124:391–398.

Barlow, L. A., Chien, C.-B., and Northcutt, R. G. 1996. Embryonic taste buds develop in the absence of innervation. *Development* 122:1103–1111.

Barlow, L. A., and Northcutt, R. G. 1994. Analysis of the embryonic lineage of vertebrate taste buds. *Chem Senses* 19:715–724.

Barlow, L. A., and Northcutt, R. G. 1995. Embryonic origin of amphibian taste buds. *Dev Biol* 169:273–285.

Barlow, L. A., and Northcutt, R. G. 1997. Taste buds develop autonomously from endoderm without induction by cephalic neural crest or paraxial mesoderm. *Development* 124:949–957.

Baroffio, A., Dupin, E., and Le Douarin, N. M. 1988. Clone-forming ability and differentiation potential of migratory neural crest cells. *Proc Natl Acad Sci USA* 85:5325–5339.

Baroffio, A., Dupin, E., and Le Douarin, N. M. 1991. Common precursors for neural and mesectodermal derivatives in the cephalic neural crest. *Development* 112:301–305.

Bartelmez, G. W. 1922. The origin of the otic and optic primordia. *J Comp Neurol* 34:201–232.

Bartelmez, G. W. 1960. Neural crest from the forebrain in mammals. *Anat Rec* 138:269–281.

Bartelmez, G. W. 1962. The proliferation of neural crest from forebrain levels in the rat. *Contr Embryol* 37:3–12.

Bartelmez, G. W., and Blount, M. P. 1954. The formation of neural crest from the primary optic vesicle in man. *Contr Embryol* 35:35–71.

Basler, K., Edlund, T., Jessell, T. M., and Yamada, T. 1993. Control of cell pattern in the neural tube: Regulation of cell differentiation by *dorsalin*-1, a novel TGFß family member. *Cell* 73:687–702.

Beall, A. C., and Rosenquist, T. H. 1990. Smooth muscle cells of neural crest origin from the aortico-pulmonary septum in the avian embryo. *Anat Rec* 226:360–366.

Béard, J. 1892. The transient ganglion-cells and their nerves in *Raja batis*. *Anat Anz* 7:191–206.

Béard, J. 1896. The history of transient nervous apparatus in certain Ichthyopsida. An account of the development and degeneration of ganglion-cells and nerve fibres. *Zool Jahrb Abt Morphol* 9:1–106.

Beckwith, J. B., and Perrin, E. V. 1963. *In situ* neuroblastomas: A contribution to the natural history of neural crest tumors. *Am J Pathol* 43:1089–1104.

Beddington, R. S. P. 1981. An autoradiographic analysis of the potency of embryonic ectoderm in the 8th day postimplantation mouse embryo. *J Embryol Exp Morphol* 64:87–104.

Beddington, R. S. P., and Robertson, E. J. 1998. Anterior patterning in mouse. *Trends Genet* 14:277–284.

Been, W., and Lieuw, K. S. 1978. Harelip and cleft palate conditions in chick embryos following local destruction of the cephalic neural crest. A preliminary note. *Acta Morphol Neerl-Scand* 16:245–255.

Beighton, P., and Beighton, G. 1986. *The Man Behind the Syndrome*. Berlin: Springer-Verlag.

Belensky, W. M., and Medina, J. E. 1980. First branchial cleft anomalies. *Laryngoscope* 90:28–39.

Bellairs, R. 1987. The primitive streak and the neural crest: Comparable regions of cell migration? In: Maderson, P. F. A., ed. *Developmental and Evolutionary Aspects of the Neural Crest.* New York: John Wiley & Sons. pp. 123–145.

Bemis, W. E., and Grande, L. 1992. Early development of the actinopterygian head. 1. External development and staging of the paddlefish, *Polyodon spathula. J Morphol* 213:47–83.

Bennett, J. H., Hunt, P., and Thorogood, P. V. 1995. Bone morphogenetic protein-2 and -4 expression during murine orofacial development. *Arch Oral Biol* 40:847–854.

Bergsma, D., ed. 1975. *Morphogenesis and Malformation of Face and Brain. Birth Defects: Original Article Series* 11 (7). New York: Alan R. Liss, Inc.

Bergsma, D. 1979. *Birth Defects Compendium.* New York: Alan R. Liss, Inc.

Berrill, N. J. 1955. *The Origin of Vertebrates.* Oxford: The Clarendon Press.

Berrill, N. J. 1987a. Early chordate evolution. Part 1. Amphioxus, The riddle of the sands. *Int J Invert Reprod Dev* 11:1–14.

Berrill, N. J. 1987b. Early chordate evolution. Part 2. Amphioxus and Ascidians. To settle or not to settle. *Int J Invert Reprod Dev* 11:15–28.

Bertani-Dziedzic, L. M., Krstulovic, A. M., Ciriello, S., and Gitlow, S. E. 1979. Routine reversed-phase high-performance liquid chromatographic measurements of urinary vanillylmandelic acid in patients with neural crest tumors. *J Chromatogr* 164:345–354.

Besson, W. T. III, Kirby, M. L., Van Mierop, L. H. S., and Teabeaut, J. R. II. 1986. Effects of the size of lesions of the cardiac neural crest at various embryonic ages on incidence and type of cardiac defects. *Circulation* 73:360–364.

Bevan, S. G., Southey, M. C., Armes, J. C., Venter, D. J., and Newgreen, D. F. 1996. Spatiotemporally exact cDNA libraries from quail embryos — a resource for studying neural crest development and neurocristopathies. *Genomics* 38:206–214.

Bijtel, J. 1931. Ueber die Entwicklung des Schwanzes bei Amphibien. *Wilhelm Roux Arch EntwMech Org* 125:448–486.

Blankenship, T. N., Peterson, P. E., and Hendrickx, A. G. 1996. Emigration of neural crest cells from macaque optic vesicles is correlated with discontinuities in its basement membrane. *J Anat* 188:473–483.

Blitz, I. L., and Cho, K. W. Y. 1995. Anterior neurectoderm is progressively induced during gastrulation. The role of the *Xenopus* homeobox gene *orthodenticle. Development* 121:993–1004.

Bockman, D. E., and Kirby, M. L. 1984. Dependence of thymus development on derivatives of the neural crest. *Science* 223:498–500.

Bockman, D. E., and Kirby, M. L. 1985. Neural crest interactions in the development of the immune system. *J Immunol* 135:766s–768s.

Bockman, D. E., Redmond, M. E., and Kirby, M. L. 1989. Alteration of early vascular development after ablation of cranial neural crest. *Anat Rec* 225:209–217.

Bockman, D. E., Redmond, M. E., Waldo, K., Davis, H., and Kirby, M. L. 1987. Effect of neural crest ablation on development of the heart and arch arteries in the chick. *Amer J Anat* 180:332–341.

Boisseau, S., and Simonneau, M. 1989. Mammalian neuronal differentiation: Early expression of a neuronal phenotype from mouse neural crest cells in a chemically defined culture medium. *Development* 106:665–674.

Bolande, R. P. 1974. The neurocristopathies: A unifying concept of disease arising in neural crest maldevelopment. *Hum Pathol* 5:409–429.

Bolande, R. P. 1981. Neurofibromatosis: The quintessential neurocristopathy: Pathogenetic concepts and relationships. *Adv Neurol* 29:67–75.

Bolande, R. P. 1984. Models and concepts derived from human teratogenesis and oncogenesis in early life. *J Histochem Cytochem* 32:878–884.

Bolande, R. P., Seliger, W. G., and Markwald, R. R. 1980. A histochemical analysis of polyanionic compounds found in the extracellular matrix encountered by migrating cephalic neural crest cells. *Anat Rec* 196:401–412.

Bolender, D. L., Seliger, W. G., Markwald, R. R., and Brauer, P. R. 1980. Structural analysis of extracellular matrix prior to the migration from cephalic neural crest cells. *Scanning Electron Microsc* 2:285–296.

Bols, N. C., and Lee, L. E. J. 1991. Technology and uses of cell culture from the tissues and organs of bony fish. *Cytotechnology* 6:163–187.

Bonstein, L., Elias, S., and Frank, D. 1998. Paraxial-fated mesoderm is required for neural crest induction in *Xenopus* embryos. *Dev Biol* 193:156–168.

Borcea, M. I. 1909. Sur l'origine du coeur, des cellules vasculaires migratrices et des cellules pigmentaires chez les Téléostéens. *C R Acad Sci Paris* 149:688–689.

Borstlap, W. A., Neidbuch, K. L., Freihofe, H. P., and Kulipers, A. M. 1990. Early secondary bone grafting of alveolar cleft defects: A comparison between chin and rib grafts. *J Craniomaxillofac Surg* 18:201–205.

Bosman, F. T., and Louwerens, J.-W. K. 1981. APUD cells in teratomas. *Am J Pathol* 104:174–180.

Boucaut, J.-C., Darribère, T., Poole, T. J., Aoyama, H., Yamada, K. M., and Thiery, J.-P. 1984. Biologically active synthetic peptides as probes of embryonic development: A competitive peptide inhibitor of fibronectin function inhibits gastrulation in amphibian embryos and neural crest cell migration in avian embryos. *J Cell Biol* 99:1822–1830.

Bowler, P. J. 1996. *Life's Splendid Drama. Evolutionary Biology and the Reconstruction of Life's Ancestry 1860–1940.* Chicago: University of Chicago Press.

Bracegirdle, B. 1986. *A History of Microtechnique. The Evolution of the Microtome and the Development of Tissue Preparation.* 2nd ed. Lincolnwood: Science Heritage Ltd.

Bradley, L., Wainstock, D., and Sive, H. 1996. Positive and negative signals modulate formation of the *Xenopus* cement gland. *Development* 122:2739–2750.

Brannan, C. I., Perkins, A. S., Vogel, K. S., Ratner, N., Nordlund, M. L., Reid, S. W., Buchberg, A. M., Jenkins, N. A., Parada, L. F., and Copeland, N. G. 1994. Targeted disruption of the neurofibromatosis type-1 gene leads to developmental abnormalities in heart and various neural crest-derived tissues. *Genes Dev* 8:1019–1029.

Brauer, P. R., Bolender, D. L., and Markwald, R. R. 1985. The distribution and spatial organization of the extracellular matrix encountered by mesencephalic neural crest cells. *Anat Rec* 211:57–68.

Brauer, P. R., and Markwald, R. R. 1987. Attachment of neural crest cells to endogenous extracellular matrices. *Anat Rec* 219:275–285.

Brauer, P. R., and Yee, J. A. 1993. Cranial neural crest cells synthesize and secrete a latent form of transforming growth factor ß that can be activated by neural crest cell proteolysis. *Dev Biol* 155:281–285.

Braun, C. B., and Northcutt, R. G. 1997. The lateral line system of hagfishes (Craniata:Myxinoidea). *Acta Zool (Stockh)* 78:247–268.

Brewster, R., Lee, J., and Ruiz i Altaba, A. 1998. *Gli,Zic* factors pattern the neural plate by defining domains of cell differentiation. *Nature* 393:579–583.

Briggs, D. E. G., Clarkson, E. N. K., and Aldridge, R. J. 1983. The conodont animal. *Lethaia* 16:1–14.

Bronner-Fraser, M. 1985. Effects of different fragments of the fibronectin molecule on latex bead translocation along neural crest migratory pathways. *Dev Biol* 108:131–145.

Bronner-Fraser, M. 1986. An antibody to a receptor for fibronectin and laminin perturbs cranial neural crest development *in vivo*. *Dev Biol* 117:528–536.

Bronner-Fraser, M. 1987. Perturbation of cranial neural crest migration by the HNK-1 antibody. *Dev Biol* 123:321–331.

Bronner-Fraser, M. 1993a. Neural development: Crest destiny. *Curr Biol* 3:201–203.

Bronner-Fraser, M. 1993b. Segregation of cell lineage in the neural crest. *Curr Opin Genet Dev* 3:641–647.

Bronner-Fraser, M. 1993c. Mechanisms of neural crest cell migration. *BioEssays* 15:221–230.

Bronner-Fraser, M. 1995. Origins and developmental potential of the neural crest. *Exp Cell Res* 218:405–417.

Bronner-Fraser, M., and Fraser, S. E. 1988. Cell lineage analysis reveals multipotentiality of some avian neural crest cells. *Nature* 335:161–164.

Bronner-Fraser, M., and Fraser, S. E. 1989. Developmental potential of avian trunk neural crest cells *in situ*. *Neuron* 3:755–766.

Bronner-Fraser, M., and Fraser, S. E. 1991. Cell lineage analysis of the avian neural crest. *Development* (suppl 2):17–22.

Bronner-Fraser, M., and Fraser, S. E. 1997. Differentiation of the vertebrate neural tube. *Curr Opin Cell Biol* 7:885–891.

Bronner-Fraser, M., Sieber-Blum, M., and Cohen, A. M. 1980. Clonal analysis of the avian neural crest: Migration and maturation of mixed neural crest clones injected into host chicken embryos. *J Comp Neurol* 193:423–434.

Bronner-Fraser, M., and Stern, C. 1991. Effects of mesodermal tissues on avian neural crest cell migration. *Dev Biol* 143:213–217.

Bronner-Fraser, M., Stern, C., and Fraser, S. E. 1991. Analysis of neural crest cell lineage and migration. *J Craniofac Genet Dev Biol* 11:214–222.

Bronner-Fraser, M., Wolf, J. J., and Murray, B. A. 1992. Effects of antibodies against N-cadherin and N-CAM on the cranial neural crest and neural tube. *Dev Biol* 153:291–301.

Brun, R. B. 1985. Neural fold and neural crest movement in the Mexican salamander *Ambystoma mexicanum*. *J Exp Zool* 234:57–61.

Brun, R. B. 1987. The *eyeless* mutant Mexican salamander (*Ambystoma mexicanum*): Evidence for an imbalanced antero-posterior morphogenetic system. *J Neurogenetics* 4:29–46.

Budarf, M. L., Collins, J., Gong, W., Roe, B., Wang, Z., Bailey, L. C., Sellinger, B., Michaud, D., Driscoll, D. A., and Emanual, B. S. 1995. Cloning a balanced translocation associated with DiGeorge syndrome and identification of a disrupted candidate gene. *Nat Genet* 10:269–278.

Buxton, P., Hunt, P., Ferretti, P., and Thorogood, P. 1997. A role for midline closure in the re-establishment of dorsoventral pattern following dorsal hindbrain ablation. *Dev Biol* 183:150–165.

Bytinsky-Salz, H. 1936. Lo sviluppo della coda negli anfibi. 1. Esperimenti di isolamento e di difetto. *Atti Accad Naz Lincei Rc* Serie 6, 24:34–40.

Bytinsky-Salz, H. 1937. Trapianti di organizzatore nelle uova di *Lampedra*. *Archo Ital Anat Embriol* 39:177–228.

Bytinsky-Salz, H. 1938. Chromatophorenstudien. II. Struktur und Determination des adepidermalen Melanophorennetzes bei *Bombina*. *Arch Exp Zellforsch* 22:132–170.

Caffery, B. E., and Josephson, J. E. 1988. Ocular side effects of isotretinoin therapy. *J Am Optom Assoc* 59:221–224.

Campbell, S. 1989. Melanogenesis of avian neural crest cells *in vivo* is influenced by external cues in the periorbital mesenchyme. *Development* 106:717–726.

Carpenter, E. M., and Hollyday, M. 1992a. The location and distribution of neural crest-derived Schwann cells in developing peripheral nerves in the chick forelimb. *Dev Biol* 150:144–159.

Carpenter, E. M., and Hollyday, M. 1992b. The distribution of neural crest-derived Schwann cells from subsets of brachial spinal segments into the peripheral nerves innervating the chick forelimb. *Dev Biol* 150:160–170.

Carr, V. M. 1984. Dorsal root ganglia development in chicks following partial ablation of the neural crest. *J Neurosci* 4:2434–2444.

Cassin, C. 1975. Crête neurale et capacité morphogénétique du stomodeum chez *Pleurodeles waltlii* (Amphibien Urodèle). *Rev Stomato-Odont N Fr* 118:149–162.

Cassin, C., and Capuron, A. 1972. Obtention d'ouvertures buccales et de bouches complietes par implantation, dans le blastocèle, de tissus embryonnaires de *Pleurodeles waltlii* Michah (Amphibien Urodèle). *C R Hebd Seances Acad Sci Paris Ser D* 275:2953–2956.

Cassin, C., and Capuron, A. 1977. Evolution de la capacité morphogénétique de la région stomodéale chez l'embryon de *Pleurodeles waltlii* Michah (Amphibien Urodèle). Etude par transplantation intrablastocélienne et par culture in vitro. *Wilhelm Roux Arch Entw-Mech Org* 181:107–112.

Cassin, C., and Capuron, A. 1979. Buccal organogenesis in *Pleurodeles waltlii* Michah (urodele amphibian). Study by intrablastocelic transplantation and *in vitro* culture. *J Biol Buccale* 7:61–76.

Celestino da Costa, A. 1920. Note sur la crête ganglionnaire cranienne chez le *Cobuye*. *C R Seances Soc Biol* 83:1651–1657.

Chan, W. Y., and Lee, K. K. H. 1992. The incorporation and dispersion of cells and latex beads on microinjection into the amniotic cavity of the mouse embryo at the early-somite stage. *Anat Embryol* 185:225–238.

Chan, W. Y., and Tam, P. P. L. 1986. The histogenetic potential of neural plate cells of early somite-stage mouse embryos. *J Embryol Exp Morphol* 96:183–193.

Chan, W. Y., and Tam, P. P. L. 1988. A morphological and experimental study of the mesencephalic neural crest cells in the mouse embryo using wheat germ agglutinin-gold conjugate as the cell marker. *Development* 102:427–442.

Chandebois, R., and Faber, J. 1983. *Automation in Animal Development. A New Theory Derived from the Concept of Cell Sociology.* Basel: Karger AG. Monographs in Developmental Biology, Volume 16.

Chang, C., and Hemmati-Brivanlou, A. 1998. Neural crest induction by *Xwnt7B* in *Xenopus*. *Dev Biol* 194:129–134.

Chapman, D. L., and Papaioannou, V. E. 1998. Three neural tubes in mouse embryos with mutations in the T-box gene *Tbx6*. *Nature* 391:695–697.

Chareonvit, S., Osumi-Yamashita, N., Ikeda, M.-A., and Eto, K. 1997. Murine forebrain and midbrain crest cells generate different characteristic derivatives *in vitro*. *Dev Growth Differ* 39:493–503.

Charlebois, T. S., Spencer, D. H., Tarkington, S. K., Henry, J. J., and Grainger, R. M. 1990. Isolation of a chick cytokeratin cDNA clone indicative of regional specialization in early embryonic ectoderm. *Development* 108:33–45.

Cheah, K. S. E., Lau, E. T., Au, P. K. C., and Tam, P. P. L. 1991. Expression of the mouse a1(II) collagen gene is not restricted to cartilage during development. *Development* 111:945–952.

Chen, J.-y., Dzik, J., Edgecombe, G. D., Ramsköld, L., and Zhou, G.-Q. 1995. A possible Early Cambrian chordate. *Nature* 377:720–722.

Chen, Y., and Grunz, H. 1997. The final determination of *Xenopus* ectoderm depends on intrinsic and external positional information. *Int J Dev Biol* 41:525–528.

Chen, Y., Kostetskii, I., Zile, M. H., and Solursh, M. 1995. Comparative study of *Msx-1* expression in early normal and vitamin-A deficient avian embryos. *J Exp Zool* 272:299–310.

Chiakulas, J. J. 1957. The specificity and differential fusion of cartilage derived from mesoderm and mesectoderm. *J Exp Zool* 136:287–300.

Chiarugi, G. 1894. *Contribuzioni allo Studio dello Sviluppo dei Nervi Encefalici nei mammiferi*. Florence.

Chibon, P. 1964. Analyse par la méthode de marquage nucléaire à la thymidine tritiée des dérivés de la crête neurale céphalique chez l'Urodèle *Pleurodeles waltlii. C R Acad Sci.* 259:3624–3627.

Chibon, P. 1966. Analyse expérimentale de la régionalisation et des capacités morphogénètiques de la crête neurale chez l'amphibien urodéle *Pleurodeles waltlii* Michah. *Mem Soc Zool Fr* 36:1–107.

Chibon, P. 1967. Marquage nucléaire par la thymidine tritée des dérivés de la crête neurale chez l'amphibien urodèle *Pleurodeles waltlii* Michah. *J Embryol Exp Morphol* 18:343–358.

Chibon, P. 1970. Capacité de régulation des excédents dans la crête neurale d'amphibien. *J Embryol Exp Morphol* 24:479–496.

Chibon, P. 1974. Un systeme morphogenetique remarquable: La crête neurale des vertébrés. *Année Biol* 13:459–480.

Chisaka, O., and Capecchi, M. R. 1991. Regionally restricted developmental defects resulting from targeted disruption of the mouse homeobox gene *Hox*-1.5. *Nature* 350:473–479.

Churchill, F. B. 1986. Weismann, Hydromedusae, and the biogenetic imperative: A reconsideration. In: Horder, T. J., Witkowski., J. A., and Wylie, C. C., eds. *A History of Embryology.* Cambridge: Cambridge University Press. British Society for Developmental Biology Symposium 8. pp. 7–33.

Ciment, G., Glimelius, B., Nelson, B. M., and Weston, J. A. 1986. Reversal of a developmental restriction in neural crest-derived cells of avian embryos by a phorbol ester drug. *Dev Biol* 118:392–398.

Ciment, G., and Weston, J. A. 1985. Segregation of developmental abilities in neural crest-derived cells: Identification of partially restricted intermediate cell types in the branchial arches of avian embryos. *Dev Biol* 111:73–83.

Clark, J. A. 1976. Are wattles of auricular or branchial origin? *Br J Plast Surg* 29:238–244.

Clement, J. H., Fettes, P., Knochek, S., Lef, J., and Knochel, W. 1995. Bone morphogenetic protein 2 in the early development of *Xenopus laevis. Mech Dev* 52:357–370.

Clouthier, D. E., Hosoda, K., Richardson, J. A., Williams, S. C., Yanagisawa, H., Kuwaki, T., Kumada, M., Hammer, R. E., and Yanagisawa, M. 1998. Cranial and cardiac neural crest defects in endothelin-A receptor-deficient mice. *Development* 125:613–824.

Coates, M. I. 1994. The origin of vertebrate limbs. *Development* (suppl):169–180.

Cocchia, D., Lauriola, L., Stolfi, V. M., Tallini, G., and Michetti, F. 1983. S-100 antigen labels neoplastic cells in liposarcoma and cartilaginous tumours. *Virchows Arch Pathol Anat* 402:139–145.

Cogliatti, S. B. 1986. *Diplomyelia*: Caudal duplication of the neural tube in mice. *Teratology* 34:343–352.

Cohen, A. M., and Konigsberg, I. R. 1975. A clonal approach to the problem of neural crest determination. *Dev Biol* 46:262–282.

Cohen, M. M. Jr. 1982. *The Child with Multiple Birth Defects*. New York: Raven Press.

Cohen, M. M. Jr. 1987. The Elephant Man did not have neurofibromatosis. *Proc Greenwood Genet Ctr* 6:187–192.

Cohen, M. M. Jr. 1988a. Understanding Proteus syndrome, unmasking the Elephant Man and stemming elephant fever. *Neurofibromatosis* 1:260–280.

Cohen, M. M. Jr. 1988b. Further diagnostic thoughts about the Elephant Man. *Am J Med Genet* 29:777–782.

Cohen, M. M. Jr. 1989. Syndromology: An updated conceptual overview I–VI. *Int J Oral Maxillofac Surg* 18:216–228, 281–290, 333–346.

Cohen, M. M. Jr. 1990. Syndromology: An updated conceptual overview VII–X. *Int J Oral Maxillofac Surg* 19:26–37, 81–96.

Cohen, M. M. Jr., and Baum, B. J., eds. 1997. *Studies in Stomatology and Craniofacial Biology*. Ohmsha, Japan: IOS Press.

Cohen, M. M. Jr., and Cole, D. E. C. 1989. Origins of recognizable syndromes: Etiologic and pathogenetic mechanisms and the process of syndrome delineation. *J Pediatr* 115:161–164.

Cohen, M. M. Jr., and Hayden, P. W. 1979. A newly recognized hamartomatous syndrome. *Birth Defects Orig Art Ser* XV, 5B:291–296.

Cohen, M. M. Jr., Kirasek, J. E., Guzman, R. T., and Gorlin, R. J. 1971. Holoprosencephaly and facial dysmorphia: Nosology, etiology and pathogenesis. *Birth Defects Orig Art Ser* VII, 7:125–134.

Collazo, A., Bronner-Fraser, M., and Fraser, S. E. 1993. Vital dye labelling of *Xenopus laevis* trunk neural crest reveals multipotency and novel pathways of migration. *Development* 118:363–376.

Collazo, A., Fraser, S. E., and Mabee, P. M. 1994. A dual embryonic origin for vertebrate mechanoreceptors. *Science* 264:426–430.

Conel, J. Leroy 1931. The genital system of the Myxinoidea: A study based on notes and drawings of these organs in Bdellostoma made by Bashford Dean. In: Gudger, E. W., ed. *The Bashford Dean Memorial Volume: Archaic Fishes*. New York: The American Museum of Natural History. pp. 67–102.

Conel, J. LeRoy. 1942. The origin of the neural crest. *J Comp Neurol* 76:191–215.

Conway, S. J., Henderson, D. J., and Copp, A. J. 1997. Pax3 is required for cardiac neural crest migration in the mouse: Evidence from the *Splotch* (Sp^{2H}) mutant. *Development* 124:505–514.

Cook, M. H., and Neal, H. V. 1921. Are the taste buds of elasmobranchs endodermal in origin? *J Comp Neurol* 33:45–64

Coombs, S., Gorner, P., and Munz, H. 1989. *The Mechanosensory Lateral Line. Neurobiology and Evolution*. New York: Springer-Verlag.

Copp, A. J. 1993. Neural Tube Defects—A CIBA Foundation Symposium, London, U.K. 17–20 May 1993. *Trends Neurosci* 16:381–393.

Copp, A. J., Brook, F. A., Estibeiro, J. P., Shum, A. S. W., and Cockroft, D. L. 1990. The embryonic development of mammalian neural tube defects. *Prog Neurobiol* 35:363–403.

Copp, S. N., and Wilson, D. B. 1981. Cranial glycosaminoglycans in early embryos of the *loop-tail* (*lp*) mutant mouse. *J Craniofac Genet Dev Biol* 1:253–260.

Corbo, J. C., Erives, A., DiGregorio, A., Chang, A., and Levine, M. 1997. Dorsoventral patterning of the vertebrate neural tube is conserved in a protochordate. *Development* 124:2335–2344.

Corcoran, J. 1998. What are the molecular mechanisms of neural tube defects? *BioEssays* 20:6–8.

Corsin, J. 1966a. The development of the osteocranium of *Pleurodeles waltlii* Michahelles. *J Morphol* 119:209–216.

Corsin, J. 1966b. Quelques problèmes de morphogènèse du crane chez les urodelès. In: Problèmes Actuels de Paléontologie (Evolution des Vertébrés). *Coll Int Cent Nat Rec Scient* 163:295–299.

Corsin, J. 1975. Différenciation *in vitro* de cartilage a partir des crêtes neurales céphaliques chez *Pleurodeles waltlii* Michah. *J Embryol Exp Morphol* 33:335–342.

Corsin, J. 1977. Le materiel extracellulaire au cours du développément du chondrocrâne des amphibiens. Mise en place et constitution. *J Embryol Exp Morphol* 38:139–150.

Couly, G. F. 1981. Les neurocristopathies du bourgeon nasofrontal humain: les syndromes ethmöidiens (hypo et hyperseptoethmöidismes). *Rev Stomatol Chir Maxillofac* 82:213–225.

Couly, G. F., Coltey, P. M., and Le Douarin, N. M. 1992. The developmental fate of the cephalic mesoderm in quail chick chimeras. *Development* 114:1–15.

Couly, G. F., Coltey, P. M., and Le Douarin, N. M. 1993. The triple origin of the skull in higher vertebrates. A study in quail chick chimeras. *Development* 117:409–429.

Couly, G. F., Grapin-Botton, A., Coltey, P., and Le Douarin, N. M. 1996. The regeneration of the cephalic neural crest, a problem revisited: The regenerating cells originate from the contralateral or from the anterior and posterior neural fold. *Development* 122:3393–3407.

Couly, G. F., and Le Douarin, N. M. 1985. Mapping of the neural early primordium in quail–chick chimeras. I. Developmental relationships between placodes, facial ectoderm and prosencephalon. *Dev Biol* 110:422–439.

Couly, G. F., and Le Douarin, N. M. 1987. Mapping of the early neural primordium in quail–chick chimeras. II. The prosencephalic neural plate and neural folds: Implications for the genesis of cephalic human congenital abnormalities. *Dev Biol* 120:198–214.

Couly, G. F., and Le Douarin, N. M. 1990. Head morphogenesis in embryonic avian chimeras: Evidence for a segmental pattern in the ectoderm corresponding to the neuromeres. Development 108:543–558.

Cousley, R. R. J., and Wilson, D. J. 1992. Hemifacial microsomia. Developmental consequences of perturbation of the auriculofacial cartilage model. *Am J Med Genet* 42:461–466.

Crawford, A. H. 1986. Neurofibromatosis in children. *Acta Orthop Scand Suppl* 57;218:1–60.

Croucher, S. J., and Tickle, C. 1989. Characterization of epithelial domains in the nasal passages of chick embryos: Spatial and temporal mapping of a range of extracellular matrix and cell surface molecules during development of the nasal placode. *Development* 106:493–509.

Crowe, F. W., Schull, J., and Neel, J. V. 1956. *A Clinical, Pathological and Genetic Study of Multiple Neurofibromatosis*. Springfield, Ill.: C. C. Thomas.

Cruz, Y. P., Yousef, A., and Selwood, L. 1996. Fate-map analysis of the epiblast of the dasyurid marsupial *Sminthopsis macroura* (Gould). *Reprod Fertil Dev* 8:779–788.

Currie, P. D. 1996. Zebrafish genetics: Mutant cornucopia. *Curr Biol* 6:1548–1552.

Cusimano, T., Fagone, A., and Reverberi, G. 1962. On the origin of the larval mouth in the Anurans. *Acta Embryol Morphol Exp* 5:82–103.

Cusimano-Carollo, T. 1962. Sulle capacità organo-formative delle pieghe neurali degli Anuri: richerche su *Discoglossus pictus* Otth. *Atti Accad Naz Lincei Rc* 33:354–358.

Cusimano-Carollo, T. 1963. Investigation on the ability of the neural folds to induce a mouth in the *Discoglossus pictus* embryo. *Acta Embryol Morphol Exp* 6:158–168.

Cusimano-Carollo, T. 1967. La piega neurale trasverse e la formazione della bocca nelle larve di *Discoglossus pictus. Atti Accad Naz Lincei Rc* 43:252–258.

Cusimano-Carollo, T. 1969. Phenomena of induction by the transverse neural fold during the formation of the mouth in *Discoglossus pictus. Acta Embryol Exp* 1:97–110.

Cusimano-Carollo, T. 1972. On the mechanism of the formation of the larval mouth in *Discoglossus pictus. Acta Embryol Exp* 4:289–332.

Dahl, E., Koseki, H., and Balling, R. 1997. *Pax* genes and organogenesis. *BioEssays* 19:755–765.

Damas, H. 1944. Recherches sur le développement de *Lampetra fluviatilis* L. Contribution à l'étude de la céphalogénèse des vertébrés. *Arch Biol Paris* 55:1–284.

Damas, H. 1951. Observations sur le développement des ganglions crâniens chez *Lampetra fluviatilis* (L). *Arch Biol Paris* 62:55–95.

Darland, T., and Leblanc, G.-G. 1996. Immortalized Hensen's node cells secrete a factor that regulates avian neural crest cell fate *in vitro. Dev Biol* 176:62–75.

Davidson, D. 1995. The function and evolution of *Msx* genes: Pointers and paradoxes. *Trends Genet* 11:405–411.

Dean, A. D., and Frost-Mason, S. K. 1995. Effects of melanocyte-stimulating hormone on wild-type and white axolotl neural crest cells. *Biochem Biophys Res Commun* 210:239–245.

Dean, B. 1896. The early development of *Amia. Q J Microsc Sci* 38:413–444.

Dean, B. 1899. On the embryology of *Bdellostoma stouti.* A general account of myxinoid development from the egg and segmentation to hatching. In: *Festschr F C van Kuppfer.* Jena: G. Fischer, pp. 221–277.

de Beer, G. R. 1937. *The Development of the Vertebrate Skull.* Oxford: Oxford University Press. Reissued 1985, Chicago: University of Chicago Press.

de Beer, G. R. 1947. The differentiation of neural crest cells into visceral cartilages and odontoblasts in *Amblystoma*, and a re-examination of the germ-layer theory. *Proc R Soc Lond B Biol Sci* 134:377–398.

de Beer, G. R. 1958. *Embryos and Ancestors.* 3rd ed. Oxford and London: Oxford University Press.

de Beer, G. R. 1971. *Homology: An Unsolved Problem.* London: Oxford University Press. Oxford Biology Reader No. 11.

Delannet, M., and Duband, J.-L. 1992. Transforming growth factor-ß control of cell-substrate adhesion during avian neural crest cell emigration *in vitro. Development* 116:275–287.

Dencker, L. 1986. Accumulation of retinoids in embryonic neural and neural crest cells as part of the mechanism of teratogenesis. *Ups J Med Sci* 91:295–298.

Dencker, L., Annerwall, E., Busch, C., and Ericksson, U. 1990. Localization of specific retinoid-binding sites and expression of cellular retinoic-acid-binding protein (CRABP) in the early mouse embryo. *Development* 110:343–352.

Dencker, L., d'Argy, R., Danielsson, B. R. G., Ghantous, H., and Sperber, G. O. 1987. Saturable accumulation of retinoic acid in neural and neural crest derived cells in early embryonic development. *Dev Pharmacol Ther* 10:212–223.

Deol, M. S. 1970. The relationship between abnormalities of pigmentation and the inner ear. *Proc R Soc Lond B Biol Sci* 175:201–217.

De Robertis, E. M., Fainsod, A., Gont, L. K., and Steinbeisser, H. 1994. The evolution of vertebrate gastrulation. *Development* (suppl):117–124.

de Selys-Longchapms, M. 1910. Gastrulation et formation des feuillets chez *Petromyzon planeri*. *Arch Biol* 25:1–75.

Detwiler, S. R. 1934. An experimental study of spinal nerve segmentation in *Ambylstoma* with reference to the plurisegmental contribution to the brachial plexus. *J Exp Zool* 67:395–443.

Detwiler, S. R. 1937. Observations upon the migration of neural crest cells and upon the development of the spinal ganglia and vertebral arches in *Amblystoma*. *Am J Anat* 61:63–94.

Detwiler, S. R., and Kehoe, K. 1939. Further observations on the origin of the sheath cells of Schwann. *J Exp Zool* 81:415–433.

De Villers, C. 1947. Récherches sur la crâne dermique des téléostéens. *Ann Paleont* 33:1–94.

Dias, M. S., and Schoenwolf, G. C. 1990. Formation of ectopic neuroepithelium in chick blastoderms: Age-related capacities for induction and self-differentiation following transplantation of quail Hensen's nodes. *Anat Rec* 229:437–448.

Dickinson, M. E., Sellek, M. A. J., McMahon, A. P., and Bronner-Fraser, M. 1995. Dorsalization of the neural tube by the non-neural ectoderm. *Development* 121:2099–2106.

Dickman, E. D., Thaller, C., and Smith, S. M. 1997. Temporally-regulated retinoic acid depletion produces specific neural crest, ocular and nervous system defects. *Development* 124:3111–3121.

DiMario, F. J., Dunham, B., Burleson, J. A., Moskovitz, J, and Cassidy, S. B. 1994. An evaluation of autonomic nervous system function in patients with Prader-Willi syndrome. *Pediatrics* 93:76–81.

Dirksen, M.-L., Mathers, P., and Jamrich, M. 1993. Expression of a *Xenopus Distal-less* homeobox gene involved in forebrain and cranio-facial development. *Mech Dev* 41:121–128.

Dirksen, M.-L., Morasso, M. I., Sargent, T. D., and Jamrich, M. 1994. Differential expression of a *Distal-less* homeobox gene *Xdll-2* in ectodermal cell lineages. *Mech Dev* 46:63–70.

Di Simone, R. E., and Berman, A. T. 1989. Gene linkage in neurofibromatosis. *Clin Orthop* 245:49–52.

Di Simone, R. E., Berman, A. T., and Schwentker, E. P. 1988. The orthopedic manifestation of neurofibromatosis. A clinical experience and review of the literature. *Clin Orthop* 230:277–283.

Dollé, P., Price, M., and Duboule, D. 1992. Expression of the murine *Dlx-1* homeobox gene during facial, ocular and limb development. *Differentiation* 49:93–99.

Doniach, T. 1995. Basic FGF as an inducer of anteroposterior neural pattern. *Cell* 83:1967–1070.

Donoghue, P. C. J. 1998. Growth and patterning in the conodont skeleton. *Philos Trans R Soc Lond B Biol Sci* 353:633–666.

Dorris, F. 1936. Differentiation of pigment cells in tissue cultures of chick neural crest. *Proc Soc Exp Biol Med* 34:448–449.

Dorris, F. 1938. The production of pigment *in vitro* by chick neural crest. *Wilhelm Roux Arch EntwMech Org* 138:323–335.

Dorris, F. 1939. The production of pigment by chick neural crest in grafts to the 3-day limb bud. *J Exp Zool* 80:315–345.

Drews, U., Kocher-Becker, U., and Drews, U. 1972. The induction of visceral cartilage from cranial neural crest by pharyngeal endoderm in hanging drop cultures and the locomotory behaviour of the neural crest cells during cartilage differentiation. *Wilhelm Roux Arch EntwMech Org* 171:17–37.

Drysdale, T. A., and Elinson, R. P. 1991. Development of the *Xenopus laevis* hatching gland and its relationship to surface ectoderm patterning. *Development* 111:469–478.

Drysdale, T. A., and Elinson, R. P. 1992. Cell migration and induction in the development of the surface ectodermal pattern of the *Xenopus laevis* tadpole. *Dev Growth Differ* 34:51–59.

Drysdale, T. A., and Elinson, R. P. 1993. Inductive events in the patterning of the *Xenopus laevis* hatching and cement glands, two cell types which delimit head boundaries. *Dev Biol* 158:245–253.

Duband, J.-L., Monier, F., Delannet, M., and Newgreen, D. 1995. Epithelium-mesenchyme transition during neural crest development. *Acta Anat* 154:63–78.

Duband, J.-L., Rocher, S., Chen, W.-T., Yamada, K. M., and Thiery, J.-P. 1986. Cell adhesion and migration in the early vertebrate embryo: Location and possible role of the putative fibronectin receptor complex. *J Cell Biol* 102:160–178.

Duband, J.-L., and Thiery, J. P. 1987. Distribution of laminin and collagens during avian neural crest development. *Development* 101:461–478.

Du Brul, E. L. 1964. Evolution of the temporomandibular joint. In: *The Temporomandibular Joint.* Sarnat, B. G., ed. Springfield, Ill.: C.C. Thomas. pp. 3–27.

Duellman, W. E., and Trueb, L. 1986. *Biology of Amphibians.* New York: McGraw-Hill Book Co.

Duff, R. S., Langtimm, C. J., Richardson, M. K., and Sieber-Blum, M. 1991. *In vitro* clonal analysis of progenitor cell patterns in dorsal root and sympathetic ganglia of the quail embryo. *Dev Biol* 147:451–459.

Dufresne, C., and Richtsmeier, J. T. 1995. Interaction of craniofacial dysmorphology, growth and prediction of surgical outcome. *J Craniofac Surg* 6:270–281.

Dunlop, L.-L. T., and Hall, B. K. 1995. Relationships between cellular condensation, preosteoblast formation and epithelial-mesenchymal interactions in initiation of osteogenesis. *Int J Dev Biol* 39:357–371.

Dunn, M. K., Mercola, M., and Moore, D. D. 1995. Cyclopamine, a steroidal alkaloid, disrupts development of cranial neural crest cells in *Xenopus. Dev Dynam* 202:255–270.

Dupin, E., Sextier-Sainte-Claire Deville, F., Nataf, V., and Le Douarin, N. M. 1993. The ontogeny of the neural crest. *C R Acad Sci* 316:1072–1081.

DuShane, G. P. 1935. An experimental study of the origin of pigment cells in Amphibia. *J Exp Zool* 72:1–31.

DuShane, G. P. 1943. The embryology of vertebrate pigment cells. Part I. Amphibia. *Q Rev Biol* 18:108–127.

DuShane, G. P. 1944. The embryology of vertebrate pigment cells. Part II. Birds. *Q Rev Biol* 19:98–117.

Ebendal, T. 1995. Cell movement in neurogenesis—An interview with Professor Jacobson, Carl Olaf. *Int J Dev Biol* 39:705–711.

Eckhoff, C., and Willhite, C. C. 1997. Embryonic delivered doses of isotretinoin (13 cis-retinoic acid) and its metabolites in hamsters. *Toxicol Appl Pharmacol* 146:79–87.

Edelman, G. M. 1983. Cell adhesion molecules. *Science* 219:450–457.

Edwards, L. F. 1929. The origin of the pharyngeal teeth of the carp (*Cyprinus carpo, Linnaeus*). *Ohio J Sci* 29:93–130.

Eichele, G. 1997. Retinoids: From hindbrain patterning to Parkinson disease. *Trends Genet* 13:343–345.

Eisen, J. S., and Weston, J. A. 1993. Development of the neural crest in the zebrafish. *Dev Biol* 159:50–59.

Ekanayake, S., and Hall, B. K. 1994. Formation of cartilaginous nodules and heterogeneity in clones of HH 17 mandibular ectomesenchyme from the embryonic chick. Acta Anat 151:171–179.

Ekanayake, S., and Hall, B. K. 1997. The *in vivo* and *in vitro* effects of bone morphogenetic protein-2 on the development of the chick mandible. *Int J Dev Biol* 41:67–81.

Ekker, M., and Akimenko, M.-A. 1991. Le poisson zébre (*Danio rerio*), un modèle en biologie du développement. *Méd Sci* 7:553–560.

Ekker, M., Akimenko, M.-A., Allende, M. L., Smith, R., Drouin, G., Langille, R. M., Weinberg, E. S., and Westerfield, M. 1997. Relationships among *msx* gene structure and function in zebrafish and other vertebrates. *Mol Biol Evol* 14:1008–1022.

Elinson, R. P. 1990. Direct development in frogs: Wiping the recapitulationist slate clean. *Semin Dev Biol* 1:263–270.

Ellies, D. L., Langille, R. M., Martin, C. C., Akimenko, M.-A., and Ekker, M. 1997. Specific craniofacial cartilage dysmorphogenesis coincides with a loss of *Dlx* gene expression in retinoic acid-treated zebrafish embryos. *Mech Dev* 61:23–36.

End, D., Pevzner, L., Lloyd, A., and Guroff, G. 1983. Identification of nerve growth factor receptors in primary cultures of chick neural crest cells. *Dev Brain Res* 7:131–136.

Epperlein, H.-H. 1974. The ectomesenchymal-endodermal interaction-system (EEIS) of *Triturus alpestris* in tissue culture. 1. Observations on attachment, migration, and differentiation of neural crest cells. *Differentiation* 2:151–168.

Epperlein, H.-H. 1988. Die Kontrolle der Wanderung und Differenzierung von Neuralleistezellen durch die Extrazelluläre matrix. *Naturwissenschaften* 75:443–450.

Epperlein, H.-H., and Claviez, M. 1982. Formation of pigment cell patterns in *Triturus alpestris* embryos. *Dev Biol* 91:497–502.

Epperlein, H.-H., Halfter, W., and Tucker, R. P. 1988. The distribution of fibronectin and tenascin along the migratory pathways of the neural crest in the trunk of amphibian embryos. *Development* 103:743–756.

Epperlein, H.-H., and Lehman, R. 1975. The ectomesenchymal-endodermal interaction system (EEIS) of *Triturus alpestris* in tissue culture. 2. Observations on differentiation of visceral cartilage. *Differentiation* 4:159–174.

Epperlein, H.-H., and Löfberg, J. 1990. The development of the larval pigment patterns in *Triturus, Alpestris* and *Ambystoma mexicanum. Adv Anat Embryol Cell Biol* 118:1–101.

Epperlein, H.-H., and Löfberg, J. 1993. The development of the neural crest in amphibians. *Ann Anat* 175:483–499.

Epperlein, H.-H., and Löfberg, J. 1996. What insights into the phenomena of cell fate determination and cell migration has the study of the urodele neural crest provided? *Int J Dev Biol* 40:695–707.

Epperlein, H.-H., Löfberg, J., and Olsson, L. 1996. Neural crest cell migration and pigment pattern formation in urodele amphibians. *Int J Dev Biol* 40:229–238.

Epstein, J. A. 1996. *Pax3*, neural crest and cardiovascular development. *Trends Cardiovasc Med* 6:255–261.

Erickson, C. A. 1986. Morphogenesis of the neural crest. In: Browder, L. W., ed. *Developmental Biology Vol. 2*. New York: Plenum Publishing Corp. pp. 481–543.
Erickson, C. A. 1987. Behavior of neural crest cells on embryonic basal laminae. *Dev Biol* 120:38–49.
Erickson, C. A. 1993a. Morphogenesis of the avian trunk neural crest—use of morphological techniques in elucidating the process. *Microsc Res Tech* 26:329–351.
Erickson, C. A. 1993b. From the crest to the periphery—control of pigment cell migration and lineage segregation. *Pigment Cell Res* 6:336–347
Erickson, C. A., Duong, T. D., and Tosney, K. W. 1992. Descriptive and experimental analysis of the dispersion of neural crest cells along the dorsolateral path and their entry into ectoderm in the chick embryos. *Dev Biol* 151:251–272.
Erickson, C. A., and Goins, T. L. 1995. Avian neural crest cells can migrate in the dorsolateral path only if they are specified as melanocytes. *Development* 121:915–924.
Erickson, C. A., Loring, J. F., and Lester, S. M. 1989. Migratory pathways of HNK-1 immunoreactive neural crest cells in the rat embryo. *Dev Biol* 134:112–118.
Erickson, C. A., and Perris, R. 1993. The role of cell-cell and cell-matrix interactions in the morphogenesis of the neural crest. *Dev Biol* 159:60–74.
Erickson, C. A., and Turley, E. A. 1987. The effects of epidermal growth factor on neural crest cells in tissue culture. *Exp Cell Res* 169:267–269.
Erickson, C. A., and Weston, J. A. 1983. An SEM analysis of neural crest migration in the mouse. *J Embryol Exp Morphol* 74:97–118.
Erickson, H. P., and Bourdon, M. A. 1989. Tenascin: An extracellular matrix protein prominent in specialized embryonic tissues and tumors. *Annu Rev Cell Biol* 5:71–92.
Etheridge, A. L. 1968. Determination of the mesonephric kidney. *J Exp Zool* 169:357–370.
Etheridge, A. L. 1972. Suppression of kidney formation by neural crest cells. *Wilhelm Roux Arch EntwMech Org* 169:268–270.
Evanson, J. E., and Milos, N. C. 1996. A monoclonal antibody against neural crest-stage *Xenopus laevis* lectin perturbs craniofacial development of *Xenopus*. *J Craniofac Genet Dev Biol* 16:74–93.
Ewart, J. L., Cohen, M. F., Meyer, R. A., Huang, G. Y., Wessels, A., Gourdie, R. G., Chin, A. J., Park, S. M. J., Lazatin, B. O., Villabon, S., and Lo, C. W. 1997. Heart and neural tube defects in transgenic mice overexpressing the Cx43 gap junction gene. *Development* 124:1281–1292.
Eyal-Giladi, H., and Zinberg, N. 1964. The development of the chondrocranium of *Pleurodeles waltlii*. *J Morph* 114:527–548.
Fagone, A. 1959. Ricerche sperimentali sulla formazione della bocca in *Discoglossus pictus*. *Acta Embryol Morphol Exp* 2:133–150.
Fagone, A. 1960. Ulteriori richerche sperimentali sulla formazione della bocca in *Discoglossus pictus*. *Atti Accad Naz Lincei Rc* 28:249–253.
Fang, H., and Elinson, R. P. 1996. Patterns of Distal-less gene expression and inductive interactions in the head of the direct developing frog *Eleutherodactylus coqui*. *Dev Biol* 179:160–172.
Fang, J., and Hall, B. K. 1995. Differential expression of neural cell adhesion molecule (N-CAM) during osteogenesis and secondary chondrogenesis in the embryonic chick. *Int J Dev Biol* 39:519–528.
Fang, J., and Hall, B. K. 1996. *In vitro* differentiation potential of the periosteal cells from a membrane bone, the quadratojugal of the embryonic chick. *Dev Biol* 180:701–712.

Fang, J., and Hall, B. K. 1997. Chondrogenic cell differentiation from membrane bone periostea. *Anat Embryol* 196:349–362.

Fauquet, M., Smith, J., Ziller, C., and Le Douarin, N. M. 1981. Differentiation of autonomic neuron precursors *in vitro*: Cholinergic and adrenergic traits in cultured neural crest cells. *J Neurosci* 1:478–492.

Ferguson, M. W. J. 1978. The teratogenic effects of 5-fluoro-2'-deoxyuridine (FUDR) on the Wistar rat foetus, with particular reference to cleft palate. *J Anat* 126:37–49.

Ferguson, M. W. J. 1981. The value of the American alligator (*Alligator mississippiensis*) as a model for research in craniofacial development. *J Craniofac Genet Dev Biol* 1:123–144.

Ferguson, M. W. J. 1984. Craniofacial development in *Alligator mississippiensis*. *Symp Zool Soc Lond* 52:223–274.

Ferguson, M. W. J. 1985. Reproductive biology and embryology of the crocodilians. In: Gans, C., Billett, F. and Maderson, P. F. A., eds. *Biology of the Reptilia, Vol. 14. Development A*. New York: John Wiley & Sons. pp. 329–492.

Ferguson, M. W. J., and Honig, L. S. 1984. Epithelial-mesenchymal interactions during vertebrate palatogenesis. In: Zimmermann, E. F., ed. *Palate Development: Normal and Abnormal Cellular and Molecular Aspects*. New York: Academic Press. pp. 138–165.

Filogamo, G., Corvetti, G., and Daneo, L. S. 1990. Differentiation of cardiac conducting cells from the neural crest. *J Auton Nerv Syst* 30:S55–S58.

Fishman, M. C., and Chien, K. R. 1997. Fashioning the vertebrate heart: Earliest embryonic decisions. *Development* 124:2099–2117.

Fleming, A., and Copp, A. J. 1998. Embryonic folate metabolism and mouse neural tube defects. *Science* 280:2107–2109.

Fleming, T. P., and George, M. A. 1986. Fluorescent latex microparticles: A non-invasive short-term cell lineage marker suitable for use in the mouse early embryo. *Wilhelm Roux's Arch Dev Biol* 196:1–11.

Flint, O. P. 1980. Cell behaviour and cleft palate in the mutant mouse, *amputated*. *J Embryol Exp Morphol* 58:131–142.

Fontaine, J. 1979. Multistep migration of calcitonin cell precursors during ontogeny of the mouse pharynx. *Gen Comp Endocrinol* 37:81–92.

Foret, J. E. 1970. Regeneration of larval urodele limbs containing homoplastic transplants. *J Exp Zool* 175:297–322.

Forey, P. L. 1995. Agnathans recent and fossil, and the origin of jawed vertebrates. *Rev Fish Biol Fisheries* 5:267–303.

Forey, P., and Janvier, P. 1993. Agnathans and the origin of jawed vertebrates. *Nature* 361:129–134.

Forey, P., and Janvier, P. 1994. Evolution of the early vertebrates. *Am Sci* 82:554–565.

Franz, T. 1992. Neural tube defects without neural crest defects in *Splotch* mice. *Teratology* 46:599–604.

Franz, T., and Kothary, R. 1993. Characterization of the neural crest defect in *Splotch* (Sp^{2H}) mutant mice using a LacZ transgene. *Dev Brain Res* 72:99–105.

Franz, T., Kothary, R., Surani, M. A. H., Halata, Z., and Grim, M. 1993. The *Splotch* mutation interferes with muscle development in the limbs. *Anat Embryol* 187:153–160.

Fraser, S. E., and Bronner-Fraser, M. 1991. Migrating neural crest cells in the trunk of the avian embryo are multipotent. *Development* 112:913–920.

Frederickson, R. G., and Low, F. N. 1971. The fine structure of perinotochordal microfibrils in control and enzyme-treated chick embryos. *Am J Anat* 130:347–376.

French, V., Ingham, P., Cooke, J., and Smith, J. 1988. Mechanisms of segmentation. *Development* 104(suppl):1–248.

Frohman, M. A., Boyle, M., and Martin, G. R. 1990. Isolation of the mouse *Hox-2.9* gene; analysis of embryonic expression suggests that positional information along the anterior-posterior axis is specified by mesoderm. *Development* 110:589–607.

Frost, D. E. 1985. *Amphibian Species of the World: A Taxonomic and Geographic Reference.* Washington, D. C.: Assoc. Systematics Collections.

Frost, S. K., and Malacinski, G. M. 1980. The developmental genetics of pigment mutants in the Mexican axolotl: A review. *Dev Genet* 1:271–294.

Fukiishi, Y., and Morriss-Kay, G. M. 1992. Migration of cranial neural crest cells to the pharyngeal arches and heart in rat embryos. *Cell Tissue Res* 268:1–8.

Fyfe, D. M., and Hall, B. K. 1979. Lack of association between avian cartilages of different embryological origins when maintained *in vitro. Am J Anat* 154:485–496.

Fyfe, D. M., and Hall, B. K. 1983. The origin of the ectomesenchymal condensations which precede the development of the bony scleral ossicles in the eyes of embryonic chicks. *J Embryol Exp Morphol* 73:69–86.

Gabbott, S. E., Aldridge, R. J., and Theron, J. N. 1995. A giant conodont with preserved muscle tissue from the Upper Ordovician of South Africa. *Nature* 374:800–803.

Gajovic, S., and Kostovic-Knezevic, L. 1995. Ventral ectodermal ridge and ventral ectodermal groove: Two distinct morphological features in the developing rat embryo tail. *Anat Embryol* 192:181–187.

Gale, E., Prince, U., Lumsden, A., Clarke, J., Holder, N., and Maden, M. 1996. Late effects of retinoic acid on neural crest and aspects of rhombomere identity. *Development* 122:783–793.

Galili, N., Epstein, J. A., Leconte, I., Nayak, S., and Buck, C. A. 1998. *Gscl*, a gene within the minimal DiGeorge critical region, is expressed in primordial germ cells and the developing pons. *Dev Dyn* 212:86–93.

Gallo, V., Bertolotto, A., and Levi, G. 1987. The proteoglycan chondroitin sulfate is present in a subpopulation of cultured astrocytes and in their precursors. *Dev Biol* 123:282–285.

Gans, C. 1987. The neural crest: A spectacular invention. In: Maderson, P. F. A., ed. *Developmental and Evolutionary Aspects of the Neural Crest.* New York: John Wiley & Sons. pp. 361–379.

Gans, C. 1989. Stages in the origin of vertebrates: Analysis by means of scenarios. *Biol Rev Camb Philos Soc* 64:221–268.

Gans, C. 1993. Evolutionary origin of the vertebrate skull. In: Hanken, J., and Hall, B. K., eds. *The Vertebrate Skull. Volume II. Patterns of Structural and Systematic Diversity.* Chicago: University of Chicago Press. pp. 1–35.

Gans, C., and Northcutt, R. G. 1983. Neural crest and the origin of vertebrates: A new head. *Science* 220:268–274.

Gans, C., and Northcutt, R. G. 1985. Neural crest: The implications for comparative anatomy. *Fortschr Zool* 30:507–514.

Garcia-Martinez, V., Alvarez, I. S., and Schoenwolf, G. C. 1993. Locations of the ectodermal and nonectodermal subdivisions of the epiblast at stages 3 and 4 of avian gastrulation and neurulation. *J Exp Zool* 267:431–446.

Gardner, C. A., and Barald, K. F. 1991. The cellular environment controls the expression of *engrailed*-like protein in the cranial neuroepithelium of quail-chick chimeric embryos. *Development* 113:1037–1048.

Gavalas, A., Studer, M., Lumsden, A., Rijli, F. M., Krumlauf, R., and Chambon, P. 1998. *Hoxa1* and *Hoxb1* synergize in patterning the hindbrain, cranial nerves and second pharyngeal arch. *Development* 125:1123–1136.

Gee, H. 1996. *Before the Backbone: Views on the Origin of the Vertebrates.* London: Chapman & Hall.

Gendron-Maguire, M., Mallo, M., Zhang, M., and Gridley, T. 1993. *Hoxa-2* mutant mice exhibit homeotic transformation of skeletal elements derived from cranial neural crest. *Cell* 75:1317–1331.

Gilbert, S. F. 1997. *Developmental Biology.* 5th ed. New York: Sinauer Inc.

Gimlich, R. L., and Braun, J. 1985. Improved fluorescent compounds for tracing cell lineage. *Dev Biol* 109:509–514.

Gipouloux, J.-D., and Girard, C. 1986. Effects of the removal of neural crest anlage upon endodermal morphogenesis and primordial germ cell migration in toad embryos. *Wilhelm Roux's Arch Dev Biol* 195:355–358.

Gipouloux, J.-D., Girard, C., and Delbos, M. 1992. The neural crest enhances the attractive potency of the somitic mesoderm towards the primordial germ cells of the toad embryo. *Eur Arch Biol* 103:37–40.

Goh, K. L., Yang, J. T., and Hynes, R. O. 1997. Mesodermal defects and cranial neural crest apoptosis in α5 integrin-null embryos. *Development* 124:4309–4319.

Goldstein, R. S., Avivi, C., and Geffe, R. 1995. Initial axial level-dependent differences in size of avian dorsal root ganglia are imposed by the sclerotome. *Dev Biol* 168:214–222.

Gont, L. K., Steinbesser, H., Blumberg, B., and De Robertis, E. M. 1993. Tail formation as a continuation of gastrulation: The multiple cell populations of the *Xenopus* tailbud derive from the late blastopore lip. *Development* 119:991–1004.

Gooday, D., and Thorogood, P. 1985. Contact behaviour exhibited by migrating neural crest cells in confrontation culture with somitic cells. *Cell Tissue Res* 241:165–169.

Goodrich, E. S. 1930. *Studies on the Structure and Development of Vertebrates.* London: Macmillan & Co. Reprinted, 1958, New York: Dover Publications, Inc.; 1986, Chicago: University of Chicago Press.

Gorbman, A. 1983. Early development of the hagfish pituitary gland: Evidence for the endodermal origin of the adenohypophysis. *Am Zool* 23:639–654.

Gorbman, A., and Tamarin, A. 1985. Early development of oral, olfactory and adenohypophyseal structures of agnathans and its evolutionary implications. In: Foreman, R. E., Gorbman, A., Dodd, J. M., and Olsson, R., eds. *Evolutionary Biology of Primitive Fishes.* New York: Plenum Press. pp. 165–185.

Gorlin, R. J. 1978. Syndromes associated with jaw deformity. In: Whitaker, L. A., ed. *Symposium on Reconstruction of Jaw Deformities.* St. Louis: C. V. Mosby. pp. 76–84.

Gorlin, R. J., Cohen, M. M. Jr., and Levin, L. S. 1988. *Syndromes of the Head and Neck.* 3rd ed. Oxford and London: Oxford University Press.

Goronowitsch, N. 1892. Die axiale und die laterale Kopfmetamerie der Vögelembryonen. Die Rolle der sog. 'Ganglienleisten' im Aufbaue der Nervenstämme. *Anat Anz* 7:454–464.

Goronowitsch, N. 1893a. Untersuchungen über die Entwicklung der sogenannten 'Ganglienleisten' im Kopfe der Vögelembryonen. *Morphol Jb* 20:187–259.

Goronowitsch, N. 1893b. Weiters über die ektodermal Entstehung von Skeletanlagen im Kopfe der Wirbeltiere. *Morphol Jb* 20:425–428.

Gorski, J. P. 1998. Is all bone the same? Distinctive distributions and properties of non-collagenous matrix proteins in lamellar vs. woven bone imply the existence of different underlying osteogenic mechanisms. *Crit Rev Oral Biol Med* 9:201–223.

Gould, A. 1997. Functions of mammalian Polycomb group and trithorax group related genes. *Curr Opin Genet Dev* 7:488–494.

Gould, G. M., and Pyle, W. L. 1896. *Anomalies and Curiosities of Medicine*. New York: Bell Publishing Co.

Gould, S. J. 1977. *Ontogeny and Phylogeny*. Cambridge: Belknap Press of Harvard University Press.

Graham, A., Francis-West, P. H., Brickell, P., and Lumsden, A. 1994. The signalling molecule BMP-4 mediates apoptosis in the rhombencephalic neural crest. *Nature* 372:684–686.

Graham, A., Heyman, I., and Lumsden, A. 1993. Even-numbered rhombomeres control the apoptotic elimination of neural crest cells from odd-numbered rhombomeres in the chick hindbrain. *Development* 119:233–245.

Graham-Smith, W. 1978. On the lateral lines and dermal bones in the parietal region of some crossopterygian and dipnoan fishes. *Philos Trans R Soc Lond B Biol Sci* 282:1–39.

Grainger, R. M. 1992. Embryonic lens induction: Shedding light on vertebrate tissue determination. *Trends Genet* 8:349–355.

Granström, G., Jacobsson, C., and Magnusson, B. C. 1991. Enzyme histochemical analysis of craniofacial malformations induced by retinoids. *Scand J Plast Reconst Hand Surg* 25:133–141.

Grant, D. S., Kleinman, H. K., Leblond, C. P., Inoue, S., Chung, A. E., and Martin, G. R. 1985. The basement-membrane-like matrix of the mouse EHS tumor. II. Immunohistochemical quantitation of six of its components. *Am J Anat* 174:387–398.

Grant, J. H., Maggioprice, L., Reutebuch, J., and Cunningham, M. L. 1997. Retinoic acid exposure of the mouse on embryonic day 9 selectively spares derivatives of the frontonasal neural crest. *J Craniofac Genet Dev* Biol 17:1–8.

Grapìn-Botton, A., Bonnin, M.-A., and Le Douarin, N. M. 1997. *Hox* gene induction in the neural tube depends on three parameters: Competence, signal supply and paralogue group. *Development* 124:849–859.

Grapìn-Botton, A., Bonnin, M.-A., McNaughton, L. A., Krumlauf, R., and Le Douarin, N. M. 1995. Plasticity of transposed rhombomeres: *Hox* gene induction is correlated with phenotypic modifications. *Development* 121:2707–2721.

Graveson, A. C. 1993. Neural crest: Contributions to the development of the vertebrate head. *Am Zool* 33:424–433.

Graveson, A. C., and Armstrong, J. B. 1987. Differentiation of cartilage from cranial neural crest in the axolotl (*Ambystoma mexicanum*). *Differentiation* 35:16–20.

Graveson, A. C., and Armstrong, J. B. 1990. The premature death (*p*) mutation of *Ambystoma mexicanum* affects a subpopulation of neural crest cells. *Differ Ontog Neoplasia* 45:71–75.

Graveson, A. C., and Armstrong, J. B. 1996. Premature death (*p*) mutation of *Ambystoma mexicanum* affects the ability of ectoderm to respond to neural induction. *J Exp Zool* 274:248–254.

Graveson, A. C., Hall, B. K., and Armstrong, J. B. 1995. The relationship between migration and chondrogenic potential of trunk neural crest cells in *Ambystoma mexicanum. Wilhem Roux's Arch Dev Biol* 204:477–483.

Graveson, A. C., Smith, M. M., and Hall, B. K. 1997. Neural crest potential for tooth development in a urodele amphibian: Developmental and evolutionary significance. *Dev Biol* 188:34–42.

Green, E., ed. 1966. *Biology of the Laboratory Mouse*. New York: McGraw Hill.

Green, M. C. 1970. The developmental effect of *congenital hydrocephalus* (*ch*) in the mouse. *Dev Biol* 23:585–608.

Griffith, C. M., Wiley, M. J., and Sanders, E. J. 1992. The vertebrate tail bud: Three germ layers from one tissue. *Anat Embryol* 185:101–113.

Groves, A. K., and Anderson, D. J. 1996. Role of environmental signals and transcriptional regulators in neural crest development. *Dev Genet* 18:64–72.

Grüneberg, H. 1947. *Animal Genetics and Medicine*. London: Hamish Hamilton Medical Books.

Grüneberg, H. 1956. A ventral ectodermal ridge of the tail in mouse embryos. *Nature* 177:787–788.

Grüneberg, H., and des Wickramaratne, G. A. 1974. A re-examination of two skeletal mutants of the mouse, *vestigial tail* (*vt*) and *congenital hydrocephalus* (*ch*). *J Embryol Exp Morphol* 31:207–222.

Grunz, H., Schüren, C., and Richter, K. 1995. The role of vertical and planar signals during the early steps of neural induction. *Int J Dev Biol* 39:539–543.

Grunz, H., and Tacke, L. 1986. The inducing capacity of the presumptive endoderm of *Xenopus laevis* studies by transfilter experiments. *Wilhelm Roux Arch EntwMech Org* 195:467–473.

Guillory, G., and Bronner-Fraser, M. 1986. An *in vitro* assay for neural crest cell migration through the somites. *J Embryol Exp Morphol* 98:85–97.

Gurdon, J. B. 1987. Embryonic induction—molecular prospects. *Development* 99:285–306.

Gurdon, J. B. 1992. The generation of diversity and pattern in animal development. *Cell* 68:185–187.

Guthrie, S., Muchamore, I., Kuroiwa, A., Marshall, H., Krumlauf, R., and Lumsden, A. 1992. Neuroectodermal autonomy of *Hox-2.9* expression revealed by rhombomere transpositions. *Nature* 356:157–159.

Gvirtzmann, G., Goldstein, R. S., and Kalcheim, C. 1992. A positive correlation between permissiveness of mesoderm to neural crest migration and early DRG growth. *J Neurobiol* 23:205–216.

Habeck, J. O. 1990. Islands of cartilage and bone at the margins of the carotid bodies in rats. *Anat Anz* 171:277–280.

Haeckel, E. 1874. Die Gastræa-Theorie, die phylogenetische Classification des Thierreichs und die Homologie der Keimblätter. *Jena Z Naturwiss* 8:1–55.

Halford, S., Wilson, D. I., Daw, S. C. M., Roberts, C., Wadey, R., Kamath, S., Wickremasinghe, A., Burn, J., Goodship, J., Mattei, M.-G., Moormon, A. F. M., and Scambler, P. J. 1993. Isolation of a gene expressed during early embryogenesis from the region of 22q11 commonly deleted in DiGeorge syndrome. *Hum Mol Genet* 2:1577–1582.

Halfter, W., Chiquet-Ehrismann, R., and Tucker, R. P. 1989. The effect of tenascin and embryonic basal lamina on the behavior and morphology of neural crest cells in vitro. *Dev Biol* 132:14–25.

Hall, B. K. 1975. Evolutionary consequences of skeletal development. *Am Zool* 15:329–350.

Hall, B. K. 1978. *Developmental and Cellular Skeletal Biology*. New York: Academic Press.

Hall, B. K. 1980. Tissue interactions and the initiation of osteogenesis and chondrogenesis in the neural crest-derived mandibular skeleton of the embryonic mouse as seen in isolated murine tissues and in recombinations of murine and avian tissues. *J Embryol Exp Morphol* 58:251–264.

Hall, B. K. 1982. Mandibular morphogenesis and craniofacial malformations. *J Craniofac Genet Dev Biol* 2:309–322.

Hall, B. K. 1983a. Epithelial-mesenchymal interactions in cartilage and bone development. In: Sawyer, R. H., and Fallon, J. F., eds. *Epithelial-mesenchymal Interactions in Development.* New York: Praeger Publishers. pp. 189–214.

Hall, B. K. 1983b. Tissue interactions and chondrogenesis. In: Hall, B. K., ed. *Cartilage. Volume 2. Development, Differentiation and Growth.* Orlando: Academic Press. pp. 187–222.

Hall, B. K. 1983c. Epigenetic control in development and evolution. In: Goodwin, B. C., Holder, N. J., and Wylie, C. C., eds. *Development and Evolution.* Cambridge: Cambridge University Press. pp. 353–379.

Hall, B. K. 1985. Critical periods during development as assessed by thallium-induced inhibition of growth of embryonic chick tibiae *in vitro. Teratology* 31:353–361.

Hall, B. K. 1986. The role of movement and tissue interactions in the development and growth of bone and secondary cartilage in the clavicle of the embryonic chick. *J Embryol Exp Morphol* 93:133–152.

Hall, B. K. 1987a. Tissue interactions in the development and evolution of the vertebrate head. In: Maderson, P. F. A., ed. *Developmental and Evolutionary Aspects of the Neural Crest.* New York: John Wiley & Sons. pp. 215–259.

Hall, B. K. 1987b. Development of the mandibular skeleton in the embryonic chick as evaluated using the DNA-inhibiting agent 5-fluoro-2'-deoxyuridine (FUDR). *J Craniofac Genet Dev Biol* 7:145–160.

Hall, B. K. 1988a. Embryonic induction and the development of bone. *Am Sci* 76:174–181.

Hall, B. K. 1988b. Patterning of connective tissues in the head: Discussion report. *Development* 103(suppl):171–174.

Hall, B. K. 1990a. Genetic and epigenetic control of vertebrate development. Netherlands *J Zool* 40:362–361.

Hall, B. K. 1990b. Evolutionary issues in craniofacial biology. *Cleft Palate J* 27:95–100.

Hall, B. K. 1991a. Cellular interactions during cartilage and bone development. Proceedings of the NIDR-NIEH sponsored Conference on Research Advances in Prenatal Craniofacial Development. *J Craniofac Genet Dev Biol* 11:238–250.

Hall, B. K. 1991b. Evolution of connective and skeletal tissues. In: Hinchliffe, J. R., Hurle, J. M., and Summerbell, D., eds. *Developmental Patterning of the Vertebrate Limb.* New York: Plenum Press. NATO ASI Series A: Life Sciences Vol. 205. pp. 303–311.

Hall, B. K. 1994a. Embryonic bone formation with special reference to epithelial-mesenchymal interactions and growth factors. In: Hall, B. K., ed. *Bone Volume 8: Mechanisms of Bone Development and Growth.* Boca Raton: CRC Press. pp. 137–192.

Hall, B. K., ed. 1994b. *Homology: The Hierarchical Basis of Comparative Biology.* Boca Raton: Academic Press.

Hall, B. K., 1994c. Introduction. In: Hall, B. K., ed. *Homology: The Hierarchical Basis of Comparative Biology.* Boca Raton: Academic Press. pp. 1–19.

Hall, B. K. 1994d. Biology and mechanisms of tissue interactions in developing systems. In: Hodges, G. M., and Rowlatt, C., eds. *Developmental Biology and Cancer.* Boca Raton: CRC Press. pp. 161–185.

Hall, B. K. 1995. Homology and embryonic development. *Evol Biol* 28:1–37.

Hall, B. K. 1997. Bone, Embryonic development. In: *Encyclopedia of Human Biology, Vol. 2*. 2nd ed. San Diego: Academic Press. pp. 105–114.

Hall, B. K. 1998a. *Evolutionary Developmental Biology.* 2nd ed. London: Chapman & Hall.

Hall, B. K. 1998b. Germ layers and the germ-layer theory revisited: Primary and secondary germ layers, neural crest as a fourth germ layer, homology, demise of the germ-layer theory. *Evol Biol* 30:121–186.

Hall, B. K. 1999a. Developmental and cellular origins of the amphibian skeleton. In: Heatwole, H., and Davies, M., eds. *Amphibian Biology, Volume 6, Osteology.* Chipping Norton, NSW: Surrey Beatty & Sons. (in press).

Hall, B. K. 1999b. The evolution of the neural crest in vertebrates. In: Jacobson, C.- O., Olsson, L., and Laurent, T., eds. *Regulatory Processes in Development: The Legacy of Sven Hörstadius*. London: The Portland Press. (in press).

Hall, B. K. 1999c. Epithelial-mesenchymal interactions. In: Tuan, R. S., and Lo, C. W., eds. *Developmental Biology Protocols,* Vol. II. Totowa, NJ: Humana Press Inc. (in press).

Hall, B. K., and Ekanayake, S. 1991. Effects of growth factors on the differentiation of neural crest cells and neural crest cell-derivatives. *Int J Dev Biol* 35:367–386.

Hall, B. K., and Hanken, J. 1985. Foreword. In: de Beer, G. R. *The Development of the Vertebrate Skull*. Chicago: University of Chicago Press. pp. vii–xxviii.

Hall, B. K., and Hörstadius, S. 1988. *The Neural Crest.* Oxford: Oxford Press.

Hall, B. K., and Miyake, T. 1992. The membranous skeleton: The role of cell condensations in vertebrate skeletogenesis. *Anat Embryol* 186:107–124.

Hall, B. K., and Miyake, T. 1995a. How do embryos tell time? In: McNamara, K. J., ed. *Evolution Through Heterochrony.* Chichester: John Wiley & Sons. pp. 1–20.

Hall, B. K., and Miyake, T. 1995b. Divide, accumulate, differentiate: Cell condensation in skeletal development revisited. *Int J Dev Biol* 39:881–893.

Hall, B. K., and Miyake, T. 1999. Craniofacial development in avian and rodent embryos. In: Tuan, R. S., and Lo, C. W., eds. *Developmental Biology Protocols,* Vol. II. Totowa, NJ: Humana Press, Inc. (in press).

Hall, B. K., and Wake, M. H., eds. 1999. *The Origin and Evolution of Larval Forms.* San Diego: Academic Press.

Hallet, M.-M., and Ferrand, R. 1984. Quail melanoblast migration in two breeds of fowl and in their hybrids: Evidence for a dominant genic control of the mesodermal pigment cell pattern through the tissue environment. *J Exp Zool* 230:229–238.

Halstead, B. L. 1987. Evolutionary aspects of neural crest-derived skeletogenic cells in the earliest vertebrates. In: Maderson, P. F. A., ed. *Developmental and Evolutionary Aspects of the Neural Crest.* New York: John Wiley & Sons. pp. 339–358.

Hammerschmidt, M., Serbedzija, G. N., and McMahon, A. P. 1996. Genetic analysis of dorsoventral pattern formation in the zebrafish: Requirement of a BMP-like ventralizing activity and its dorsal repressor. *Genes Devel* 10:2452–2461.

Hammond, W. S., and Yntema, C. L. 1947. Depletions in the thoracolumbar sympathetic system following removal of neural crest in the chick. *J Comp Neurol* 86:237–265.

Hanken, J., and Hall, B. K. 1983. Evolution of the skeleton. *Nat Hist* 92(4):28–39.

Hanken, J., and Hall, B. K., eds. 1993. *The Vertebrate Skull. Volume I–3.* Chicago: University of Chicago Press.

Hanken, K., Jennings, D. H., and Olsson, L. 1997. Mechanistic basis of life-history evolution in anuran amphibians; Direct development. *Am Zool* 37:160–171.

Hanken, J., Klymkowsky, M. W., Summers, C. H., Seufert, D. W., and Ingebrigtsen, N. 1992. Cranial ontogeny in the direct-developing frog, *Eleutherodactylus coqui* (Anura: Leptodactylidae), analyzed using whole-mount immunohistochemistry. *J Morph* 211:95–118.

Hanken, J., and Thorogood, P. V. 1993. Evolution and development of the vertebrate skull: The role of pattern formation. *Trends Ecol Evol* 8:9–14.

Hardesty, R. A., and Marsh, J. L. 1990. Craniofacial onlay bone grafting—a prospective evaluation of graft morphology, orientation and embryonic origin. *Plast Reconst Surg* 85:5–14.

Hardisty, M. W. 1979. *Biology of the Cyclostomes*. London: Chapman & Hall.

Harris, M. J., and Juriloff, D. M. 1989. Test of the isoallele hypothesis at the mouse *First arch* (*far*) locus. *J Hered* 80:127–131.

Harrison, F. 1986. Autoradiographic study of the origin of basement membrane components in the avian embryo. In: Slavkin, H. C., ed. *Progress in Developmental Biology, Part B*. New York: Alan R. Liss, Inc. pp. 377–380.

Harrison, F. 1993. Cellular origin of the basement membrane in embryonic chicken/quail chimeras. *Int J Dev Biol* 37:337–347.

Harrison, R. G. 1904. Neue Versuche und Beobachtungen über die Entwicklung der peripheren Nerven der Wirbeltiere. *Sber Niederrhein Ges Nat-u Heilk* 35:55–62.

Harrison, R. G. 1907a. Experiments in transplanting limbs and their bearing upon the problems of the development of nerves. *J Exp Zool* 4:239–281.

Harrison, R. G. 1907b. Observations on the living developing nerve fiber. *Anat Rec* 1:116–118.

Harrison, R. G. 1910. The outgrowth of the nerve fiber as a mode of protoplasmic movement. *J Exp Zool* 9:787–846.

Harrison, R. G. 1924. Neuroblast versus sheath cell in the development of peripheral nerves. *J Comp Neurol* 37:123–206.

Harrison, R. G. 1929. Correlation in the development and growth of the eye studied by means of heteroplastic transplantation. *Wilhelm Roux Arch EntwMech Org* 120:1–55.

Harrison, R. G. 1935a. Heteroplastic grafting in embryology. *The Harvey Lectures* 29:116–157.

Harrison, R. G. 1935b. On the origin and development of the nervous system studied by the methods of experimental embryology. The Croonian Lecture. *Proc R Soc Lond B Biol Sci* 118:155–196.

Harrison, R. G. 1936. Relations of symmetry in the developing embryo. *The Coll Net* 11:217–226.

Harrison, R. G. 1938. Die Neuralleiste Erganzheft. *Anat Anz* 85:3–30.

Harrison, T. A., Stadt, H. A., Kumiski, D., and Kirby, M. L. 1995. Compensatory responses and development of the nodose ganglion following ablation of placodal precursors in the embryonic chick (*Gallus domesticus*). *Cell Tissue Res* 281:379–385.

Hart, R. C., McCue, P. A., Ragland, W. L., Winn, K. J., and Unger, E. R. 1990. Avian model for 13-cis-retinoic acid embryopathy: Demonstration of neural crest related defects. *Teratology* 41:463–472.

Harvey, A. R., Fan, Y., Beilharz, M. W., and Grounds, M. D. 1992. Survival and migration of transplanted male glia in adult female mouse brains monitored by a Y-chromosome-specific probe. *Mol Brain Res* 12:339–343.

Hassell, J. R., Greenberg, J. H., and Johnston, M. C. 1977. Inhibition of cranial neural crest cell development by vitamin A in cultured chick embryo. *J Embryol Exp Morphol* 39:267–271.

Hatada, Y., and Stern, C. D. 1994. A fate map of the epiblast of the early chick embryo. *Development* 120:2879–2889.

Hatano, M., Iitsuka, Y., Yamamoto, H., Dezawa, M., Yusa, S., Kohno, Y., and Tokuhisa, T. 1997. *Ncx*, a *Hox11* related gene, is expressed in a variety of tissues derived from neural crest cells. *Anat Embryol* 195:419–425.

Hawley, S. H. B., Wunnerberg-Stapleton, K., Hashimoto, C., Laurent, M. N., Watabe, T., Blumberg, B. W., and Cho, K. W. Y. 1995. Disruption of BMP signals in embryonic *Xenopus* ectoderm leads to direct neural induction. *Genes Devel* 9:2923–2935.

Hay, E. D. 1989. Extracellular matrix, cell skeletons, and embryonic development. *Am J Med Genet* 34:14–29.

Hayashi, A., and Arima, M. 1987. Cell culture study on neurofibromatosis. *Brain Dev* 9:588–592.

Heath, L., and Thorogood, P. 1989. Keratan sulfate expression during avian craniofacial morphogenesis. *Wilhelm Roux's Arch Dev Biol* 198:103–113.

Heath, L., Wild, A., and Thorogood, P. V. 1992. Monoclonal antibodies raised against premigratory neural crest reveal population heterogeneity during crest development. *Differentiation* 49:151–165.

Heine, V. I., Munoz, E. F., Flanders, K. C., Ellingsworth, L. R., Lam, H.-Y. P., Thompson, N. L., Roberts, S. B., and Sporn, M. B. 1987. Role of transforming growth factor-ß in the development of the mouse embryo. *J Cell Biol* 105:2861–2876.

Helff, O. M. 1940. Studies on amphibian metamorphosis. XVII. Influence of non-living annular tympanic cartilage on tympanic membrane formation. *J Exp Biol* 17:45–60.

Helson, L. 1984. Pre-Gardner's syndrome, thyroglossal cysts and undifferentiated tumor of neural crest origin. *Anticancer Res* 4:247–250.

Helwig, U., Imai, K., Schmahl, W., Thomas, B. E., Varnum, D. S., Nadeau, J. H., and Balling, R. 1995. Interaction between *Undulated* and *Patch* leads to an extreme form of spina bifida in double-mutant mice. *Nature Genet* 11:60–63.

Hemmati-Brivanlou, A., de la Torre, J. R., Holt, C., and Harland, R. M. 1991. Cephalic expression and molecular characterization of *Xenopus En-2*. *Development* 111:715–724.

Hemmati-Brivanlou, A., and Melton, D. 1997. Vertebrate embryonic cells will become nerve cells unless told otherwise. *Cell* 88:13–17.

Hemmati-Brivanlou, A., Stewart, R. M., and Harland, R. M. 1990. Region-specific neural induction of an engrailed protein by anterior notochord in *Xenopus*. *Science* 250:800–802.

Hemond, S. G., and Morest, D. K. 1991. Ganglion formation from the otic placode and the otic crest in the chick embryos: Mitosis, migration, and the basal lamina. *Anat Embryol* 184:1–13.

Henderson, D. J., and Copp, A. J. 1997. Role of the extracellular matrix in neural crest cell migration. *J Anat* 191:507–515.

Henion, P. D., Raible, D. W., Beattie, C. E., Stoesser, K. L., Weston, J. A., and Eisen, J. S. 1996. Screen for mutations affecting development of zebrafish neural crest. *Dev Genet* 18:11–17.

Henion, P. D., and Weston, J. A. 1994. Retinoic acid selectively promotes the survival and proliferation of neurogenic precursors in cultured neural crest cell populations. *Dev Biol* 161:243–250.

Henion, P. D., and Weston, J. A. 1997. Timing and pattern of cell fate restriction in the neural crest lineage. *Development* 124:4351–4359.

Hennig, W. 1966. *Phylogenetic Systematics*. Urbana: University of Illinois Press.

Henry, J. J., and Grainger, R. M. 1987. Inductive interactions in the spatial and temporal restriction of lens-forming potential in embryonic ectoderm of *Xenopus laevis*. *Dev Biol* 124:200–214.

Hertwig, O., and Hertwig, R. 1882. Die Coelomtheorie, Versuch einer Erklärung des mittleren Keimblattes. *Jenaische Zeit* 15:1–150.

Hill, D. P., and Robertson, K. A. 1997. Characterization of the cholinergic differentiation of the human neuroblastoma cell line La-N-5 after treatment with retinoic acid. *Dev Brain Res* 102:53–67.

Hill, J. P., and Watson, K. M. 1958. The early development of the brain in marsupials. *J Anat* 92:493–497.

Hill, R. E., Jones, P. F., Rees, A. R., Sime, C. M., Justice, M. J., Copeland, N. G., Jenkins, N. A., Graham, E., and Davidson, D. R. 1989. A new family of homo box-containing genes: Molecular structure, chromosomal location, and developmental expression of Hox-7.1 *Genes Dev* 3:26–37.

Hinchliffe, J. R. 1994. Evolutionary developmental biology of the tetrapod limb. *Development* (suppl):163–168.

Hinchliffe, J. R., and Johnson, D. R. 1980. *The Development of the Vertebrate Limb.* London: Oxford University Press.

Hinrichs, S. H., Nerenberg, M., Reynolds, R. K., Khoury, G., and Jay G. 1987. A transgenic mouse model for human neurofibromatosis. *Science* 237:1340–1343.

Hirano, S. 1986. Observations on the migration and differentiation of neural crest cells in somite extirpated salamander larvae. *Arch Hist Japon* 49:309–320.

Hirano, S., and Shirai, T. 1984. Morphogenetic studies on the neural crest of *Hynobius* larvae using vital staining and india ink labelling methods. *Arch Histol Japon* 47:57–70.

Hirata, M., Ito, K., and Tsuneki, K. 1997. Migration and colonization patterns of HNK-1-immunoreactive neural crest cells in lamprey and swordtail embryos. *Zool Sci* 14:305–312.

His, W. 1868. *Untersuchungen über die erste Anlage des Wirbeltierleibes. Die erste Entwicklung des Hühnchens im Ei.* Leipzig: F. C. W. Vogel.

His, W. 1870. Beschreibung eines Mikrotoms. *Arch Mikr Anat* 6:229–232.

His, W. 1874. *Unserer Körperform und das Physiologische Problem ihrer Entstehung.* Leipzig: Engelmann.

His, W. 1895a. Die anatomischer Nomenclature. Nomina anatomica. *Arch Ant Physiol Leipzig Anat Abth* (suppl):1–180.

His, W. 1895b. Anatomische Forschungen über Johann Sebastian Bach's Gebeine und Antlitz. *Abh Math-Phys Cl Königl Sächs Gesellsch Wissench* 22:379–420.

Ho, L., Symes, K., Yordan, C., Gudas, L. J., and Mercola, M. 1994. Localization of PDGFA and PDGFRa mRNA in *Xenopus* embryos suggests signalling from neural ectoderm and pharyngeal endoderm to neural crest cells. *Mech Dev* 48:165–174.

Hodges, G. M., and Rowlatt, C., eds. 1994. *Developmental Biology and Cancer.* Boca Raton: CRC Press.

Hoefnagel, C. A., Voûte, P. A., Kraker, J. de., and Marcuse, H. R. 1987. Radionuclide diagnosis and therapy of neural crest tumors using iodine ^{131}metaiodobenzyl-guanidine. *J Nucl Med* 28:308–314.

Holland, L. Z., and Holland, N. D. 1996. Expression of *AmphiHox-1* and *AmphiPax-1* in amphioxus embryos treated with retinoic acid: Insights into evolution and patterning of the chordate nerve cord and pharynx. *Development* 122:1829–1838.

Holland, L. Z., Kene, M., Williams, N. A., and Holland, N. D. 1997. Sequence and embryonic expression of the amphioxus *engrailed* gene (*AmphiEn*): The metameric pattern of transcription resembles that of its segment-polarity homolog in *Drosophila. Development* 124:1723–1732.

Holland, N. D. 1996. Homology, homeobox genes, and the early evolution of the vertebrates. In: Ghiselin, M. T., and Pinna, G., eds. *New Perspectives on the History of Life: Essays on Systematic Biology as Historical Narrative. Mem Calif Acad Sci* 20:63–70.

Holland, N. D., Holland, L. Z., Honma, Y., and Fujii, T. 1993. *Engrailed* expression during development of a lamprey, *Lampetra japonica*: A possible clue to homologies between Agnathan and gnathostome muscles of the mandibular arch. *Dev Growth Differ* 35:153–160.

Holland, N. D., Panganiban, G., Henyey, E. L., and Holland, L. Z. 1996. Sequence and developmental expression of *AmphiDll*, an amphioxus *Distal-less* gene transcribed in the ectoderm, epidermis and nervous system: Insights into evolution of craniate forebrain and neural crest. *Development* 122:2911–2920.

Holland, P. W. H. 1992. Homeobox genes in vertebrate evolution. *BioEssays* 14:267–273.

Holland, P. W. H. 1996. Molecular biology of lancelets: Insights into development and evolution. *Isr J Zool* 42:S247–S272.

Holland, P. W. H., and Garcia-Fernàndez, J. 1996. *Hox* genes and chordate evolution. *Dev Biol* 173:382–395.

Holland, P. W. H., and Graham, A. 1995. Evolution of regional identity in the vertebrate nervous system. *Perspect Dev Neurobiol* 3:17–27.

Holland, P. W. H., Holland, L. Z., Williams, N. A., and Holland, N. D. 1992. An amphioxus homeobox gene: Sequence conservation, spatial expression during development and insights into vertebrate evolution. *Development* 116:653–661.

Holley, S. A., and Ferguson, E. L. 1997. Fish are like flies are like frogs: Conservation of dorsal-ventral patterning mechanisms. *BioEssays* 19:281–284.

Holley, S. A., Jackson, P. D., Sasai, Y., Lu, B., De Robertis, E. M., Hoffmann, F. M., and Ferguson, E. L. 1995. A conserved system for dorsal-ventral patterning in insects and vertebrates involving *Sog* and *Chordin*. *Nature* 376:249–253.

Hollinshead, H. 1940. Chromaffin tissue and paraganglia. *Q Rev Biol* 15:156–171.

Holmdahl, D. E., 1925a. Experimentelle Untersuchungen über die Lage der Grenze zwischen primärer und sekundärer Körperentwicklung beim Huhn. *Anat Anz* 59:393–396.

Holmdahl, D. E., 1925b. Die erst Entwicklung des Körpels bei den Vögeln und Säugetoeren inkl. dem Menschen, besonders mit Rücksicht auf die Bildung des Rückenmarks, des Zoloms und der entodermalen Kloake nebst einem Exkürs über die Entstehung der Spina bifida in der Lumbosakralregion. I. *Gegenbaurs Morphol Jahrb* 54:333–384.

Holmdahl, D. E., 1925c. Die erst Entwicklung des Körpels bei den Vögeln und Säugetoeren inkl. dem Menschen, besonders mit Rücksicht auf die Bildung des Rückenmarks, des Zoloms und der entodermalen Kloake nebst einem Exkürs über die Entstehung der Spina bifida in der Lumbosakralregion. II-V. *Gegenbaurs Morphol Jahrb* 55:112–208.

Holmdahl, D. E. 1928. Die Enstehung und weitere Entwicklung der Neuralleiste (Ganglienleiste) bei Vogeln und Saugetieren. *Z Mikrosk-Anat Forsch* 14:99–298.

Holmdahl, D. E. 1934. Neuralleiste und Ganglienleiste beim Menschen. *Z Mikrosk-Anat Forsch* 36:137–178.

Holmgren, N. 1940. Studies on the head in fishes. Embryological, morphological, and phylogenetic researches. Part I: Development of the skull in sharks and rays. *Acta Zool* 21:51–267.

Holmgren, N. 1943. Studies on the head in fishes. Embryological, morphological, and phylogenetic researches. Part IV: General morphology of the head in fish. *Acta Zool* 24:1–188.

Holtfreter, J. 1929. Uber die Aufzucht isolierter Teile des Amphibienkeims. I. Methode einer Gewebezüchtung *in vivo. Wilhelm Roux Arch EntwMech Org* 117:421–510.

Holtfreter, J. 1933. Der Einfluss von Wirtsalter und verschiedenen Organbezirken aud die Differenzierung von angelagerten Gastrulaektoderm.*Wilhelm Roux Arch EntwMech Org 127*:619–775.

Holtfreter, J. 1935a. Morphologische Beeinflussung von Urodelenektoderm bei xenoplastischer Transplantation. *Wilhelm Roux Arch EntwMech Org* 133:367–426.

Holtfreter, J. 1935b. Über das Verhalten von Anurenektoderm in Urodelenkeimen. *Wilhelm Roux Arch EntwMech Org* 133:427–494.

Holtfreter, J. 1936. Regionale Induktionen in xenoplastisch zusammengesetzten Explantaten. *Wilhelm Roux Arch EntwMech Org* 134:466–550.

Holtfreter, J. 1968. Mesenchyme and epithelia in inductive and morphogenetic processes. In: Fleischmajer, R., and Billingham, R. E., eds. *Epithelial-Mesenchymal Interactions.* Baltimore: Williams & Wilkins. pp. 1–30.

Holtfreter, J., and Hamburger, V. 1955. Amphibians. In: Willier, B. H., Weiss, P. A., and Hamburger, V., eds. *Analysis of Development.* Philadelphia: W. B. Saunders Co. Reprinted 1971, New York: Hafner Publishing Co. pp. 230–296.

Hood, L. C., and Rosenquist, T. H. 1992. Coronary artery development in the chick: Origin and deployment of smooth muscle cells, and the effects of neural crest ablation. *Anat Rec* 234:291–300.

Hope, D. G., and Mulvihill, J. J. 1981. Malignancy in neurofibromatosis. In: Riccardi, V. M., and Mulvihill, J. J., eds. *Neurofibromatosis (von Recklinghausen Disease).* New York: Raven Press. pp. 33–56.

Hörstadius, S. 1928. Über die determination des Keimes bei Echinodermen. *Acta Zool* 9:1–191.

Hörstadius, S. 1939. The mechanics of sea urchin development studies by operative methods. *Biol Rev Camb Philos Soc* 14:132–179.

Hörstadius, S. 1950. *The Neural Crest: Its Properties and Derivatives in the Light of Experimental Research.* Oxford: Oxford University Press.

Hörstadius, S. 1973. *Experimental Embryology of Echinoderms.* Oxford: Clarendon Press.

Hörstadius, S., and Sellman, S. 1941. Experimental studies on the determination of the chondrocranium in Amblystoma mexicanum. *Ark Zool* 33A(13):1–8

Hörstadius, S., and Sellman, S. 1946. Experimentelle untersuchungen über die Determination des Knorpeligen Kopfskelettes bei Urodelen. *Nova Acta R Soc Scient Ups* Ser. 4, 13:1–170.

Hosoda, K., Hammer, R. E., Richardson, J. A., Baynish, A. G., Cheung, J. C., Giaid, A., and Yanagisawa, M. 1994. Targeted and natural (*Piebald-lethal*) mutations of endothelin-B receptor gene produce megacolon associated with spotted coat color in mice. *Cell* 79:1267–1276.

Hou, L., and Takeuchi, T. 1992. Differentiation of reptilian neural crest cells *in vitro. In Vitro Cell Dev Biol Anim* 28:348–354.

Hou, L., and Takeuchi, T. 1994. Neural crest development in reptilian embryos: Studies with monoclonal antibody, HNK-1. *Zool Sci* 11:423–431.

Howard, M. J., and Bronner-Fraser, M. 1985. The influence of neural tube-derived factors in differentiation of neural crest cells in vitro. I. Histochemical study of the appearance of adrenergic cells. *J Neurosci* 5:3302–3309.

Howell, M., and Ford, P. 1980. *The True History of the Elephant Man.* Bungay, Suffolk: Richard Clay Ltd.

Howes, R. I. Jr. 1977. Root formation in ectopically transplanted teeth of the frog, *Rana pipiens*, I. Tooth morphogenesis. *Acta Anat* 97:151–165.
Howes, R. I. Jr. 1978a. Root formation in ectopically transplanted teeth of the frog, *Rana pipiens*, II. Comparative aspects of the root tissues. *Acta Anat* 100:461–470.
Howes, R. I. Jr. 1978b. Regeneration of ankylosed teeth in the adult frog premaxilla. *Acta Anat* 101:179–186.
Howes, R. I. Jr., and Eakers, E. C. 1984. Augmentation of tooth and jaw regeneration in the frog with a digital transplant. *J Dental Res* 63:670–674.
Huang, R., Zhi, Q., Ordahl, C. P., and Christ, B. 1997. The fate of the first somite. *Anat Embryol* 195:435–449.
Hughes, A. 1957. The development of the primary sensory system in *Xenopus laevis* (Daudin). *J Anat* 91:323–338.
Hunt, P., Ferretti, P., Krumlauf, R., and Thorogood, P. 1995. Restoration of normal *Hox* code and branchial arch morphogenesis after extensive deletion of hindbrain neural crest. *Dev Biol* 168:584–597.
Hunt, P., Gulisano, M., Cook, M., Sham, M.-H., Faiella, A., Wilkinson, D., Boncinelli, E., and Krumlauf, R. 1991a. A distinct *Hox* code for the branchial region of the vertebrate head. *Nature* 353:861–864.
Hunt, P., and Krumlauf, R. 1992. *Hox* codes and positional specification in vertebrate embryonic axes. *Ann Rev Cell* 8:227–256.
Hunt, P., Whiting, J., Muchamore, I., Marshall, H., and Krumlauf, R. 1991b. Homeobox genes and models for patterning the hindbrain and branchial arches. *Development* (suppl 1):187–196.
Hunt, P., Whiting, J., Nonchev, S., Sham, M.-H., Marshall, H., Graham, A., Cook, M., Allemann, R., Rigby, P. W. J., Gulisano, M., Faiella, A., Boncinelli, E., and Krumlauf, R. 1991c. The branchial *Hox* code and its implications for gene regulation, patterning of the nervous system and head evolution. *Development* (suppl 2):63–77.
Huson, S. M. 1987. The First European Symposium on Neurofibromatosis. *J Med Genet* 24:572–573.
Huszar, D., Sharpe, A., Hashmi, S., Bouchard, B., Houghton, A., and Jaenisch, R. 1991. Generation of pigmented stripes in albino mice by retroviral marking of neural crest melanoblasts. *Development* 113:653–660.
Huszar, D., Sharpe, A., and Jaenisch, R. 1991. Migration and proliferation of cultured neural crest cells in W mutant neural crest chimeras. *Development* 112:131–141.
Huxley, T. H. 1849. On the anatomy and the affinities of the family of the Medusæ. *Philos Trans R Soc Lond B Biol Sci* 139:413–434.
Huxley, T. H. 1871. *A Manual of the Anatomy of Vertebrated Animals*. London.
Hyman, L. H. 1940. *The Invertebrates; Protozoa through Ctenophora*. New York: McGraw Hill.
Hyman, L. H. 1951. *The Invertebrates; Platyhelminthes and Rhynchocoela*. New York: McGraw Hill.
Ikeya, M., Lee, S. M. K., Johnson, J. E., McMahon, A. P., and Takada, S. 1997. *Wnt* signalling required for expansion of neural crest and CNS progenitors. *Nature* 389:966–970.
Imai, H., Osumi-Yamashita, N., Ninomiya, Y., and Eto, K. 1996. Contribution of early-migrating midbrain crest cells to the dental mesenchyme of mandibular molar teeth in rat embryos. *Dev Biol* 176:151–165.
Innes, P. B. 1985. The ultrastructure of early cephalic neural crest cell migration in the mouse. *Anat Embryol* 172:33–38.

Ishibe, T., Cremer, M. A., and Yoo, T. J. 1989. Type II collagen distribution in the ear of the guinea pig fetus. *Ann Otol Rhinol Laryngol* 98:648–654.

Ito, K., and Morita, T. 1995. Role of retinoic acid in mouse neural crest cell development *in vitro. Dev Dyn* 204:211–218.

Ito, K., Morita, T., and Sieber-Blum, M. 1993. *In vitro* clonal analysis of mouse neural crest development. *Dev Biol* 157:517–525.

Ito, K., and Sieber-Blum, M. 1991. *In vitro* clonal analysis of quail cardiac neural crest development. *Dev Biol* 148:95–106.

Ito, K., and Sieber-Blum, M. 1993. Pluripotentiality and developmentally restricted neural crest-derived cells in posterior visceral arches. *Dev Biol* 156:191–200.

Ito, K., and Takeuchi, T. 1984. The differentiation *in vitro* of the neural crest cells of the mouse embryo. *J Embryol Exp Morphol* 84:49–62.

Jablonski, N. G. 1992. Sun, skin colour and spina bifida: An exploration of the relationship between ultraviolet light and neural tube defects. *Proc Australasian Soc Human Biol* 5:455–462.

Jacobs-Cohen, R. J., Payette, R. F., Gershon, M. D., and Rothman, T. P. 1987. Inability of neural crest cells to colonize the presumptive aganglionic bowel of *Ls/Ls* mutant mice: Requirement for a permissive microenvironment. *J Comp Neurol* 255:425–438.

Jacobson, A. G. 1966. Inductive processes in embryonic development. *Science* 152:25–34.

Jacobson, A. G. 1987. Determination and morphogenesis of axial structures: Mesodermal metamerism, shaping of the neural plate and tube, and segregation and functions of the neural crest. In: Maderson, P. F. A., ed. *Developmental and Evolutionary Aspects of the Neural Crest*pp. New York: John Wiley & Sons. pp. 147–180.

Jacobson, A. G., and Duncan, J. T. 1968. Heart induction in salamanders. *J Exp Zool* 167:79–103.

Jacobson, A. G., and Meier, S. 1984. Morphogenesis of the head of a newt: Mesodermal segments, neuromeres, and distribution of neural crest. *Dev Biol* 106:181–193.

Jacobson, A. G., and Moury, J. D. 1995. Tissue boundaries and cell behavior during neurulation. *Dev Biol* 171:98–110.

Jacobson, A. G., and Sater, A. K. 1988. Features of embryonic induction. *Development* 104:341–359.

Jacobson, M. 1991. *Developmental Neurobiology.* 3rd ed. New York: Plenum Press.

Jacobson, M., and Moody, S. A. 1984. Quantitative lineage analysis of the frog's nervous system. I. Lineages of Rohon-Béard neurons and primary motoneurons. *J Neurosci* 4:1361–1369.

Jaenisch, R. 1985. Mammalian neural crest cells participate in normal embryonic development on microinjection into post-implantation mouse embryos. *Nature* 318:181–183.

Jaffe, R., Santamaria, M., Yunis, E. J., Tannery, N. H., Agostini, R. M. Jr., Medina, J., and Goodman, M. 1984. The neuroectodermal tumor of bone. *Am J Surg Path* 8:885–898.

Jahn, A. F., and Ganti, K. 1987. Major auricular malformations due to Accutane (isotretinoin). *Laryngoscope* 97:832–835.

Jamrich, M., and Sato, S. 1989. Differential gene expression in the anterior neural plate during gastrulation of *Xenopus laevis. Development* 105:779–786.

Janvier, P. 1996a. *Early Vertebrates.* Oxford: Clarendon Press.

Janvier, P. 1996b. Fishy fragments tip the scales. *Nature* 383:757–758.

Jarvis, B. L., Johnston, M. C., and Sulik, K. K. 1990. Congenital malformations of the external, middle and inner ear produced by isotretinoin exposure in mouse embryos. *Otolaryngol Head Neck Surg* 102:391–401.

Jaskoll, T., Greenberg, G., and Melnick, M. 1991. Neural tube and neural crest: A new view with time-lapse high-definition photomicroscopy. *Am J Med Genet* 41:333–345.

Jeffery, W. R. 1997. Evolution of ascidian development. *Bioscience* 47:417–425.

Jeffs, P., Jaques, K., and Osmond, M. 1992. Cell death in cranial neural crest development. *Anat Embryol* 185:583–588.

Jensen, N. A., Rodriguez, M. L., Garvey, J. S., Miller, C. A., and Hood, L. 1993. Transgenic mouse model for neurocristopathy. Schwannomas and facial bone tumors. *Proc Natl Acad Sci USA* 90:3192–3196.

Jessell, T. M., and Melton, D. A. 1992. Diffusible factors in vertebrate embryonic induction. *Cell* 68:257–270.

Jesuthasan, S. 1996. Contact inhibition/collapse and path finding of neural crest cells in the zebrafish trunk. *Development* 122:381–389.

Jiang, R., Lan, Y., Norton, C. R., Sundberg, J. P., and Gridley, T. 1998. The *Slug* gene is not essential for mesoderm or neural crest development in mice. *Dev Biol* 198:277–285.

Johansen, K., and Strahan, R. 1963. The respiratory system of *Myxine glutinosa* L. In: Brodal, A., and Fänge, R., eds. *The Biology of Myxine*. Oslo: Grödahl and Sons. pp. 352–371.

Johnels, A. G. 1948. On the development and morphology of the skeleton of the head of *Petromyzon*. *Acta Zool Stockh* 29:139–279.

Johnson, D. R. 1986. *The Genetics of the Skeleton. Animal Models of Skeletal Development*. Oxford: Clarendon Press.

Johnston, M. C. 1966. A radioautographic study of the migration and fate of cranial neural crest cells in the chick embryo. *Anat Rec* 156:143–156.

Johnston, M. C. 1975. The neural crest in abnormalities of the face and brain. *Birth Defects Orig Art Ser* 11:1–18.

Johnston, M. C., and Bronsky, P. T. 1991. Animal models for human craniofacial malformations. *J Craniofac Genet Dev Biol* 11:277–291.

Johnston, M. C., and Bronsky, P. T. 1995. Prenatal craniofacial development: New insights on normal and abnormal mechanisms. *Crit Rev Oral Biol Med* 6:25–79.

Johnston, M. C., and Listgarten, M. A. 1972. Observations on the migration, interaction and early differentiation of orofacial tissues. In: Slavkin, H. C., and Bavetta, L. A., eds. *Developmental Aspects of Oral Biology*. New York: Academic Press. pp. 53–80.

Johnston, M. C., Morriss, G. M., Kushner, D., and Bingle, G. J. 1977. Abnormal organogenesis of facial structures. In: Wilson, J. G., and Fraser, F. C., eds. *Handbook of Teratology, Vol. 2*. New York: Plenum Press. pp. 421–451.

Johnston, M. C., Noden, D. M., Hazelton, R. D., Coulombre, J. L., and Coulombre, A. J. 1979. Origins of avian ocular and periocular tissues. *Exp Eye Res* 29:27–45.

Johnston, M. C., and Sulik, K. K. 1979. Some abnormal patterns of development in the craniofacial region. *Birth Defects Orig Art Ser* 15:23–42.

Johnston, M. C., Sulik, K. K., and Webster, W. S. 1985. Altered mouse neural crest development resulting from maternally-administered 13-cis-retinoic acid (Accutane®). *J Dental Res* 64:270.

Johnston, M. C., Sulik, K. K., and Webster, W. S. 1986. Administration of 13-cis-retinoic acid to pregnant mice provides insights into the pathogenesis of human craniofacial malformations. *J Dental Res* 65:847.

Johnston, M. C., Vig, K. W. L., and Ambrose, L. J. H. 1981. Neurocristopathy as a unifying concept: Clinical correlations. In: Mulvihill, J. J., and Riccardi, V. M., eds. *Neurofi-*

bromatosis: Genetics, Cell Biology and Biochemistry. New York: Raven Press. pp. 97–104.

Jones, C. M., Lyons, K. M., and Hogan, B. L. M. 1991. Involvement of bone morphogenetic protein-4 (BMP-4) and Vgr-1 in morphogenesis and neurogenesis in the mouse. *Development* 111:531–542.

Jones, D. S. 1937. The origin of the sympathetic trunks in the chick embryo. *Anat Rec* 70:45–62.

Jones, D. S. 1939. Studies on the origin of sheath cells and sympathetic ganglia in the chick. *Anat Rec* 73:343–357.

Jones, D. S. 1941. Further studies on the origin of sympathetic ganglia in the chick embryo. *Anat Rec* 79:7–13.

Jones, D. S. 1942. The origin of the vagi and the parasympathetic ganglion cells of the viscera of the chick. *Anat Rec* 82:185–195.

Jones, M. C. 1990. The neurocristopathies: Reinterpretation based upon the mechanism of abnormal morphogenesis. *Cleft Palate J* 27:136–140.

Joyner, A. L. 1996. *Engrailed, Wnt* and *Pax* genes regulate midbrain-hindbrain development. *Trends Genet* 12:15–20.

Jungalwala, F. B., Chou, D. K. H., Suzuki, Y., and Maxwell, G. D. 1992. Temporal expression of HNK1-reactive sulfoglucuronyl glycolipid in cultured quail trunk neural crest cells—comparison with other developmentally regulated glycolipids. *J Neurochem* 58:1045–1051.

Juriloff, D. M., Harris, M. J., and Froster-Iskenius, U. 1987. Hemifacial deficiency induced by a shift in dominance of the mouse mutation *far*: A possible genetic model for hemifacial microsomia. *J Craniofac Genet Dev Biol* 7:27–44.

Kaan, H. W. 1938. Further studies on the auditory vesicle and cartilaginous capsule of *Amblystoma punctatum. J Exp Zool* 78:159–183.

Kalcheim, C., and Le Douarin, N. M. 1986. Requirement of a neural tube signal for the differentiation of neural crest cells into dorsal root ganglia. *Dev Biol* 116:451–466.

Kalcheim, C., and Leviel, V. 1988. Stimulation of collagen production *in vitro* by ascorbic acid released from explants of migrating avian neural crest. *Cell Differ* 22:107–114.

Kallen, B. 1953. Notes on the development of the neural crest in the head of *Mus musculus. J Embryol Exp Morphol* 1:393–398.

Kalter, H. 1980. A compendium of the genetically induced congenital malformations of the house mouse. *Teratology* 21:397–429.

Kameda, Y. 1995. Evidence to support the distal vagal ganglion as the origin of C cells of the ultimobranchial gland in the chick. *J Comp Neurol* 359:1–14.

Kanki, J. P., and Ho, R. K. 1997. The development of the posterior body in zebrafish. *Development* 124:881–893.

Kanzler, B., Kuschert, S. J., Liu, Y.-H., and Mallo, M. 1998. *Hoxa-2* restricts the chondrogenic domain and inhibits bone formation during development of the branchial area. *Development* 125:2587–2597.

Kappetein, A. P., Gittenbe, A. C., Zwinerdam, A. H., Rohmer, J., Poelmann, R. E., and Huysmans, H. A. 1991. The neural crest as a possible pathogenetic factor in coarctation of the aorta and bicuspid aortic valve. *J Thor Surg* 102:830–836.

Kapron-Bras, C. M., and Trasler, D. G. 1985. Reduction in the frequency of neural tube defects in *Splotch* mice by retinoic acid. *Teratology* 32:87–92.

Kasperk, C., Wergedal, J., Strong, D., Farley, J., Wangerin, K., Gropp, H., Ziegler, R., and Baylink, D. J. 1995. Human bone cell phenotypes differ depending on their skeletal site of origin. *J Clin Endocrinol Metab* 80:2511–2517.

Kastschenko, N. 1888. Zur Entwicklungsgeschichte der Selachierembryos. *Anat Anz* 3:445–467.

Kavka, A. I., and Barald, K. F. 1994. The role of neurofibromatosis-1 (NF1) in the developing nervous system and neural crest. *Mol Biol Cell* 5(suppl):P2321.

Kay, E. D. 1986. The phenotypic interdependence of the musculoskeletal characters of the mandibular arch in mice. *J Embryol Exp Morphol* 98:123–136.

Kay, E. D. 1987. Craniofacial dysmorphogenesis following hypervitaminosis A in mice. *Teratology* 35:105–117.

Kay, E. D., and Kay, C. H. 1989. Dysmorphogenesis of the mandible, zygoma and middle ear ossicles in hemifacial microsomia and mandibulofacial dysostosis. *Am J Med Genet* 32:27–31.

Keith, J. 1977. Effects of excess vitamin A on the cranial neural crest in the chick embryo. *Ann R Coll Surg* 59:479–483.

Kelsh, R. N., Brand, M., Jiang, Y.-J., Heisenberg, C.-P., Lin, S., Haffter, P., Odenthal, J., Mullins, M. C., van Eeden, F. J. M., Furutani-Seiki, M., Granato, M., Hammerschmidt, M., Kane, D. A., Warga, R. M., Beuchle, D., Vogelsang, L., and Nüsslein-Volhard, C. 1996. Zebrafish pigmentation mutations and the processes of neural crest development. *Development* 123:369–389.

Kemp, A. 1990. Involvement of the neural crest in development of the Australian lungfish *Neoceratodus forsterei* (Krefft, 1870). *Mem Queensland Mus* 28:101–102.

Kemshead, J. T., Goldman, A., Fritschy, J., Malpas, J. S., and Pritchard, J. 1983. Use of monoclonal antibodies in the differential diagnosis of neuroblastoma and lymphoblastic disorders. *Lancet* i:12–15.

Kengaku, M., and Okamoto, H. 1993. Basic fibroblast growth factor induces differentiation of neural tube and neural crest lineages of cultured ectodermal cells from *Xenopus* gastrulae. *Development* 119:1067–1078.

Kerr, J. G. 1919. *Text-Book of Embryology. Volume II. Vertebrata with the Exception of Mammalia.* London: Macmillan and Co.

Kerr, R. S. E., and Newgreen, D. F. 1997. Isolation and characterization of chondroitin sulfate proteoglycans from embryonic quail that influence neural crest cell behavior. *Dev Biol* 192:108–124.

Kessel, M. 1992. Respecification of vertebral identities by retinoic acid. *Development* 115:487–501.

Kessel, M., and Gruss, P. 1990. Murine developmental control genes. *Science* 249:374–379.

Kessel, M., and Gruss, P. 1991. Homeotic transformations of murine vertebrae and concomitant alteration of *Hox* codes induced by retinoic acid. *Cell* 67:89–104.

Khare, M. K., and Choudhury, S. 1985. Establishment of the neural differentiation pattern in the prospective prosencephalic ectoderm of the chick embryo (*Gallus domesticus*). *Dev Growth Differ* 27:83–93.

Kil, S. H., Lallier, T., and Bronner-Fraser, M. 1996. Inhibition of cranial neural crest adhesion *in vitro* and migration *in vivo* using integrin antisense oligonucleotides. *Dev Biol* 179:91–101.

Kilcoyne, R. F., Cope, R., Cunningham, W., Nardella, F. A., Denman, S., Franz, T. J., and Hanifin, J. 1986. Minimal spinal hyperostosis with low-dose isotretinoin therapy. *Invest Radiol* 21:41–44.

Kimura, C., Takeda, N., Suzuki, M., Oshimura, M., Aizawa, S., and Matsuo, I. 1997. *Cis*-acting elements conserved between mouse and pufferfish *Otx2* genes govern the expression in mesencephalic neural crest cells. *Development* 124:3929–3941.

Kindblom, L.-G., Lodding, P., and Angervall, L. 1983. Clear-cell sarcoma of tendons and aponeuroses. An immunohistochemical and electron microscopic analysis indicating neural crest origin. *Virchows Arch Path Anat Histopath* 401:109–128.

Kingsley, D. M. 1994. The TGF-ß superfamily: New members, new receptors, and new genetic tests of function in different organisms. *Genes Dev* 8:133–146.

Kinutani, M., Tan, K., Desaki, J., Coltey, M., Kitaoka, K., Nagano, Y., Takashima, Y., and Le Douarin, N. M. 1989. Avian spinal cord chimeras. Further studies on the neurological syndrome affecting the chimeras after birth. *Cell Differ Dev* 26:145–162.

Kirby, M. L. 1988a. Nodose placode provides ectomesenchyme to the developing chick heart in the absence of cardiac neural crest. *Cell Tissue Res* 252:17–22.

Kirby, M. L. 1988b. Nodose placode contributes autonomic neurons to the heart in the absence of cardiac neural crest. *J Neurosci* 8:1089–1095.

Kirby, M. L. 1989. Plasticity and predetermination of mesencephalic and trunk neural crest transplanted into the region of the cardiac neural crest. *Dev Biol* 134:402–412.

Kirby, M. L. 1993. Cellular and molecular contributions of the cardiac neural crest to cardiovascular development. *Trends Cardiovasc Med* 3:18–23.

Kirby, M. L., Aronstam, R. S., and Buccafusco, J. J. 1985. Changes in cholinergic parameters associated with failure of conotruncal septation in embryonic chick hearts after neural crest ablation. *Circ Res* 56:392–401.

Kirby, M. L., and Bockman, D. E. 1984. Neural crest and normal development: A new perspective. *Anat Rec* 209:1–6.

Kirby, M. L., Creazzo, T. L., and Christiansen, J. L. 1989. Chronotropic responses of chick atria to field stimulation after various neural crest ablations. *Circ Res* 65:1547–1554.

Kirby, M. L., Gale, T. F., and Stewart, D. E. 1983. Neural crest cells contribute to normal aorticopulmonary septation. *Science* 220:1059–1061.

Kirby, M. L., Kumiski, D. H., Myers, T., Cerjan, C., and Mishima, N. 1993. Back transplantation of chick cardiac neural crest cells cultured in LIF rescues heart development. *Dev Dyn* 198:296–311.

Kirby, M. L., and Stewart, D. E. 1983. Neural crest origin of cardiac ganglion cells in the chick embryo: Identification and extirpation. *Dev Biol* 97:433–443.

Kirby, M. L., and Stewart, D. E. 1984. Adrenergic innervation of the developing chick heart: Neural crest ablations to produce sympathetically aneural hearts. *Am J Anat* 171:295–305.

Kirby, M. L., Turnage, K. L. III., and Hays, B. M. 1985. Characterization of conotruncal malformations following ablation of "cardiac" neural crest. *Anat Rec* 213:87–93.

Kirby, M. L., and Waldo, K. L. 1990. Role of neural crest in congenital heart disease. *Circulation* 82:332–340.

Kirby, M. L., and Waldo, K. L. 1995. Neural crest and cardiovascular patterning. *Circ Res* 77:211–215.

Kirchgessner, A. L., Adlersberg, M. A., and Gershon, M. D. 1992. Colonization of the developing pancreas by neural precursors from the bowel. *Dev Dyn* 194:142–154.

Kjaer, I. 1998. Neuro-osteology. *Crit Rev Oral Biol Med* 9: 224–244.

Klima, M., and Bangma, G. C. 1987. Unpublished drawings of marsupial embryos from the Hill collection and some problems of marsupial ontogeny. *Z Säugetierkunde* 52:201–211.

Kochhar, D. M., and Christian, M. S. 1997. Tretinoin: A review of the nonclinical developmental toxicology experience. *J Am Acad Dermatol* 36:S47–S59.

Kochhar, D. M., Jiang, H., Penner, J. D., Beard, R. L., and Chandraratna, A. S. 1996. Differential teratogenic response of mouse embryos to receptor selective analogs of retinoic acid. *Chem Biol Interact* 100:1–12.

Kollar, E. J. 1983. Epithelial-mesenchymal interactions in the mammalian integument: Tooth development as a model for instructive induction. In: Sawyer, R. H., and Fallon, J. F., eds. *Epithelial-Mesenchymal Interactions in Development.* New York: Praeger Publishing Co. pp. 27–50.

Kollar, E. J., and Baird, G. R. 1969. The influence of the dental papilla on the development of tooth shape in embryonic mouse tooth germs. *J Embryol Exp Morphol* 21:131–148.

Koltzoff, N. K. 1901. Entwicklungsgeschichte des Kopfes von *Petromyzon planeri. Bull Soc Imp Nat Moscow* 15:259–589.

Köntges, G., and Lumsden, A. 1996. Rhombencephalic neural crest segmentation is preserved throughout craniofacial ontogeny. *Development* 122:3229–3242.

Koole, R. 1994a. Ectomesenchymal mandibular symphysis bone graft—an improvement in alveolar cleft grafting. *Cleft Palate Craniofac J* 31:217–223.

Koole, R. 1994b. *The Bone Graft in the Alveolar Cleft.* Ph.D. diss., University of Utrecht. pp. 1–180.

Koole, R., Bosker, H., and van der Dussen, F. N. 1989. Late secondary autogenous bone grafting in cleft patients comparing mandibular (ectomesenchymal) and iliac crest (mesenchymal) grafts. *J Craniomaxillofac Surg* 17:28–30.

Korade, Z., and Frank, E. 1996. Restriction in cell fate of developing spinal cord cells transplanted to neural crest pathways. *J Neurosci* 16:7638–7648.

Kostovic-Knezevic, L., Gajovic, S., and Svajger, A. 1991. Morphogenetic features in the tail region of the rat embryo. *Int J Dev Biol* 35:191–195.

Koumans, J. T. M., and Sire, J.-Y. 1996. An *in vitro*, serum-free organ culture technique for the study of development and growth of the dermal skeleton in fish. *In Vitro Cell Dev Biol Anim* 32:612–626.

Kraemer, M. 1974. La morphogénèse du chondrocrâne de *Discoglossus pictus* Otth (Amphibiens, anoure). *Bull Biol* 108:211–228.

Krauss, S., Maden, M., Holder, N., and Wilson, S. W. 1992. Zebrafish *Pax* [b] is involved in the formation of the midbrain-hindbrain boundary. *Nature* 360:87–89.

Krotoski, D., and Bronner-Fraser, M. 1986. Mapping of neural crest pathways in *Xenopus laevis.* In: Slavkin ,H. C., ed. *Progress in Developmental Biology, Part B.* New York: Alan R. Liss, Inc. pp. 229–233.

Krotoski, D., and Bronner-Fraser, M. 1990. Distribution of integrins and their ligands in the trunk of *Xenopus laevis* during neural crest cell migration. *J Exp Zool* 253:139–150.

Krotoski, D. M., Domingo, C., and Bronner-Fraser, M. 1986. Distribution of a putative cell surface receptor for fibronectin and laminin in the avian embryo. *J Cell Biol* 103:1061–1071.

Krotoski, D. M., Fraser, S. E., and Bronner-Fraser, M. 1988. Mapping of neural crest pathways in *Xenopus laevis* using inter- and intra-specific cell markers. *Dev Biol* 127:119–132.

Krull, C. E., Collazo, A., Fraser, S. E., and Bronner-Fraser, M. 1995. Segmental migration of trunk neural crest: Time-lapse analysis reveals a role for PNA-binding molecules. *Development* 121:3733–3743.

Krull, C. E., Lansford, R., Gale, N. W., Collazo, A., Marcelle, C., Yancopoulos, G. D., Fraser, S. E., and Bronner-Fraser, M. 1997. Interactions of Eph-related receptors and ligands confer rostrocaudal pattern to trunk neural crest migration. *Curr Biol* 1:571–580.

Krumlauf, R. 1993. *Hox* genes and pattern formation in the branchial region of the vertebrate head. *Trends Genet* 9:106–112.

Krumlauf, R., Hunt, P., Graham, A., and Wilkinson, D. 1991. Patterning regional identity: Spatially-restricted and dynamic expression patterns of *Krox-20* and *Hox* genes in the developing nervous system. *Semin Dev Biol* 2:375–384.

Kubota, Y., Morita, T., and Ito, K. 1996. New monoclonal antibody (4Egr) identifies mouse neural crest cells. *Dev Dyn* 206:368–378.

Kuntz, A. 1922. Experimental studies on the histogenesis of the sympathetic nervous system. *J Comp Neurol* 34:1–36.

Kuntz, A., and Batson, O. V. 1920. Experimental observations on the histogenesis of the sympathetic trunks in the chick. *J Comp Neurol* 32:335–345.

Kuratani, S. C. 1991. Alternate expression of the HNK-1 epitope in rhombomeres of the chick embryo. *Dev Biol* 144:215–219.

Kuratani, S. C. 1997. Spatial distribution of postotic crest cells defines the head/trunk interface of the vertebrate body: Embryological interpretation of peripheral nerve morphology and evolution of the vertebrate head. *Anat Embryol* 195:1–13.

Kuratani, S. C., and Aizawa, S. 1995. Patterning of the cranial nerves in the chick embryo is dependent on cranial mesoderm and rhombomeric metamerism. *Dev Growth Differ* 37:717–731.

Kuratani, S. C., and Bockman, D. E. 1991. Capacity of neural crest cells from various axial levels to participate in thymic development. *Cell Tissue Res* 263:99–106.

Kuratani, S. C., and Eichele, G. 1993. Rhombomere transplantation repatterns the segmental organization of cranial nerves and reveals cell-autonomous expression of a homeodomain protein. *Development* 117:105–117.

Kuratani, S. C., and Kirby, M. L. 1991. Initial migration and distribution of the cardiac neural crest in the avian embryo: An introduction to the concept of the circumpharyngeal crest. *Am J Anat* 191:215–227.

Kuratani, S. C., and Kirby, M. L. 1992. Migration and distribution of circumpharyngeal crest cells in the chick embryo: Formation of the circumpharyngeal ridge and E/c8$^+$ crest cells in the vertebrate head region. *Anat Rec* 234:263–280.

Kuratani, S. C., Miyagawa-Tomita, S., and Kirby, M. L. 1991. Development of cranial nerves in the chick embryo with special reference to the alterations of cardiac branches after ablation of the cardiac neural crest. *Anat Embryol* 183:501–514.

Kuratani, S. C., Ueki, T., Hirano, S., and Aizawa, S. 1998. Rostral truncation of a cyclostome, *Lampetra japonica*, induced by All-*trans* retinoic acid defines the head/trunk interface of the vertebrate body. *Dev Dyn* 211:35–51.

Kutsky, R. J. 1973. *Handbook of Vitamins and Hormones*. New York: Van Nostrand Reinhold Co.

La Bonne, C., and Bronner-Fraser, M. 1998. Neural crest induction in *Xenopus*: evidence for a two-signal model. *Development* 125:2403–2414.

Lacalli, T. C. 1995. Dorsoventral axis inversion. *Nature* 373:110–111.

Lacalli, T. C. 1996. Dorsoventral axis inversion: A phylogenetic perspective. *BioEssays* 18:251–254.

Lacalli, T. C., Holland, N. D., and West, J. E. 1994. Landmarks in the anterior central nervous system of amphioxus larvae. *Philos Trans R Soc Lond B Biol Sci* 344:165–185.

LaFlamme, S. E., and Dawid, I. B. 1990. Differential keratin gene expression during the differentiation of the cement glands of *Xenopus laevis*. *Dev Biol* 137:414–418.

Lahav, R., Ziller, C., Dupin, E., and Le Douarin, N. M. 1996. Endothelin 3 promotes neural crest cell proliferation and mediates a vast increase in melanocyte number in culture. *Proc Natl Acad Sci USA* 93:3892–3897.

Lallier, T., Leblanc, G., Artinger, K. B., and Bronner-Fraser, M. 1992. Cranial and trunk neural crest cells use different mechanisms for attachment to extracellular matrices. *Development* 116:531–541.

Lamb, T. M., Knecht, A. K., Smith, W. C., Stachel, S. E., Economides, A. N., Stahl, N., Yancopolous, G. D., and Harland, R. M. 1993. Neural induction by the secreted polypeptide Noggin. *Science* 262:713–718.

Lamborghini, J. E. 1980. Rohon-Béard cells and other large neurons in *Xenopus* originate during gastrulation. *J Comp Neurol* 189:323–333.

Lamers, C. H. J., Rombout, J. W. H. M., and Timmermans, L. P. M. 1981. An experimental study on neural crest migration in *Barbus conchonius* (Cyprinidae, Teleostei) with special reference to the origin of the enteroendocrine cells. *J Embryol Exp Morphol* 62:309–323.

Lammer, E. J., Chen, D. T., Hoar, R. M., Agnish, N. D., Benke, P. J., Braun, J. T., Curry, C. J., Fernhoff, P. M., Grix, A. W. Jr., Lott, I. T., Ruchard, J. M., and Sun, S. C. 1985. Retinoic acid embryopathy. *New Engl J Med* 313:837–841.

Landacre, F. L. 1921. The fate of the neural crest in the head of the Urodeles. *J Comp Neurol* 33:1–43.

Landis, S. C., and Patterson, P. H. 1981. Neural crest cell lineages. *Trends Neurosci* 4:172–174.

Landolt, R. M., Vaughan, L., Winterhalter, K. H., and Zimmermann, D. R. 1995. Versican is selectively expressed in embryonic tissues that act as barriers to neural crest cell migration and axon outgrowth. *Development* 121:2303–2312.

Landry, C. F., Youson, J. H., and Brown, I. R. 1990. Expression of the Beta-S100 gene in brain and craniofacial cartilage of the embryonic rat. *Dev Neurosci* 12:225–234.

Langille, R. M. 1994. Differentiation of craniofacial mesenchyme In: Hall, B. K., ed. *Bone, Vol. 9 Differentiation and Morphogenesis of Bone.* Boca Raton: CRC Press. pp. 1–64.

Langille, R. M., and Hall, B. K. 1986. Evidence of cranial neural crest cell contribution to the skeleton of the sea lamprey, *Petromyzon marinus.* In: Slavkin, H. C., ed. *New Discoveries and Technologies in Developmental Biology. Part B.* New York: Alan R. Liss, Inc. pp. 263–266.

Langille, R. M., and Hall, B. K. 1987. Development of the head skeleton of the Japanese medaka, *Oryzias latipes* (Teleostei). *J Morphol* 193:135–158.

Langille, R. M., and Hall, B. K. 1988a. Role of the neural crest in development of the cartilaginous cranial and visceral skeleton of the medaka, *Oryzias latipes* (Teleostei). *Anat Embryol* 177:297–305.

Langille, R. M., and Hall, B. K. 1988b. Role of the neural crest in development of the trabeculae and branchial arches in embryonic sea lamprey, *Petromyzon marinus* (L). *Development* 102:301–310.

Langille, R. M., and Hall, B. K. 1989. Neural crest-derived branchial arches link lampreys and gnathostomes. *Fortsch Zool Prog Zool* 35:210–212.

Langille, R. M., and Hall, B. K. 1993a. Pattern formation and the neural crest. In: Hanken, J., and Hall, B. K., eds. *The Skull, Vol. 1, Development.* Chicago: University of Chicago Press. pp. 77–111.

Langille, R. M., and Hall, B. K. 1993b. In vitro calcification of cartilage from the lamprey, *Petromyzon marinus* (L). *Acta Zool* 74:31–41.

Lankester, E. R. 1873. On the primitive cell-layers of the embryo as the basis of genealogical classification of animals, and on the origin of vascular and lymph systems. *Ann Mag Nat Hist Series* 4:11:321–338.

Lankester, E. R. 1877. Notes on the embryology and classification of the animal kingdom: Comprising a revision of speculations relative to the origin and significance of the germ layers. *Q J Microsc Sci* 17:399–454.

Lannoo, M. J., and Smith, S. C. 1989. The lateral line. In: Armstrong, J. B., and Malacinski, G. M., eds. *Developmental Biology of the Axolotl.* Oxford: Oxford University Press. pp. 176–184.

Laudel, T. P., and Lim, T.-M. 1993. Development of the dorsal root ganglion in a teleost, *Oreochromis mossambicus* (Peters). *J Comp Neurol* 327:141–150.

Lauder, J. M., and Zimmermann, E. F. 1988. Sites of serotonin uptake in epithelia of the developing mouse palate, oral cavity, and face: Possible role in morphogenesis. *J Craniofac Genet Dev Biol* 8:265–278.

Lawson, K. A., Meneses, J. J., and Pedersen, R. A. 1991. Clonal analysis of epiblast fate during germ layer formation in the mouse embryo. *Development* 113:891–911.

Leblanc, G.-G., and Holbert, T. E. 1994. Coculture with Hensen's node respecifies neural crest cell fate. *Mol Biol Cell* 5(suppl):106a.

Leblanc, G.-G., and Holbert, T. E. 1996. Hensen's node regulates avian neural crest differentiation *in vitro. J Neurobiol* 29:249–261.

Leblanc, G.-G., Holbert, T. E., and Darland, T. 1995. Role of the transforming growth factor-ß family in the expression of cranial neural crest-specific phenotypes. *J Neurobiol* 26:497–510.

Leblond, C. P., and Inoue, S. 1989. Structure, composition and assembly of basement membrane. *Am J Anat* 185:367–390.

Le Douarin, N. M. 1969. Particularités du noyau interphasique chez la Caille Japonaise (*Coturnix coturnix japonica*). Utilisation de ces particularites comme 'marquage biologique' dans les recherches sur les interactions tissulaires et les migrations cellulaires au cours de l'ontogenèse. *Bull Biol Fr Belg* 103:435–452.

Le Douarin, N. M. 1971a. Etude ultrastructurale comparative du noyau interphasique chez la caille (*Coturnix coturnix japonica*) et le Poulet (*Gallus gallus*) par la méthode de coloration régressive à l'EDTA. *C R Hebd Seances Acad Sci* 272:2334–2337.

Le Douarin, N. M. 1971b. Caractéristiques ultrastructurales du noyau interphasique chez la Caille et chez le Poulet et utilisation de cellules de Caille comme 'marquers biologiques' en embryologie expérimentale. *Ann Embryol Morphol* 4:125–135.

Le Douarin, N. M. 1974. Cell recognition based on natural morphological nuclear markers. *Med Biol* 52:281–319.

Le Douarin, N. M. 1982. *The Neural Crest.* Cambridge: Cambridge University Press.

Le Douarin, N. M. 1986. Cell line segregation during peripheral nervous system ontogeny. *Science* 231:1515–1522.

Le Douarin, N. M. 1988a. The Claude Bernard Lecture 1987. Embryonic chimeras: A tool for studying the development of the nervous and immune systems. *Proc R Soc Lond B Biol Sci* 235:1–17.

Le Douarin, N. M. 1988b. Recherches sur la différenciation de la crête neurale: Influence de facteurs de croissance. *Ann Endocrinol (Paris)* 49:256–269.

Le Douarin, N. M. Dupin, E., Baroffio, A., and Dulac, C. 1992. New insights into the development of neural crest derivatives. *Int Rev Cytol* 138:269–314.

Le Douarin, N. M., Fontaine-Perus, J., and Couly, G. 1986. Cephalic ectodermal placodes and neurogenesis. *Trends Neurosci* 9:175–180.

Le Douarin, N. M., Grapin-Botton, A., and Catala, M. 1996. Patterning of the neural primordium in the avian embryo. *Semin Cell Dev Biol* 1:157–167.

Le Douarin, N. M., Ziller, C., and Couly, G. F. 1993. Patterning of neural crest derivatives in the avian embryo: *in vivo* and *in vitro* studies. *Dev Biol* 159:24–49.

Lee, Y. M., Osumi-Yamashita, N., Ninomiya, Y., Moon, C. K., Eriksson, U., and Eto, K. 1995. Retinoic acid stage-dependently alters the migration pattern and identity of hindbrain neural crest cells. *Development* 121:825–837.

Le Lièvre, C. 1971a. Recherches sur l'origine embryologique des arcs viscéraux chez l'embryon d'Oiseau par la méthode des greffes interspécifiques entre Caille et Poulet. *C R Seances Soc Biol* 165:395–400.

Le Lièvre, C. 1971b. Recherche sur l'origine embryologique du squelette viscéral chez l'embryon d'Oiseau. *C R Ass Anat* 152:575–583.

Le Lièvre, C. 1974. Rôle des cellules mésectodermiques issues des crêtes neurales céphaliques dans la formation des arcs branchiaux et du squelette viscéral. *J Embryol Exp Morphol* 31:453–477.

Le Lièvre, C. 1976. Contribution des crêtes neurales à la genèse des structures céphaliques et cervicales chez les Oiseaux. Thèse d'Etat, Nantes, France.

Le Lièvre, C. 1978. Participation of neural crest derived cells in the genesis of the skull in birds. *J Embryol Exp Morphol* 47:17–37.

Le Lièvre, C., and Le Douarin, N. M. 1974. Origine ectodermique du derme de la face et du cou, montrée par des combinaisons interspécifiques chez l'embryon d'Oiseau. *C R Hebd Seances Acad Sci* 278:517–520.

Le Lièvre, C., and Le Douarin, N. M. 1975. Mesenchymal derivatives of the neural crest: Analysis of chimaeric quail and chick embryos. *J Embryol Exp Morphol* 34:125–154.

León, Y., Miner, C., Represa, J., and Giraldez, F. 1992. Mybp75 oncoprotein is expressed in developing otic and epibranchial placodes. *Dev Biol* 153:407–410.

Leu, F.-J., and Damjanov, I. 1988. Protease treatment combined with immunohistochemistry reveals heterogeneity of normal and neoplastic basement membranes. *J Histochem Cytochem* 36:213–220.

Levak-Svajger, B., and Svajger, A. 1974. Investigations on the origin of the definitive endoderm in the rat embryo. *J Embryol Exp Morphol* 32:445–459.

Levi, G., Crossin, K. L., and Edelman, G. M. 1987. Expression, sequences and distribution of two primary cell adhesion molecules during embryonic development of *Xenopus laevis*. *J Cell Biol* 105:2359–2372.

Levy, B. M., Detwiler, S. R., and Copenhaver, W. M. 1956. The production of developmental abnormalities of the oral structures in *Amblystoma punctatum*. *J Dental Res* 36:659–662.

Lewis, W. H. 1907. On the origin and differentiation of the otic vesicle in amphibian embryos. *Anat Rec* 1:141–145.

Liem, K. F. Jr., Tremml, G., and Jessell, T. M. 1997. A role for the roof plate and its resident TGF-ß-related proteins in neuronal patterning in the dorsal spinal cord. *Cell* 91:127–138.

Liem, K. F., Jr., Tremml, G., Roelink, H., and Jessell, T. M. 1995. Dorsal differentiation of neural plate cells induced by BMP-mediated signals from epidermal ectoderm. *Cell* 82:969–979.

Lim, T. M., Lunn, E. R., Keynes, R. J., and Stern, C. D. 1987. The differing effects of occipital and trunk somites on neural development in the chick embryo. *Development* 100:525–533.

Liwnicz, B. H. 1982. Mitogenic lectin receptors of nervous system tumors. Study of gliomas, neural crest tumors and meningiomas *in vitro* using phytohemagglutin and concanavalin A. *J Neuropath Exp Neurol* 41:281–297.

Lizard-Nacol, S., Lizard, G., Justrabo, E., and Turccarei, C. 1989. Immunologic characterization of Ewing's sarcoma using mesenchymal and neural markers. *Am J Pathol* 135:847–856.

Lo, L.-C., Birren, S. J., and Anderson, D. J. 1991. *V-myc* immortalization of early rat neural crest cells yields a clonal cell line which generates both glial and adrenergic progenitor cells. *Dev Biol* 145:139–153.

Lo, L.-C., Johnson, J. E., Wuenschell, C. W., Saito, T., and Anderson, D. J. 1991. Mammalian *achaete-scute* homolog 1 is transiently expressed by spatially restricted subsets of early neuroepithelial and neural crest cells. *Genes Dev* 5:1524–1537.

Lodewyk, H. S., Van Mierop, L. H. S., and Kutsche, L. M. 1986. Cardiovascular anomalies in DiGeorge syndrome and importance of neural crest as a possible pathogenetic factor. *Am J Cardiol* 58:133–137.

Löfberg, J., Nynäs-McCoy, A., Olsson, C., Jönsson, L., and Perris, R. 1985. Stimulation of initial neural crest cell migration in the axolotl embryo by tissue grafts and extracellular matrix transplanted on microcarriers. *Dev Biol* 107:442–459.

Löfberg, J., Perris, R., and Epperlein, H.-H. 1989. Timing in the regulation of neural crest cell migration: Retarded "maturation" of regional extracellular matrix inhibits pigment cell migration in embryos of the white axolotl mutant. *Dev Biol* 131:168–181.

Lopashov, G. V. 1944. Origins of pigment cells and visceral cartilage in teleosts. *C R Acad Sci USSR* 44:169–172.

Loring, J. F., and Erickson, C. A. 1987. Neural crest cell migratory pathways in the trunk of the chick embryo. *Dev Biol* 121:220–236.

Loring, J. F., Glimelius, B., and Weston, J. A. 1982. Extracellular matrix materials influence quail neural crest cell differentiation *in vitro. Dev Biol* 90:165–174.

Luckenbill-Edds, L., and Carrington, J. 1988. Effect of hyaluronic acid on the emergence of neural crest cells from the neural tube of the quail, *Coturnix coturnix japonica. Cell Tissue Res* 252:573–579.

Lufkin, T., Dierich, A., LeMeur, M., Mark, M., and Chambon, P. 1991. Disruption of the *Hox-1.6* homeobox gene results in defects in a region corresponding to its rostral domain of expression. *Cell* 66:1105–1119.

Lufkin, T., Mark, M., Hart, C. P., Dollé, P., LeMeur, M., and Chambon, P. 1992. Homeotic transformation of the occipital bones of the skull by ectopic expression of a homeobox gene. *Nature* 359:839–841.

Luider, T. M., Bravenboer, N., Meijers, C., van der Kamp, A. W. M., Tibboel, D., and Poelmann, R. E. 1993. The distribution and characterization of HNK-1 antigens in the developing avian heart. *Anat Embryol* 188:307–316.

Luider, T. M., Peters-van der Sanden, M. J. H., Molemaar, J. C., Tibboel, D., van der Kamp, A. W. M., and Meijers, C. 1992. Characterization of HNK-1 antigens during the formation of the avian enteric nervous system. *Development* 115:561–572.

Lumsden, A. 1987. The neural crest contribution to tooth development in the mammalian embryo. In: Maderson, P. F. A., ed. *Developmental and Evolutionary Aspects of the Neural Crest.* New York: John Wiley & Sons. pp. 261–300.

Lumsden, A. 1988. Spatial organization of the epithelium and the role of neural crest cells in the initiation of the mammalian tooth germ. *Development* 102(suppl):155–169.

Lumsden, A., and Graham, A. 1996. Death in the neural crest: Implications for pattern formation. *Semin Cell Dev Biol* 1:169–174.

Lumsden, A., and Krumlauf, R. 1996. Patterning the vertebrate neuraxis. *Science* 274:1109–1115.

Lumsden, A., Sprawson, N., and Graham, A. 1991. Segmental origin and migration of neural crest cells in the hindbrain region of the chick embryo. *Development* 113:1281–1291.

Lunn, E. R., Scourfield, J., Keynes, R. J., and Stern, C. D. 1987. The neural tube origin of ventral root sheath cells in the chick embryo. *Development* 101:247–254.

Luther, A. 1924. Entwicklungsmechanische Untersuchungen am Labyrinth einiger Anuren. *Conn Biol Soc Sci Fenn* 2:1–24.

Maas, R., and Bei, M. 1997. The genetic control of early tooth development. *Crit Rev Oral Biol Med* 8:4–39.

MacKenzie, A., Ferguson, M. W. J., and Sharpe, P. T. 1991. *Hox-7* expression during murine craniofacial development. *Development* 113:601–611.

MacKenzie, A., Leeming, G. L., Jowett, A. K., Ferguson, M. W. J., and Sharpe, P. T. 1991. The homeobox gene Hox 7.1 has specific regional and temporal expression patterns during early murine craniofacial embryogenesis, especially tooth development *in vivo* and *in vitro*. *Development* 111:269–285.

Mackie, G. O. 1995. On the 'visceral nervous system' of *Ciona*. *J Mar Biol Assoc UK* 75:141–151.

Maclean, N., and Hall, B. K. 1987. *Cell Commitment and Differentiation.* Cambridge: Cambridge University Press.

Maden, M., Gale, E., Kostetskii, I., and Zile, M. 1996. Vitamin A-deficient quail embryos have half a hindbrain and other neural defects. *Curr Biol* 6:417–426.

Maden, M., Horton, C., Graham, A., Leonard, L., Pizzey, J., Siegenthaler, G., Lumsden, A., and Eriksson, U. 1992. Domains of cellular retinoic-acid binding protein I (CRABP1) expression in the hindbrain and neural crest of the mouse embryo. *Mech Dev* 37:13–23.

Maderson, P. F. A. 1975. Embryonic tissue interactions as the basis for morphological change in evolution. *Am Zool* 15:315–328.

Maderson, P. F. A. 1983. An evolutionary view of epithelial-mesenchymal interactions. In: Sawyer, R. H., and Fallon, J. F., eds. *Epithelial-mesenchymal Interactions in Development.* New York: Praeger Press. pp. 215–422.

Maderson, P. F. A. 1987. *Developmental and Evolutionary Aspects of the Neural Crest.* New York: John Wiley & Sons.

Mahmood, R., Flanders, K. C., and Morriss-Kay, G. M. 1992. Interactions between retinoids and TGFßs in mouse embryogenesis. *Development* 115:67–74.

Mahmood, R., Kiefer, P., Guthrie, S., Dickson, C., and Mason, I. 1995. Multiple roles for FGF-3 during cranial neural development in the chicken. *Development* 121:1399–1410.

Mahmood, R., Mason, I. J., and Morriss-Kay, G. M. 1996. Expression of *Fgf-3* in relation to hindbrain segmentation, otic pit position, and pharyngeal arch morphology in normal and retinoic-acid-exposed mouse embryos. *Anat Embryol* 194:13–22.

Maisey, J. G. 1986. Heads and tails: A chordate phylogeny. *Cladistics* 2:201–256.

Malicki, J., Schier, A. F., Solnicakrezel, L., Stemple, D. L., Neuhauss, S. C. F., Stainier, D. Y. R., Abdelilah, S., Rangini, Z., Zwartkruis, F., and Driever, W. 1996. Mutations affecting development of the zebrafish ear. *Development* 123(suppl):275–283.

Mallatt, J. 1984. Early vertebrate evolution: Pharyngeal structure and the origin of gnathostomes. *J Zool* 204:169–183.

Mallatt, J. 1996. Ventilation and the origin of jawed vertebrates: A new mouth. *Zool J Linn Soc* 117:329–404.

Mallatt. J. 1997. Shark pharyngeal muscles and early vertebrate evolution. *Acta Zool* 78:279–294.

Mallo, M. 1997. Retinoic acid disturbs mouse middle ear development in a stage-dependent fashion. *Dev Biol* 184:175–186.

Mallo, M. 1998. Embryological and genetic aspects of middle ear development. *Int J Dev Biol* 42:11–22.

Mallo, M., and Gridley, T. 1996. Development of the mammalian ear: Coordinate regulation of formation of the tympanic ring and the external acoustic meatus. *Development* 122:173–179.

Manasek, F. J., and Cohen, A. M. 1977. Anionic glycopeptides and glycosaminoglycans synthesized by embryonic neural tube and neural crest. *Proc Natl Acad Sci USA* 74:1057–1061.

Mancilla, A., and Mayor, R. 1996. Neural crest formation in *Xenopus laevis*: Mechanisms of *Xslug* induction. *Dev Biol* 177:580–589.

Mangold, O. 1929. Experimente zur Analyse der Determination und Induktion der Medullarplatte. *Wilhelm Roux Arch EntwMech Org* 117:586–696.

Manley, N. R., and Capecchi, M. R. 1995. The role of *Hoxa-3* in mouse thymus and thyroid development. *Development* 121:1989–2003.

Manley, N. R., and Capecchi, M. R. 1997. Hox group 3 paralogous genes act synergistically in the formation of somitic and neural crest-derived structures. *Dev Biol* 192:274–288.

Manley, N. R., and Capecchi, M. R. 1998. Hox group 3 paralogs regulate the development and migration of the thymus, thyroid, and parathyroid glands. *Dev Biol* 195:1–15.

Mansouri, A., Stoykova, A., Torres, M., and Gruss, P. 1996. Dysgenesis of cephalic neural crest derivatives in *Pax7*$^{-/-}$ mutant mice. *Development* 122:831–838.

Marchant, L., Linker, C., Ruiz, P., Guerrero, N., and Mayor, R. 1998. The inductive properties of mesoderm suggest that the neural crest cells are specified by a BMP gradient. *Dev Biol* 198:319–329.

Mark, M., Lohnes, D., Mendelsohn, C., Dupé, V., Vonesch, J.-L., Kastner, P., Rijli, F., Bloch-Zupam, A., and Chambon, P. 1995. Roles of retinoic acid receptors and of *Hox* genes in the patterning of the teeth and of the jaw skeleton. *Int J Dev Biol* 39:111–121.

Mark, M., Lufkin, T., Vonesch, J.-L., Ruberte, E., Olivo, J.-C., Dollé, P., Gorry, P., Lumsden, A., and Chambon, P. 1993. Two rhombomeres are altered in *Hoxa-1* mutant mice. *Development* 119:319–338.

Marshall, A. M. 1878. The development of the cranial nerves in the chick. *Q J Microsc Sci* 18:10–40.

Marshall, A. M. 1979. The morphology of the vertebrate olfactory organ. *Q J Microsc Sci* 19:300–340.

Marshall, A. M. 1882. *The Frog: An Introduction to Anatomy, Histology, and Embryology.* London: Macmillan & Co.

Marshall, A. M. 1893. *Vertebrate Embryology: A Text-Book for Students and Practitioners.* London: Smith Elder & Co.

Marshall, H., Nonchev, S., Sham, M. H., Muchamore, I., Lumsden, A., and Krumlauf, R. 1992. Retinoic acid alters hindbrain *Hox* code and induces transformation of rhombomeres 2/3 into a 4/5 identity. *Nature* 360:737–741.

Martha, G. N., Frunchak, Y.-N., Frost, S. K., Thibaudeau, D. G., and Milos, N. C. 1990. Developmentally regulated lectin in dark versus white axolotl embryos. *Biochem Biophys Res Commun* 166:695–700.

Martinez, S., and Alvarado-Mallart, R.-M. 1990. Expression of the homeobox *chick-en* gene in chick/quail chimeras with inverted mes-metencephalic grafts. *Dev Biol* 139:432–436.

Martinsen, B. J., and Bronner-Fraser, M. 1997. Comparison of cranial and trunk neural crest/tube using differential display. *Mol Biol Cell* 1(suppl):128a.

Martins-Green, M., and Erickson, C. A. 1986. Development of neural tube basal lamina during neurulation and neural crest cell emigration in the trunk of the mouse embryo. *J Embryol Exp Morphol* 98:219–236.

Martins-Green, M., and Erickson, C. A. 1987. Basal lamina is not a barrier to neural crest cell emigration: Documentation by TEM and by immunofluorescent and immunogold labelling. *Development* 101:517–533.

Marusich, M. F., Pourmehr, K., and Weston, J. A. 1986a. A monoclonal antibody (SN1) identifies a subpopulation of avian sensory neurons whose distribution is correlated with axial levels. *Dev Biol* 118:494–504.

Marusich, M. F., Pourmehr, K., and Weston, J. A. 1986b. The development of an identified subpopulation of avian sensory neurons is regulated by interaction with the periphery. *Dev Biol* 118:505–510.

Mathew, T. 1982. Evidence supporting neural crest origin of an alveolar soft part sarcoma: An ultrastructural study. *Cancer* 50:507–514.

Matsumoto, J., Lynch, T. J., Grabowski, S., Richards, C. M., Lo, S. L., Clark, C., Kern, D., Taylor, J. D., and Tchen, T. T. 1983. Fish tumor pigment cells: Differentiation and comparison to their normal counterparts. *Am Zool* 23:569–580.

Matsumoto, J., Wada, K., and Akiyama, T. 1989. Neural crest cell differentiation and carcinogenesis: Capability of goldfish erythrophoroma cells for multiple differentiation and clonal polymorphism in their melanogenic variants. *J Invest Dermatol* 92:255S–260S.

Matsuo, I., Kuratani, S. C., Kimura, C., Takeda, N., and Aizawa, S. 1995. Mouse *Otx2* functions in the formation and patterning of rostral head. *Genes Dev* 9:2646–2658.

Mautner, V.-F., Umnus-Schnelle, S., Köppen, J., and Heise, U. 1988. The diagnosis of von Recklinghausen neurofibromatosis. *Deutsche Med Wochenschrift* 113:1149–1151.

Maxwell, G. D. 1976. Cell cycle changes during neural crest cell differentiation *in vitro*. *Dev Biol* 49:66–79.

Maxwell, G. D., and Forbes, M. E. 1987. Exogenous basement membrane-like matrix stimulates adrenergic development in avian neural crest cultures. *Development* 101:767–776.

Maxwell, G. D., and Forbes, M. E. 1988. Adrenergic development of neural crest cells grown in a defined medium under a reconstituted basement membrane-like matrix. *Neuroscience* 95:64–68.

Maxwell, G. D., and Forbes, M. E. 1990a. The phenotypic response of cultured quail trunk neural crest cells to a reconstituted basement membrane-like matrix is specific. *Dev Biol* 141:233–237.

Maxwell, G. D., and Forbes, M. E. 1990b. Stimulation of adrenergic development in trunk neural crest cultures by a reconstituted basement membrane-like matrix is inhibited by agents that elevate cAMP. *J Neurosci Res* 25:172–179.

Maxwell, G. D., and Forbes, M. E. 1991. Spectrum of in vitro differentiation of quail trunk neural crest cells isolated by cell sorting using the HNK-1 antibody and analysis of the adrenergic development of HNK-1$^+$ sorted subpopulations. *J Neurobiol* 22:276–286.

Maxwell, G. D., Forbes, M. E., and Christie, D. S. 1988. Analysis of the development of cellular subsets present in the neural crest using cell sorting and cell culture. *Neuron* 1:557–568.

Mayor, R., Guerrero, N., and Martínez, C. 1997. Role of FGF and Noggin in neural crest induction. *Dev Biol* 189:1–12.

Mayor, R., Morgan, R., and Sargent, M. G. 1995. Induction of the prospective neural crest of *Xenopus. Development* 121:767–777.

McCarthy, R. A., and Hay, E. D. 1991. Collagen I, laminin, and tenascin: Ultrastructure and correlation with avian neural crest formation. *Int J Dev Biol* 35:437–452.

McClearn, D., and Noden, D. M. 1988. Ontogeny of architectural complexity in embryonic quail visceral arch muscles. *Am J Anat* 183:277–293.

McGrew, L. L., Lai, C.-J., and Moon, R. T. 1995. Specification of the anteroposterior neural axis through synergistic interaction of the wnt signaling cascade with noggin and follistatin. *Dev Biol* 172:337–342.

McKee, G. J., and Ferguson,. M. W. J. 1984. The effects of mesencephalic neural crest cell extirpation on the development of chicken embryos. *J Anat* 139:491–512.

McLeod, M. J., Harris, M. J., Chernoff, G. F., and Miller, J. R. 1980. First arch malformation: A new craniofacial mutant in the mouse. *J Hered* 71:331–335.

McMahon, A. P., and Bradley, A. 1990. The *Wnt-1* (*int-1*) proto-oncogene is required for development of a large region of the mouse brain. *Cell* 62:1073–1085.

Medvedeva, I. M. 1975. *The Olfactory Organ in Amphibians and Its Phylogenetic Significance.* Leningrad: Nauka.

Medvedeva, I. M. 1986a. On the origin of nasolacrimal duct in Tetrapoda. In: Rocek, Z., ed. *Studies in Herpetology.* Prague: Charles University. pp. 37–40.

Medvedeva, I. M. 1986b. Nasolachrymal canal of Ambystomatidae in the light of its origin in other terrestrial vertebrates. In: Vorobyeva, E. I., and Lebedkina, N. S., eds. *Morphology and Evolution of Animals.* Moscow: Nauka. pp. 138–156.

Meier, S. 1978a. Development of the embryonic chick otic placode. I. Light microscopic analysis. *Anat Rec* 191:447–458.

Meier, S. 1978b. Development of the embryonic chick otic placode. II. Electron microscopic analysis. *Anat Rec* 191:459–478.

Meier, S., and Packard, D. S. Jr. 1984. Morphogenesis of the cranial segments and distribution of neural crest in the embryo of the snapping turtle, *Chelydra serpentina. Dev Biol* 102:309–323.

Meinke, D. K. 1986. Morphology and evolution of the dermal skeleton in lungfishes. *J Morphol* 1(suppl):133–149.

Melnick, M., Bixler, D., and Shields, E. D. 1980. *Etiology of Cleft Lip and Cleft Palate.* New York: Alan R. Liss, Inc.

Melville, R. V., and Smith, J. D. D. 1987. *Official Lists and Indexes of Names and Works in Zoology.* London: International Trust for Zoological Nomenclature, British Museum (Natural History).

Menoud, P. A., Debrot, S., and Schowing, J. 1989. Mouse neural crest cells secrete both urokinase-type and tissue-type plasminogen activators *in vitro. Development* 106:685–690.

Mérida-Velasco, J. A., Sánchez-Montesinos, I., Espín-Ferra, J., Garcia-Garcia, J. D., and Roldán-Schilling, V. 1996. Grafts of the third branchial arch in chick embryos. *Acta Anat* 155:73–80.
Merrilees, M. J. 1975. Tissue interactions: Morphogenesis of the lateral line system and labyrinth of vertebrates. *J Exp Zool* 192:113–118.
Merrilees, M. J., and Crossman, E. J. 1973. Surface pits in the family Esocifae. II. Epidermal–dermal interactions and evidence of aplasia of the lateral line sensory system. *J Morphol* 141:321–332.
Mes-Hartree, M., and Armstrong, J. B. 1980. Evidence that the premature death mutation (p) in the Mexican axolotl (*Ambystoma mexicanum*) is not an autonomous cell lethal. *J Embryol Exp Morphol* 60:295–302.
Metscher, B. D., Northcutt, R. G., Gardner, D. M., and Bryant, S. V. 1997. Homeobox genes in axolotl lateral line placodes and neuromasts. *Dev Genes Evol* 207:287–295.
Mii, Y., Miyauchi, Y., Hohnoki, K., Maruyama, H., Tsutsumi, M., Dohmae, K., Tamai, S., Konishi, Y., and Yamanouchi, T. 1989. Neural crest cell origin of clear cell sarcoma of tendons and aponeuroses. Ultrastructural and enzyme cytochemical study of human and nude mouse-transplanted tumours. *Virchows Arch A Pathol Anat Histopathol* 415:51–60.
Mikhailov, A. T., and Gorgolyuk, N. A. 1988. Concanavalin A induces neural tissue and cartilage in amphibian early gastrula ectoderm. *Cell Differ* 22:145–154.
Milos, N. C., Ma, Y., Varma, P. V., Bering, M. P., Mohamed, Z., Pilarski, L. M., and Frunchar, Y. N. 1990. Localization of endogenous galactoside-binding lectins during morphogenesis of *Xenopus laevis*. *Anat Embryol* 182:319–327.
Milos, N. C., Meadows, G., Evanson, J. E., Pinchbeck, J. B, Bawa, N., Young, K. J., Palmer, N. G., Murdoch, C. A., and Carmel, D. 1998. Expression of the endogenous galactoside-binding lectin of *Xenopus laevis* during cranial neural crest development: Lectin localization is similar to that of members of the N-CAM and cadherin families of cell adhesion molecules. *J Craniofac Genet Dev Biol* 18:11–29.
Milos, N. C., and Wilson, H. C. 1986. Cell surface carbohydrate involvement in controlling the adhesion and morphology of neural crest cells and melanophores of *Xenopus laevis*. *J Exp Zool* 238:211–224.
Milos, N. C., Wilson, H. C., Ma, Y.-L., Mohanraj, T. M., and Frunchak, Y.-N. 1987. Studies on cell adhesion of *Xenopus laevis* melanophores: Modulation of cell-cell and cell-substratum adhesion in vitro by endogenous *Xenopus* galactoside-binding lectins. *Pigment Cell Res* 1:188–196.
Mina, M., Gluhak, J., Upholt, W. B., Kollar, E. J., and Rogers, B. 1995. Experimental analysis of *Msx-1* and *Msx-2* gene expression during chick mandibular morphogenesis. *Dev Dyn* 202:195–214.
Minkoff, R., and Kuntz, A. J. 1977. Cell proliferation during morphogenetic change: Analysis of frontonasal morphogenesis in the chick embryo employing DNA labeling indices. *J Embryol Exp Morphol* 40:101–113.
Minuth, M., and Grunz, H. 1980. The formation of mesodermal derivatives after induction with vegetalizing factor depends on secondary cell interactions. *Cell Differ* 9:229–238.
Mitani, S., and Okamoto, H. 1991. Inductive differentiation of two neural lineages reconstituted in a microculture system from *Xenopus* early gastrula cells. *Development* 112:21–31.
Mitchell, P. J., Timmons, P. M., Hébert, J. M., Rigby, P. W. J., and Tjian, R. 1991. Transcription factor AP-2 is expressed in neural crest cell lineages during mouse embryogenesis. *Genes Dev* 5:105–119.

Miya, T., Morita, K., Suzuki, A., Ueno, N., and Satoh, N. 1997. Functional analysis of an ascidian homologue of vertebrate BMP-2/BMP-4 suggests its role in the inhibition of neural fate specification. *Development* 124:5149–5159.

Miya, T., Morita, K., Ueno, N., and Satoh, N. 1996. An ascidian homologue of vertebrate BMPs-5-8 is expressed in the midline of the anterior neuroectoderm and in the midline of the ventral epidermis of the embryo. *Mech Dev* 57:181–190.

Miyagawa-Tomita, S., Waldo, K., Tomita, H., and Kirby, M. L. 1991. Temporospatial study of the migration and distribution of cardiac neural crest in quail-chick chimeras. *Am J Anat* 192:79–88.

Miyake, T., Cameron, A. M., and Hall, B. K. 1996. Stage-specific onset of condensation and matrix deposition for Meckel's and other first arch cartilages in inbred C57BL/6 mice. *J Craniofac Genet Dev Biol* 16:32–47.

Miyake, T., Cameron, A. M., and Hall, B. K. 1997a. Stage-specific expression patterns of alkaline phosphatase during development of the first arch skeleton in inbred C57BL/6 mouse embryos. *J Anat* 190:239–260.

Miyake, T., Cameron, A. C., and Hall, B. K. 1997b. Space, size and time: Variability and constancy of embryonic development in and among three inbred strains of mice. *Growth Dev Aging* 61:141–155.

Miyake, T., and Hall, B. K. 1994. Development of *in vitro* organ culture techniques for differentiation and growth of cartilages and bones from teleost fish and comparisons with *in vivo* skeletal development. *J Exp Zool* 268:22–43.

Miyake, T., McEachran, J. D., and Hall, B. K. 1992. Edgeworth's legacy of cranial muscle development with an analysis of the ventral gill arch muscles in batoid fishes (Batoidea: Chondrichthyes). *J Morphol* 212:213–256.

Moase, C. E., and Trasler, D. G. 1989. Spinal ganglia reduction in the *Splotch*-delayed mouse neural tube defect mutant. *Teratology* 40:67–75.

Moase, C. E., and Trasler, D. G. 1990. Delayed neural crest cell emigration from *Sp* and Sp^4 mouse neural tube explants. *Teratology* 42:171–182.

Moase, C. E., and Trasler, D. G. 1991. N-CAM alterations in *Splotch* neural tube defect mouse embryos. *Development* 113:1049–1058.

Moase, C. E., and Trasler, D. G. 1992. *Splotch* locus mouse mutants: Models for neural tube defects and Waardenburg syndrome type I in humans. *J Med Genet* 29:145–151.

Moens, C. B., Cordes, S. P., Giorgianni, M. W., Barsh, G. S., and Kimmel, C. B. 1998. Equivalence in the genetic control of hindbrain segmentation in fish and mouse. *Development* 125:381–391.

Moiseiwitsch, J. R. D., and Lauder, P. M. 1995. Serotonin regulates mouse cranial neural crest migration. *Proc Natl Acad Sci USA* 92:7182–7186.

Monga, M. 1997. Vitamin A and its congeners. *Semin Perinatol* 21:135–142.

Monsoro-Burq, A.-H., Duprez, D., Watanabe, Y., Bontoux, M., Vincent, C., Brickell, P., and Le Douarin, N. M. 1996. The role of bone morphogenetic protein in vertebral development. *Development* 122:3607–3616.

Montagu, A. 1971. *The Elephant Man : A Study in Human Dignity.* New York: Outerbridge & Dienstfrey.

Moro-Balbás, J. A., Gato, A., Alonso, M. I., and Barbosa, E. 1998. Local increase level of chondroitin sulfate induces changes in the rhombencephalic neural crest migration. *Int J Dev Biol* 42:207–216.

Moro-Balbás, J. A., Gato, A., Alonso Revuelta, M. I., Pastor, J. F., Represa, J. J., and Barbosa, E. 1993. Retinoic acid induces changes in the rhombencephalic neural crest cells migration, and extracellular matrix composition in chick embryos. *Teratology* 48:197–206.

Morrison-Graham, K., Bork, T., and Weston, J. A. 1990. Association between collagen and glycosaminoglycans is altered in dermal extracellular matrix of fetal *Steel* (Sl^d/Sl^d) mice. *Dev Biol* 139:308–319.

Morrison-Graham, K., Schatteman, G. C., Bork, T., Bowen-Pope, D. F., and Weston, J. A. 1992. A PDGF receptor mutation in the mouse (*Patch*) perturbs the development of a non-neural subset of neural crest-derived cells. *Development* 115:133–142.

Morrison-Graham, K., West-Johnsrud, L., and Weston, J. A. 1990. Extracellular matrix from normal but not *Steel* mutant mice enhances melanogenesis in cultured mouse neural crest cells. *Dev Biol* 139:299–307.

Morrison-Graham, K., and Weston, J. A. 1989. Mouse mutants provide new insights into the role of extracellular matrix in cell migration and differentiation. *Trends Genet* 5:116–121.

Morrison-Graham, K., and Weston, J. A. 1993. Transient *Steel* factor dependence by neural crest-derived melanocyte precursors. *Dev Biol* 159:346–352.

Morriss, G. M. 1972. Morphogenesis of the malformations induced in rat embryos by maternal hypervitaminosis A. *J Anat* 113:241–250.

Morriss, G. M. 1975. Abnormal cell migration as a possible factor in the genesis of vitamin A induced craniofacial anomalies. In: Neubert, D., and Merker, H. J., eds. *New Approaches to the Evaluation of Abnormal Embryonic Development.* Stuttgart: George Thieme. pp. 678–687.

Morriss, G. M. 1976. Vitamin A and congenital malformations. *Int J Vitam Nutr Res* 46:220–222.

Morriss, G. M. 1980. Neural tube defects: Towards prevention and understanding. *Nature* 184:121–123.

Morriss, G. M., and New, D. A. T. 1979. Effect of oxygen concentration on morphogenesis of cranial neural folds and neural crest in cultured rat embryos. *J Embryol Exp Morphol* 54:17–35.

Morriss, G. M., and Thorogood, P. V. 1978. An approach to cranial neural crest cell migration and differentiation in mammalian embryos. In: Johnson, M. H., ed. *Development of Mammals.Vol. 3*. Amsterdam: North Holland. pp. 363–412.

Morriss-Kay, G. M. ed. 1992. *Retinoids in Normal Development and Teratogenesis.* Oxford: Oxford University Press.

Morriss-Kay, G. M. 1993. Retinoic acid and craniofacial development: Molecules and morphogenesis. *BioEssays* 15:9–15.

Morriss-Kay, G. M. 1996. Craniofacial defects in AP-2 null mutant mice. *BioEssays* 18:785–788.

Morriss-Kay, G. M., and Tan, S.-S. 1987. Mapping cranial neural crest cell migration pathways in mammalian embryos. *Trends Genet* 3:257–261.

Morriss-Kay, G. M., and Tuckett, F. 1985. The role of microfilaments in cranial neurulation in rat embryos: Effects of short-term exposure to cytochalasin D. *J Embryol Exp Morphol* 88:333–348.

Morriss-Kay, G. M., and Tuckett, F. 1989. Immunohistochemical localization of chondroitin sulphate proteoglycans and the effects of chondroitinase ABC in 9- to 11-day rat embryos. *Development* 106:787–798.

Morris-Wiman, J., and Brinkley, L. L. 1990. Changes in mesenchymal cell and hyaluronate distribution correlate with in vivo elevation of the mouse mesencephalic neural folds. *Anat Rec* 226:383–395.

Moss, M. L., and Moss-Salentijn, L. 1983. Vertebrate cartilage. In: Hall, B. K., ed. *Cartilage. Vol. 1. Structure, Function and Biochemistry.* New York: Academic Press. pp. 1–30.

Moury, J. D., and Hanken, J. 1995. Early cranial neural crest migration in the direct-developing frog, *Eleutherodactylus coqui. Acta Anat* 153:243–253.

Moury, J. D., and Jacobson, A. G. 1989. Neural fold formation at newly created boundaries between neural plate and epidermis in the axolotl. *Dev Biol* 133:44–57.

Moury, J. D., and Jacobson, A. G. 1990. The origins of neural crest cells in the axolotl. *Dev Biol* 141:243–253.

Mujtaba, T., Mayer-Proschel, M., and Rao, M. S. 1998. A common neural progenitor for the CNS and PNS. *Dev Biol* 200:1–15.

Müller, E., and Ingvar, S. 1921. Über den Ursprung des sympathicus vei den Amphibien. *Uppsala Läkareförenings Förhandlingar Ny Följd* 26:1–15.

Müller, E., and Ingvar, S. 1923. Über den Ursprung des Sympathicus beim Hunchen. *Arch Mikrosk Anat EntwMech* 99:650–671.

Müller, F., and O'Rahilly, R. 1987. The development of the human brain, the closure of the caudal neuropore, and the beginning of secondary neurulation at stage 12. *Anat Embryol* 176:413–430.

Müller, F., and O'Rahilly, R. 1989. Mediobasal prosencephalic defects, including holoprosencephaly and cyclopia, in relation to the development of the human forebrain. *Am J Anat* 185:391–414.

Müller, F., and O'Rahilly, R. 1994. *The Embryonic Human Brain: An Atlas of Developmental Stages.* New York: John Wiley & Sons.

Muneoka, K., Wanek, N., and Bryant, S. V. 1986. Mouse embryos develop normally exo utero. *J Exp Zool* 239:289–293.

Murphy, M., Drago, J., and Bartlett, P. F. 1990. Fibroblast growth factor stimulates the proliferation and differentiation of neural precursor cells *in vitro. J Neurosci Res* 25:463–475.

Murphy, M., Reid, K., William, D. E., Lyman, S. D., and Bartlett, P. F. 1992. Steel factor is required for maintenance, but not differentiation, of melanocyte precursors in the neural crest. *Dev Biol* 153:396–401.

Murray, J. D., Deeming, D. C., and Ferguson, M. W. J. 1990. Size-dependent pigmentation-pattern formation in embryos of *Alligator mississippiensis*: Time of initiation of pattern generation mechanism. *Proc R Soc Lond B Biol Sci* 239:279–293.

Nakagawa, M., Thompson, R. P., Terracio, L., and Borg, T. K. 1993. Developmental anatomy of HNK-1 immunoreactivity in the embryonic rat heart: Co-distribution with early conductive tissue. *Anat Embryol* 187:445–460.

Nakagawa, S., and Takeichi, M. 1998. Neural crest emigration from the neural tube depends on regulated cadherin expression. *Development* 125:2963–2871.

Nakamura, H., and Ayer-Le Lièvre, C. 1982. Mesectodermal capabilities of the trunk neural crest of birds. *J Embryol Exp Morphol* 70:1–18.

Nakata, K., Nagai, T., Aruga, J., and Mikoshiba, K. 1997. *Xenopus Zic3*, a primary regulator both in neural and neural crest development. *Proc Natl Acad Sci USA* 94:11980–11985.

Neuhass, S. C. F., Solnica-Krezel, L., Schier, A. F., Zwartkruis, F., Stemple, D. L., Malici, J., Abdelilah, S., Stainier, D. Y. R., and Driever, W. 1996. Mutations affecting craniofacial development in zebrafish. *Development* 123:357–367.

Neumayer, L. 1906. Histogenese und Morphogenese des peripheren Nervensystems, der Spinalganglien und des Nervus sympathicus. In: Hertwig, O., ed. *Handbuch der Ver-*

gleichenden und Experimentellen Entwickelungslehre der Wirbeltiere. Vol. 2, Part 3. Jena: Gustav Fisher Verlag. pp. 513–626.

Newgreen, D. 1995. Epithelial-mesenchymal transitions, Part I. *Acta Anat* 154:1–97.

Newgreen, D. F., and Erickson, C. A. 1986. The migration of neural crest cells. *Int Rev Cytol* 103:89–145.

Newgreen, D. F., and Gibbins, I. L. 1982. Factors controlling the time of onset of the migration of neural crest cells in the fowl embryo. *Cell Tissue Res* 224:145–160.

Newgreen, D. F., and Minichiello, J. 1995. Control of epitheliomesenchymal transformation. 1. Events in the onset of neural crest cell migration are separable and inducible by protein kinase inhibitors. *Dev Biol* 170:91–101.

Newgreen, D. F., Scheel, M., and Kastner, V. 1986. Morphogenesis of sclerotome and neural crest in avian embryos. *In vivo* and *in vitro* studies on the role of notochordal extracellular material. *Cell Tissue Res* 244:299–313.

Newman, S. A., and Comper, W. D. 1990. 'Generic' physical mechanisms of morphogenesis and pattern formation. *Development* 110:1–18.

Newth, D. R. 1950. Fate of the neural crest in lampreys. *Nature* 165:284.

Newth, D. R. 1951. Experiments on the neural crest of the lamprey embryo. *J Exp Biol* 28:247–260.

Newth, D. R. 1955. Nervnii grebyen i golovnai skelet minogi. (The neural crest and head skeleton of lampreys). *Dokl Akad Nauk SSSR* 102:653–656.

Newth, D. R. 1956. On the neural crest of the lamprey embryo. *J Embryol Exp Morphol* 4:358–375.

Nguyen, V. H., Schmid, B., Trout, J., Connors, S. A., Ekker, M., and Mullins, M. C. 1998. Ventral and lateral regions of the zebrafish gastrula, including the neural crest progenitors, are established by a *bmp2b/swirl* pathway of genes. *Dev Biol* 199:93–110.

Nichols, D. H. 1981. Neural crest formation in the head of the mouse embryo as observed using a new histological technique. *J Embryol Exp Morphol* 64:105–120.

Nichols, D. H. 1985. The ultrastructure and immunohistochemistry of the basal lamina beneath escaping neural crest mesenchyme in the head of the mouse embryo. *Anat Rec* 211:138A.

Nichols, D. H. 1986a. Mesenchyme formation from the trigeminal placodes of the mouse embryo. *Am J Anat* 176:19–31.

Nichols, D. H. 1986b. Formation and distribution of neural crest mesenchyme to the first pharyngeal arch region of the mouse embryo. *Am J Anat* 176:221–231.

Nichols, D. H. 1987. Ultrastructure of neural crest formation in the midbrain/rostral hindbrain and preotic hindbrain regions of the mouse embryo. *Am J Anat* 179:143–154.

Nieto, M. A., Sargent, M. G., Wilkinson, D. G., and Cooke, J. 1994. Control of cell behavior during vertebrate development by *Slug*, a zinc finger gene. *Science* 164:835–839.

Nieto, M. A., Sechrist, J., Wilkinson, D. G., and Bronner-Fraser, M. 1995. Relationship between spatially restricted *Krox-20* gene expression in branchial neural crest and segmentation in the chick embryo hindbrain. *EMBO J* 14:1697–1710.

Nieuwkoop, P. D. 1985. Inductive interactions in early amphibian development and their general nature. *J Embryol Exp Morphol* 89(suppl):333–347.

Nieuwkoop, P. D. 1992. The formation of the mesoderm in urodelean amphibians. VI. The self-organizing capacity of the induced meso-endoderm. *Wilhelm Roux's Arch Dev Biol* 201:18–29.

Nieuwkoop, P. D., and Albers, B. 1990. The role of competence in the cranio-caudal segregation of the central nervous system. *Dev Growth Differ* 32:23–31.

Nieuwkoop, P. D., Johnen, A. G., and Albers, B. 1985. *The Epigenetic Nature of Early Chordate Development.* Cambridge: Cambridge University Press.

Nieuwkoop, P. D., and Koster, K. 1995. Vertical versus planar induction in amphibian early development. *Dev Growth Differ* 37:653–668.

Nishibatake, M., Kirby, M. L., and van Mierop, L. H. S. 1987. Pathogenesis of persistent truncus arteriosus and dextroposed aorta in the chick embryo after neural crest ablation. *Circulation* 75:255–264.

Nishihira, T., Kasai, M., Hayashi, Y., Kimura, M., Matsumura, Y., Akaishi, T., Ishiguro, S., Kataoka, S., Watanabe, H., Miura, Y., and Sato, H. 1981. Experimental studies on differentiation of cells originated from human neural crest tumors *in vitro* and *in vivo*. *Cell Mol Biol* 27:181–196.

Niu, M. C. 1947. The axial organization of the neural crest, studied with particular reference to its pigmentary component. *J Exp Zool* 105:79–114.

Noden, D. M. 1973. The migratory behavior of neural crest cells. In: Bosma, J. F., ed. *Development in the Fetus and Infant.* Bethesda: National Institutes of Health. Fourth Symposium on Oral Sensation and Perceptions. DHEW # 73-546. pp. 9–36.

Noden, D. M. 1975. An analysis of the migratory behavior of avian cephalic neural crest cells. *Dev Biol* 42:106–130.

Noden, D. M. 1978a. Interactions directing the migration and cytodifferentiation of avian neural crest cells. In: Garrod, D., ed. *The Specificity of Embryological Interactions.* London: Chapman & Hall. pp. 3–50.

Noden, D. M. 1978b. The control of avian cephalic neural crest cytodifferentiation. I. Skeletal and connective tissues. *Dev Biol* 67:296–312.

Noden, D. M. 1978c. The control of avian cephalic neural crest cytodifferentiation. II. Neural tissues. *Dev Biol* 67:313–329.

Noden, D. M. 1980. The migration and cytodifferentiation of cranial neural crest cells. In: Pratt, R. M., and Christiansen, R. L., eds. *Current Research Trends in Prenatal Craniofacial Development.* New York: Elsevier/North Holland. pp. 3–26.

Noden, D. M. 1982a. Periocular mesenchyme: Neural crest and mesodermal interactions. In: Jakobiec, F. A., ed. *Ocular Anatomy, Embryology and Teratology.* Philadelphia: Harper and Row. pp. 97–120.

Noden, D. M. 1982b. Patterns and organization of craniofacial skeletogenic and myogenic mesenchyme: A perspective. In: Dixon, A. D., and Sarnat, B. G., eds. *Factors and Mechanisms Influencing Bone Growth.* New York: Alan R. Liss, Inc. pp. 167–203.

Noden, D. M. 1983a. The role of the neural crest in patterning of avian cranial skeletal, connective and muscle tissues. *Dev Biol* 96:144–165.

Noden, D. M. 1983b. The embryonic origins of avian cephalic and cervical muscles and associated connective tissues. *Am J Anat* 168:257–726.

Noden, D. M. 1984a. The use of chimeras in analysis of craniofacial development. In: Le Douarin, N. M., and McLaren, A., eds. *Chimeras in Developmental Biology.* Orlando: Academic Press. pp. 241–280.

Noden, D. M. 1984b. Craniofacial development: New views on old problems. *Anat Rec* 208:1–13.

Noden, D. M. 1985. Embryonic patterning of craniofacial muscles. *Anat Rec* 211:139A–140A.

Noden, D. M. 1986a. Origins and patterns of craniofacial mesenchymal tissues. *J Craniofac Genet Dev Biol* 2(suppl):15–32.

Noden, D. M. 1986b. Patterning of avian craniofacial muscles. *Dev Biol* 116:347–356.

Noden, D. M. 1987. Interactions between cephalic neural crest and mesodermal populations. In: Maderson, P. F. A.ed. *Developmental and Evolutionary Aspects of the Neural Crest*. New York: John Wiley & Sons. pp. 89–119.

Noden, D. M. 1991. Origins and patterning of avian outflow tract endocardium. *Development* 111:867–876.

Noden, D. M. 1993. Spatial integration among cells forming the cranial peripheral nervous system. *J Neurobiol* 24:248–261.

Norr, S. C. 1973. *In vitro* analysis of sympathetic neuron differentiation from chick neural crest cells. *Dev Biol* 34:16–38.

Northcutt, R. G. 1992. The phylogeny of octavolateralis ontogenies: A reaffirmation of Garstang's phylogenetic hypothesis. In: Popper, A., Webster, D., and Fay, R., eds. *The Evolutionary Biology of Hearing*. New York: Springer-Verlag. pp. 21–47.

Northcutt, R. G. 1995. The forebrain of gnathostomes—In search of a morphotype. *Brain Behav Evol* 46:275–318.

Northcutt, R. G. 1996. The origin of craniates—neural crest, neurogenic placodes, and homeobox genes. *Isr J Zool* 42:S273–S313.

Northcutt, R. G., and Barlow, L. A. 1998. Amphibians provide new insights into taste-bud development. *Trends Neurosci* 21:38–43.

Northcutt, R. G., and Brändle, K. 1995. Development of branchiomeric and lateral line nerves in the axolotl. *J Comp Neurol* 355:427–454.

Northcutt, R. G., Brändle, K., and Fritzsch, B. 1995. Electroreceptors and mechanosensory lateral line organs arise from single placodes in axolotls. *Dev Biol* 168:358–373.

Northcutt, R. G., Catania, K. C., and Criley, B. B. 1994. Development of lateral line organs in the axolotl. *J Comp Neurol* 340:480–514.

Northcutt, R. G., and Gans, C. 1983. The genesis of neural crest and epidermal placodes: A reinterpretation of vertebrate origins. *Q Rev Biol* 58:1–28.

Nozue, T. 1974. Specific sensitivity of neural crest cells in mice embryos to alkylating agents and hydrocortisone acetate. *Okajimas Folia Anat Japan* 51:1–8.

Nozue, T. 1988a. Relationships between neural crest cells and mast cells in newborn mice. *Anat Anz Jena* 166:219–225.

Nozue, T. 1988b. Effects of EDTA on newborn mice with special reference to neural crest cells. *Anat Anz Jena* 166:209–218.

Nozue, T. 1988c. Relationships between neural crest cells and serotonin in newborn mice. *Anat Anz Jena* 166:227–237.

Nozue, T., and Kayano, T. 1977a. Relationship between the geographic distribution pattern of ^{14}C-dopa and the neural crest cells in the postnatal development of mice. *Acta Anat* 97:114–117.

Nozue, T., and Kayano, T. 1977b. Multiple oncogenesis of neural crest cells by steroids in suckling mice. *Experientia* 33:1640–1641.

Nozue, T., and Kayano, T. 1977c. Multiple APUD system (neural crest) tumors caused by endotoxin in suckling mice. *Experientia* 33:516–517.

Nozue, T., and Kayano, T. 1978. Effects of mitomycin C in postnatal tooth development in mice with special reference to neural crest cells. *Acta Anat* 100:85–94.

Nozue, T., and Ono, S. 1989. Exposure of newborn mice to adenosine causes neural crest dysplasia and tumor formation. *Neurofibromatosis* 2:261–273.

Nozue, T., and Tsuzaki, M. 1974a. Histochemical study on neural crest cells in mice embryos. *Okajimas Folia Anat Japan* 51:103–120.

Nozue, T., and Tsuzaki, M. 1974b. Further studies on distribution of neural crest cells in mice embryos. *Okajimas Folia Anat Japan* 51:131–160.

Nübler-Jung, K., and Arendt, D. 1994. Is ventral in insects dorsal in vertebrates? A history of embryological arguments favouring axis inversion in chordate ancestors. *Wilhem Roux's Arch Dev Biol* 203:357–366.

Oakley, R. A., Lasky, C. J., Erickson, C. A., and Tosney, K. W. 1994. Glycoconjugates mark a transient barrier to neural crest migration in the chicken embryo. *Development* 120:102–114.

Oakley, R. A., and Tosney, K. W. 1991. Peanut agglutinin and chondroitin-6-sulfate are molecular markers for tissues that act a barriers to axon advance in the avian embryo. *Dev Biol* 147:187–206.

Okada, E. W. 1955. Isolationsversuche zur Alanyse der Knorpelbildung aus Neuralleistenzellen bei Urodelenkeim. *Mem Coll Sci Kyoto Univ Ser B* 22:23–28.

Okamura, Y., Okado, H., and Takahashi, K. 1993. The ascidian embryo as a prototype of vertebrate neurogenesis. *BioEssays* 15:723–730.

Olsen, K. D., Maragos, N. E., and Weiland, L. H. 1980. First branchial cleft anomalies. *Laryngoscope* 90:423–436.

Olson, E. N., and Srivastava, D. 1996. Molecular pathways controlling heart development. *Science* 272:671–676.

Olsson, L. 1993. Pigment pattern formation in the larval salamander, *Ambystoma maculatum. J Morphol* 215:151–163.

Olsson, L. 1994. Pigment pattern formation in larval Ambystomatid salamanders — *Ambystoma talpoideum, Ambystoma barbouri,* and *Ambystoma annulatum. J Morphol* 220:123–138.

Olsson, L., and Hanken, J. 1996. Cranial neural-crest migration and chondrogenic fate in the Oriental Fire-Bellied toad, *Bombina orientalis*: Defining the ancestral pattern of head development in anuran amphibians. *J Morphol* 229:105–120.

Olsson, L., and Löfberg, J. 1992. Pigment cell migration and pattern formation in salamander larvae. In: Rensing, L., ed. *Oscillations and Morphogenesis.* New York: Marcel Dekker. pp. 453–462.

Olsson, L., Stigson, M., Perris, R., Sorrell, J. M., and Löfberg, J. 1996. Distribution of keratan sulfate and chondroitin sulfate in wild type and white mutant axolotl embryos during neural crest cell-migration. *Pigment Cell Res* 9:5–17.

Olsson, L., Svensson, K., and Perris, R. 1996. Effects of extracellular matrix molecules on sub-epidermal neural crest cell migration in wild type and white mutant axolotl embryos. *Pigment Cell Res* 9:18–27.

Opitz, J. M., and Gorlin, R. J., eds. 1988. *Neural Crest and Craniofacial Disorders: Genetic Aspects.* New York: Alan R. Liss, Inc.

Oppedal, B. R., Brandtzaeg, P., and Kemshead, J. T. 1987a. Immunohistochemical performance testing of monoclonal antibodies to neuroblastoma cells on normal adrenals, spinal and sympathetic ganglia, and neural crest tumours. *Histopathology* 11:351–362.

Oppedal, B. R., Brandtzaeg, P., and Kemshead, J. T. 1987b. Immunohistochemical differentiation of neuroblastomas from other small round cell neoplasms of childhood using a panel of mono-and polyclonal antibodies. *Histopathology* 11:363–374.

Oppenheimer, J. M. 1940. The non-specificity of the germ layers. *Q Rev Biol* 15:1–27.

O'Rahilly, R., and Gardner, E. 1979. The initial development of the human brain. *Acta Anat* 104:123–133.

Orlando, V., and Paro, R. 1995. Chromatin multiprotein complexes involved in the maintenance of transcription patterns. *Curr Opin Genet Dev* 5:174–179.

Orr, H. 1887. Contribution to the morphology of the lizard. *J Morphol* 1:311–372.

Orr-Urtreger, A., and Lonai, P. 1992. Platelet-derived growth factor-A and its receptor are expressed in separate, but adjacent cell layers of the mouse embryo. *Development* 115:1045–1058.

Ortiz-Monasterio, F. 1978. Soft tissue problems in jaw deformities. In: Whitaker, L. A., ed. *Symposium on Reconstruction of Jaw Deformities.* St. Louis: C. V. Mosby. pp. 196–203.

Osman, A., and Ruch, J. V. 1975. Topographical distribution of mitosis in odontogenic fields of the lower jaw in mice embryos. *J Biol Buccale* 3:117–132.

Osumi-Yamashita, N., and Eto, K. 1990. Mammalian cranial neural crest cells and facial development. *Dev Growth Differ* 32:451–460.

Osumi-Yamashita, N., Ninomiya, Y., Doi, H., and Eto, K. 1994. The contributions of both forebrain and midbrain crest cells to the mesenchyme in the frontonasal mass of mouse embryos. *Dev Biol* 164:409–419.

Osumi-Yamashita, N., Ninomiya, Y., Doi, H., and Eto, K. 1996. Rhombomere formation and hind-brain crest cell migration from prorhombomeric origins in mouse embryos. *Dev Growth Differ* 38:107–119.

Osumi-Yamashita, N., Ninomiya, Y., and Eto, K. 1997. Mammalian craniofacial development in vitro. *Int J Dev Biol* 41:187–194.

Oxtoby, E., and Jowett, T. 1993. Cloning of the zebrafish Krox-20 gene (Krx-20) and its expression during hindbrain development. *Nucleic Acid Res* 21:1087–1095.

Pagon, R. A., Graham, J. M. Jr., Zonana, J., and Yong, S.-L. 1981. Coloboma, congenital heart disease, and choanal atresia with multiple anomalies: CHARGE association. *J Pediatr* 99:223–227.

Pander, C. 1817. *Dissertatio inauguralis, sistens historiam metamorphoseos quam ovum incubatum prioribus quinque diebus subit.* Würzburg.

Panopoulou, G. D., Clark, M. D., Holland, L. Z., Lehrach, H., and Holland, N. D. 1998. AmphiBMP2/4, an amphioxus bone morphogenetic protein closely related to *Drosophila* decapentaplegic and vertebrate BMP2 and BMP4: Insights into evolution of dorsoventral axis specification. *Dev Dyn.* 213:130–139.

Papan, C., and Campos-Ortega, J. A. 1994. On the formation of the neural keel and neural tube in the zebrafish, *Brachydanio rerio. Wilhelm Roux's Arch Dev Biol* 203:178–186.

Paralkar, V. M., Vukicevic, S., and Reddi, A. H. 1991. Transforming growth factor ß type 1 binds to collagen IV of basement membrane matrix: Implications for development. *Dev Biol* 143:303–308.

Paralkar, V. M., Weeks, B. S., Yu, Y. M., Kleinman, H. K., and Reddi, A. H. 1992. Recombinant human bone morphogenetic protein 2B stimulates PC12 cell differentiation: Potentiation and binding to type IV collagen. *J Cell Biol* 119:1721–1728.

Parichy, D. M. 1996a. Salamander pigment patterns: How can they be used to study developmental mechanisms and their evolutionary transformation? *Int J Dev Biol* 40:871–874.

Parichy, D. M. 1996b. When neural crest and placodes collide: Interactions between melanophores and the lateral lines that generate stripes in the salamander *Ambystoma tigrinum tigrinum* (Ambystomatidae). *Dev Biol* 175:283–300.

Parichy, D. M. 1996c. Pigment patterns of larval salamanders (Ambytomatidae, Salamandridae): The role of the lateral line sensory system and the evolution of pattern-forming mechanisms. *Dev Biol* 175:265–282.

Parichy, D. M. 1998. Experimental analysis of character coupling across a complex life cycle: Pigment pattern metamorphosis in the tiger salamander, *Ambystoma tigrinum tigrinum. J Morphol* 237:53–67.

Parr, B. A., Shea, M. J., Vassileva, G., and McMahon, A. P. 1993. Mouse *Wnt* genes exhibit discrete domains of expression in the early embryonic CNS and limb buds. *Development* 119:247–261.

Pascualcastroviejo, I. 1990. Tumors of the neural crest. In: Pascualcastroviejo, I., ed. *Spinal Tumors in Children and Adolescents.* New York: Raven Press. pp. 111–128.

Pasteels, J. 1943. Proliférations et croissance dans la gastrulation et al formation de la queue des Vertébrés. *Arch Biol (Liège)* 54:1–51.

Patterson, P. H. 1990. Control of cell fate in a vertebrate neurogenic lineage. *Cell* 62:1035–1038.

Payette, R. F., Tennyson, V. M., Pomeranz, H. D., Pham, T. D., Rothman, T. P., and Gershon, M. D. 1988. Accumulation of components of basal laminae: Association with the failure of neural crest cells to colonize the presumptive aganglionic bowel of *Ls/Ls* mutant mice. *Dev Biol* 125:341–360.

Pearse, A. G. E. 1969. The cytochemistry and ultrastructure of polypeptide hormone-producing cells of the APUD series, and the embryologic, physiologic and pathologic implications of the concept. *J Histochem Cytochem* 17:303–313.

Pearse, A. G. E. 1977a. The diffuse neuroendocrine system and the 'common peptides.' In: MacIntyre, I., and Szelke, M., eds. *Molecular Endocrinology.* Amsterdam: Elsevier/North-Holland. pp. 309–323.

Pearse, A. G. E. 1977b. The APUD concept and its implications: Related endocrine peptides in brain, intestine, pituitary, placenta and anuran cutaneous glands. *Med Biol* 55:115–125.

Peeters, M. C. E., Schutte, B., Lenders, M.-H. J. N., Hekking, J. W. M., Drukker, J., and van Straaten, H. W. M. 1998. Role of differential cell proliferation in the tail bud in aberrant mouse neurulation. *Dev Dyn* 211:382–389.

Percy, R., and Potter, I. C. 1991. Aspects of the development and functional morphology of the pericardial heart and associated blood vessels of lampreys. *J Zool (Lond)* 223:49–66.

Perris, R., von Boxberg, Y., and Löfberg, J. 1988. Local embryonic matrices determine region-specific phenotypes in neural crest cells. *Science* 241:86–89.

Perris, R., and Johansson, S. 1990. Inhibition of neural crest cell migration by aggregating chondroitin sulfate proteoglycans is mediated by their hyaluronan-binding region. *Dev Biol* 137:1–12.

Perris, R., Krotoski, D., and Bronner-Fraser, M. 1991. Collagens in avian neural crest development: Distribution *in vivo* and migration-promoting ability *in vitro. Development* 113:969–984.

Perris, R., Krotowski, D., Lallier, T., Domingo, C., Sorrell, M., and Bronner-Fraser, M. 1991. Spatial and temporal changes in the distribution of proteoglycans during avian neural crest development. *Development* 111:583–599.

Perris, R., Kuon, H.-J., Glanville, R. W., and Bronner-Fraser, M. 1993. Collagen type VI in neural crest development. Distribution *in situ* and interaction with cells *in vitro. Dev Dyn* 198:135–149.

Perris, R., and Löfberg, J. 1986. Promotion of chromatophore differentiation in isolated premigratory neural crest cells by extracellular material explanted on microcarriers. *Dev Biol* 113:327–341.

Perris, R., Löfberg, J., Fällström, C., Boxberg, Y. V., Olsson, L., and Newgreen, D. F. 1990. Structural and compositional divergencies in the extracellular matrix encountered by neural crest cells in the white mutant axolotl embryo. *Development* 109:533–551.

Persaud, T. V. N., Chudley, A. E., and Skalko, R. G. 1985. *Basic Concepts in Teratology.* New York: Alan R. Liss, Inc.

Person, P. 1983. Invertebrate cartilages. In: Hall, B. K., ed. *Cartilage. Vol. 1. Structure, Function and Biochemistry.* New York: Academic Press. pp. 31–58.

Peterson, K. J. 1994. The origin and early evolution of the Craniata. In: Prothero, D. R., and Schoch, R. M., eds. *Major Features of Vertebrate Evolution. Knoxville: The Paleontology Society and the University of Tennessee. Short Courses in Paleontology No. 7.* pp. 14–37.

Peterson, P. E., Blankenship, T. H., Wilson, D. B., and Hendrickx, A. G. 1996. Analysis of hindbrain neural crest migration in the long-tailed monkey (*Macaca fascicularis*). *Anat Embryol* 194:235–246.

Peters-van der Sanden, M. J. H. 1994. The hindbrain neural crest and the development of the enteric nervous system. Thesis. Erasmus University. Rotterdam.

Peters-van der Sanden, M. J. H., Luider, T. M., van der Kamp, A. W. M., Tibboel, D., and Meijers, C. 1993. Regional differences between various axial segments of the avian neural crest regarding the formation of enteric ganglia. *Differentiation* 53:17–24.

Pettway, Z., Guillory, G., and Bronner-Fraser, M. 1990. Absence of neural crest cells from the region surrounding implanted notochord *in situ. Dev Biol* 142:335–345.

Phillips, M. T. III., Waldo, K., and Kirby, M. L. 1989. Neural crest ablation does not alter pulmonary vein development in the chick embryo. *Anat Rec* 223:292–298.

Pictet, R. L., Rall, L. B., Phelps, P., and Rutter, W. J. 1976. The neural crest and the origin of the insulin-producing and other gastrointestinal hormone-producing cells. *Science* 191:191–193.

Pictet, R. L., and Rutter, W. J. 1972. Development of the human pancreas. In: Steiner, D., and Freinkel, N., eds. *Handbook of Physiology. Sec. 7, Vol. 1. Endocrine Pancreas.* Baltimore: Williams & Wilkins. pp. 25–66.

Pierce, G. B. 1985. Carcinoma is to embryology as mutation is to genetics. *Am Zool* 25:707–712.

Pierce, G. B. 1987. The embryological basis of carcinoma. In: Maccioni, R. B., and Aréchaga, J., eds. *The Cytoskeleton in Cell Differentiation and Development.* Oxford: IRL Press. pp. 13–23.

Pinco, O., Carmeli, C., Rosenthal, A., and Kalcheim, C. 1993. Neurotrophin-3 affects proliferation and differentiation of distinct neural crest cells and is present in the early neural tube of avian embryos. *J Neurobiol* 24:1626–1641.

Piotrowski, T., Schilling, T. F., Brand, M., Jiang, Y.-J., Heisenberg, C.-P., Beuchle, D., Grandel, H., van Eeden, F. J. M., Furutani-Seiki, M., Granato, M., Haffter, P., Hammerschmidt, M., Kane, D. A., Kelsh, R. N., Mullins, M. C., Odenthal, J., Warga, R. M., and Nüsslein-Volhard, C. 1996. Jaw and branchial arch mutants in zebrafish. II: Anterior arches and cartilage differentiation. *Development* 123:345–356.

Platt, J. B. 1893. Ectodermic origin of the cartilages of the head. *Anat Anz* 8:506–509.

Platt, J. B. 1894. Ontogenetic differentiation of the ectoderm in Necturus. Second preliminary note. *Arch Mikrosk Anat EntwMech* 43:911–966.

Platt, J. B. 1896. Ontogenetic differentiation of the ectoderm in Necturus. II. On the development of the peripheral nervous system. *Q J Microsc Sci* 38:485–547.

Platt, J. B. 1897. The development of the cartilaginous skull and of the branchial and hypoglossal musculature in *Necturus. Morphol Jb* 25:377–464.

Poelmann, R. E., Gittenberger-de Groot, A. C., Mentink, M. M. T., Delpech, B., Girard, N., and Christ, B. 1990. The extracellular matrix during neural crest formation and migration in rat embryos. *Anat Embryol* 182:29–39.

Poelmann, R. E., Mikawa, T., and Gittenberger-de Groot, A. C. 1998. Neural crest cells in outflow tract septation of the embryonic chicken heart: Differentiation and apoptosis. *Dev Dyn* 212:373–384.

Pomerance, B. 1979. *The Elephant Man.* New York: Grove Press.

Pomeranz, H. D., and Gershon, M. D. 1990. Colonization of the avian hindgut by cells derived from the sacral neural crest. *Dev Biol* 137:378–394.

Pomeranz, H. D., Rothman, T. P., Chalazonitis, A., Tennyson, V. M., and Gershon, M. D. 1993. Neural crest-derived cells isolated from the gut by immunoselection develop neuronal and glial phenotypes when cultured on laminin. *Dev Biol* 156:341–361.

Ponder, B. 1990. Neurofibromatosis gene cloned. *Nature* 346:703–704.

Pons, G., O'Dea R. F., and Mikkin, B. L. 1982. Biological characterization of the C-1300 murine neuroblastoma: An in vivo neural crest tumor model. *Cancer Res* 42:3719–3723.

Poswillo, D. 1974. Otomandibular deformity: Pathogenesis as a guide to reconstruction. *J Maxillofac Surg* 2:64–72.

Poswillo, D. 1975a. Causal mechanisms of craniofacial deformity. *Br Med Bull* 31:101–106.

Poswillo, D. 1975b. The pathogenesis of the Treacher Collins syndrome (mandibulofacial dysostosis). *Br J Oral Surg* 13:1–26.

Poswillo, D. 1975c. Hemorrhage in development of the face. *Birth Defects Orig Art Ser* 16:61–81.

Poswillo, D. 1978. Pathogenesis of jaw deformity. In: Whitaker, L. A., ed. Symposium on *Reconstruction of Jaw Deformities.* St. Louis: C. V. Mosby. pp. 69–75.

Presley, R., Horder, T. J., and Slipka, J. 1996. Lancelet development as evidence of ancestral chordate structure. *Isr J Zool* 42:S97–S116.

Prieto, A. L., Crossin, K. L., Cunningham, B. A., and Edelman, G. M. 1989. Localization of mRNA for neural cell adhesion molecule (N-CAM) polypeptides in neural and non-neural tissues by *in situ* hybridization. *Proc Natl Acad Sci USA* 86:9579–9583.

Prince, V., and Lumsden, A. 1994. *Hoxa-2* expression in normal and transposed rhombomeres: Independent regulation in the neural tube and neural crest. *Development* 120:911–923.

Prince, V. E., Moens, C. B., Kimmel, C. B., and Ho, R. K. 1998. Zebrafish *hox* genes: Expression in the hindbrain region of wild-type and mutants of the segmentation gene, valentino. *Development* 125:393–406.

Puffenberger, E. G., Hosoda, K., Washington, S. S., Nakao, K. deWilt, D., Yanagiswa, M., and Chakvarti, A. 1994. A mis-sense mutation of the endothelin-B receptor gene in multigenic Hirschsprung's disease. *Cell* 79:1257–1266.

Puzdrowski, R. L. 1989. Peripheral distribution and central projections of the lateral-line nerves in goldfish, *Carassius auratus. Brain Behav Evol* 34:110–131.

Qin, F., and Kirby, M. L. 1995. *Int-2* influences the development of the nodose ganglion. *Pediatr Res* 38:485–492.

Qiu, M., Bulfone, A., Ghattas, I., Meneses, J. J., Christensen, L., Sharpe, P. T., Presley, R., Pedersen, R. A., and Rubenstein, J. L. R. 1997. Role of the Dlx homeobox genes in proximodistal patterning of the branchial arches: Mutations of *Dlx-1, Dlx-2*, and *Dlx-1* and *-2* alter morphogenesis of proximal skeletal and soft tissue structures derived from the first and second arches. *Dev Biol* 185:165–284.

Qiu, M., Bulfone, A., Martinez, S., Meneses, J. J., Shimamura, K., Pedersen, R. A., and Rubenstein, J. L. R. 1995. Role of *Dlx-2* in head development and evolution: Null mutation of *Dlx-2* results in abnormal morphogenesis of proximal first and second

branchial arch derivatives and abnormal differentiation in the forebrain. *Genes Dev* 9:2523–2538.
Quinlan, G. A., Williams, E. A., Tan, S.-S., and Tam, P. P. L. 1995. Neurectodermal fate of epiblast cells in the distal region of the mouse egg cylinder: Implications for body plan organization during early embryogenesis. *Development* 121:87–98.
Rachootin, S., and Thomson, K. S. 1981. Epigenetics, paleontology and evolution. In: Scudder, G. C. E., and Reveal, J. L., eds. *Evolution Today.* Pittsburgh: Hunt Institute for Botanical Documentation, Carnegie-Mellon University. pp. 181–193.
Raible, D. W., and Eisen, J. S. 1994. Restriction of neural crest cell fate in the trunk of the embryonic zebrafish. *Development* 120:495–503.
Raible, D. W., and Eisen, J. S. 1996. Regulative interactions in zebrafish neural crest. *Development* 122:501–507.
Raible, D. W., Wood, A., Hodson, W., Henion, P. D., Weston, J. A., and Eisen, J. S. 1992. Segregation and early dispersal of neural crest cells in the embryonic zebrafish. *Dev Dyn* 195:29–42.
Raven, C. P. 1931a. Die eigentümliche Bildungsweise des Neuralrohrs beim Axolotl und die Lage des Ganglienleistenmaterials. *Anat Anz* 71:161–166.
Raven, C. P. 1931b. Zur Entwicklung der Ganglienleiste. I. Die Kinematik der Ganglienleisten Entwicklung bei den Urodelen. *Wilhelm Roux Arch EntwMech Org* 125:210–293.
Raven, C. P. 1931c. Die Induktionsfähigkeit des Ganglienleistenmaterials von *Rana fusca.* Ein Beitrag zur Determinationsfrage. *Proc K Ned Akad Wet* 34:554–557.
Raven, C. P. 1933a. Zur Entwicklung der Ganglienleiste. II. Über das differenzierungsvermögen des Kopfganglienleistenmaterials von Urodelen. *Wilhelm Roux Arch EntwMech Org* 129:179–198.
Raven, C. P. 1933b. Zur Entwicklung der Ganglienleiste. III. Die Induktionsfähigkeit des Kopfganglienleistenmaterials von *Rana fusca. Wilhelm Roux Arch EntwMech Org* 130:517–561.
Raven, C. P. 1935. Zur Entwicklung der Ganglienleiste. IV. Untersuchungen über Zeitpunkt und Verlauf der 'materiellen Determination' des präsumptiven Kopfganglienleistenmaterials der Urodelen. *Wilhelm Roux Arch EntwMech Org* 132:509–575.
Raven, C. P. 1936. Zur Entwicklung der Ganglienleiste. V. Über die Differenzierung des Rumpfganglienleistenmaterials. *Wilhelm Roux Arch EntwMech Org* 134:122–145.
Raven, C. P. 1937. Experiments on the origin of the sheath cells and sympathetic neuroblasts in Amphibia. *J Comp Neurol* 67:220–240.
Raven, C. P. 1942. The neural crest. *Arch Neerl Zool* 6:475–478.
Raven, C. P., and Kloos, J. 1945. Induction by medial and lateral pieces of the archenteron roof with special reference to the determination of the neural crest. *Acta Neerl Morphol* 5:348–362.
Rawles, M. E. 1940. The development of melanophores from embryonic mouse tissues grown in the coelom of chick embryos. *Proc Natl Acad Sci USA* 26:673–680.
Rawles, M. E. 1944. The migration of melanoblasts after hatching into pigment-free skin grafts of the common fowl. *Physiol Zool* 17:167–183.
Rawles, M. E. 1948. Origin of melanophores and their role in development of color patterns in vertebrates. *Physiol Zool* 28:383–408.
Redekop, G., Elisevic, K., and Gilbert, J. 1990. 4th ventricular Schwannoma: Case report. *J Neurosurg* 73:777–781.
Redline, R. W., Neish, A., Holmes, L. B., and Collins, T. 1992. Homeobox genes and congenital malformations. *Lab Invest* 66:659–670.

Reedy, M. V., Faraco, C. D., and Erickson, C. A. 1998. The delayed entry of thoracic neural crest cells into the dorsolateral path is a consequence of the late emigration of melanogenic neural crest cells from the neural tube. *Dev Biol* 200:234–246.

Reichenbach, A., Schaaf, P., and Schneider, H. 1990. Primary neurulation in teleosts—evidence for epithelial genesis of central nervous tissue as in other vertebrates. *J Hirnforsch* 31:152–158.

Reif, W.-E. 1982. Evolution of dermal skeleton and dentition in vertebrates: The odontode-regulation theory. *Evol Biol* 15:287–368.

Reiss, J. O. 1990. Effect of unilateral nasal placode extirpation on anuran cranial development. *Am Zool* 30:139A.

Reissmann, E., Ernsberger, U., Francis-West, P. H., Rueger, D., Brickell, P. M., and Rohrer, H. 1996. Involvement of bone morphogenetic protein-4 and bone morphogenetic protein-7 in the differentiation of the adrenergic phenotype in sympathetic neurons. *Development* 122:2079–2088.

Remak, R. 1850–1855. *Untersuchungen über die Entwickelung der Wirbelthiere.* Berlin.

Rhinn, M., Dierich, A., Shawlot, W., Behringer, R. R., Le Meur, M., and Ang, S.-L. 1998. Sequential roles for *Otx2* in visceral endoderm and neurectoderm for forebrain and midbrain induction and specification. *Development* 125:845–856.

Riccardi, V. M. 1979. Cell–cell interaction as an epigenetic determinant in the expression of mutant neural crest cells. *Birth Defects Orig Art Ser* 15:89–98.

Riccardi, V. M. 1988. Neurofibromatosis: Challenges for applied cellular and molecular biology. *Lab Invest* 59:726–728.

Riccardi, V. M., and Eichner, J. E. 1986. *Neurofibromatosis: Phenotype, Natural History, and Pathogenesis.* Baltimore: Johns Hopkins University Press.

Riccardi, V. M., and Margos, V. A. 1981. Characteristics of skin and tumor fibroblasts from neurofibromatosis patients. In: Riccardi, V. M., and Mulvihill, J. J., eds. *Neurofibromatosis (von Recklinghausen Disease).* New York: Raven Press. pp. 191–198.

Riccardi, V. M., and Mulvihill, J. J. 1981. *Neurofibromatosis (von Recklinghausen Disease).* Genetics, Cell Biology, and Biochemistry. Advances in Neurology Vol. 29. New York: Raven Press.

Richardson, M. K., and Hornbruch, A. 1991. Quail neural crest cells cannot read positional values in the dorsal trunk feathers of the chicken embryo. *Wilhelm Roux's Arch Dev Biol* 199:397–401.

Richardson, M. K., Hornbruch, A., and Wolpert, L. 1989. Pigment pattern expression in the plumage of the quail embryo and the quail–chick chimaera. *Development* 107:805–818.

Richardson, M. K., and Sieber-Blum, M. 1993. Pluripotent neural crest cells in the developing skin of the quail embryo. *Dev Biol* 157:348–358.

Richman, J. 1994. Morphogenesis of bone. In: Hall, B. K., ed. *Bone. Vol. 8. Differentiation and Morphogenesis of Bone.* Boca Raton: CRC Press. pp. 65–118.

Rijli, F. M., Mark, M., Lakkaraju, S., Dietrich, A., Dollé, P., and Chambon, P. 1993. A homeotic transformation is generated in the rostral branchial region of the head by disruption of *Hoxa-2*, which acts as a selector gene. *Cell* 75:1333–1349.

Riopelle, R. J., Haliotis, T., Roper, J. C. 1983. Nerve growth factor receptors of human tumors of neural crest origin: Characterization of binding site heterogeneity and alteration by theophylline. *Cancer Res* 43:5184–5189.

Riopelle, R. J., and Riccardi, V. M. 1987. Neuronal growth factors from tumors of von Recklinghausen neurofibromatosis. *Can J Neurol Sci* 14:141–144.

Riou, J.-F., Shi, D.-L., Chiquet, M., and Bouchaut, J.-C. 1988. Expression of tenascin in response to neural induction in amphibian embryos. *Development* 104:511–524.

Ris, H. 1941. An experimental study on the origin of melanophores in birds. *Physiol Zool* 14:48–66.

Robertson, K., and Mason, I. 1995. Expression of *ret* in the chicken embryo suggests roles in regionalization of the vagal neural tube and somites and in development of multiple neural crest and placodal lineages. *Mech Dev* 53:329–344.

Robson, P., Wright, G. M., Youson, J. H., and Keeley, F. W. 1997. A family of non-collagen-based cartilages in the skeleton of the sea lamprey, *Petromyzon marinus. Comp Biochem Physiol* 118B:71–78

Rocek, Z., and Vesely, M. 1989. Development of the ethmoidal structures of the endocranium in the anuran *Pipa pipa. J Morphol* 200:301–320.

Rodríguez-Gallardo, L., Climent, V., Garcia-Martinez, V., Schoenwolf, G. C., and Alvarez, I. S. 1997. Targeted over-expression of FGF in chick embryos induces formation of ectopic neural cells. *Int J Dev Biol* 41:715–723.

Rogers, S. L., Cutts, J. L., Gegick, P. J., McGuire, P. G., Rosenberger, C., and Krisinski, S. 1994. Transforming growth factor-ß1 differentially regulates proliferation, morphology, and extracellular matrix expression by three neural crest-derived neuroblastoma cell lines. *Exp Cell Res* 211:252–262.

Rogers, S. L., Gegick, P. J., Alexander, S. M., and McGuire, P. G. 1992. Transforming growth factor-ß alters differentiation in cultures of avian neural crest-derived cells: Effects on cell morphology, proliferation, fibronectin expression, and melanogenesis. *Dev Biol* 151:192–203.

Rohon, V. 1884. Zur histogenese des Rückenmarks der Forelle. *Sitzungsberichte Akad Wien* 1884:39–57.

Rollhäuser-ter Horst, J. 1980. Neural crest replaced by gastrula ectoderm in amphibia. Effect on neurulation, CNS, gills and limbs. *Anat Embryol* 160:203–212.

Romer, A. S. 1972. The vertebrate as a dual animal-somatic and visceral. *Evol Biol* 6:121–156.

Rosenquist, G. C. 1981. Epiblast origin and early migration of neural crest cells in the chick embryo. *Dev Biol* 87:201–211.

Rosenquist, T. H., Beal, A. C., Modis, L., and Fishman, R. 1990. Impaired elastic matrix development in the great arteries after ablation of the cardiac neural crest. *Anat Rec* 226:347–359.

Rosenquist, T. H., Kirby, M. L., and Van Mierop, L. H. S. 1989. Solitary aortic arch artery—A result of surgical ablation of cardiac neural crest and nodose placode in the avian embryo. *Circulation* 80:1469–1475.

Roth, S. 1973. A molecular model for cell interactions. *Q Rev Biol* 48:541–563.

Rothman, T. P., and Gershon, M. D. 1984. Regionally defective colonization of the terminal bowel by the precursors of enteric neurons in lethal spotted mutant mice. *Neuroscience* 12:1293–1311.

Rothman, T. P., Gershon, M. D., Fontaine-Pèrus, J. C., Chanconie, M., and Le Douarin, N. M. 1987. The effect of back-transplants of the embryonic gut wall on growth of the neural tube. *Dev Biol* 124:331–346.

Rothman, T. P., Goldowitz, D., and Gershon, M. D. 1993. Inhibition of migration of neural crest-derived cells by the abnormal mesenchyme of the presumptive aganglionic bowel of *ls/ls* mice: Analysis with aggregation and interspecific chimeras. *Dev Biol* 159:559–573.

Rothman, T. P., Le Douarin, N. M., Fontaine-Perus, J. C., and Gershon, M. D. 1990. Developmental potential of neural crest-derived cells migrating from segments of developing quail bowel back-grafted into younger chick host embryos. *Development* 109:411–423.

Rothman, T. P., Le Douarin, N. M., Fontaine-Perus, J. C., and Gershon, M. D. 1993. Inhibition of migration of neural crest-derived cells by the abnormal mesenchyme of the presumptive bowel of *Ls/Ls* mice. *Dev Dyn* 196:217–213.

Rouleau, G. A., Wertelecki, W., Haines, J. L., Hobbs, W. J., Trofatter, J. A., Seizinger, B. R., Martuzas, R. L., Superneau, D. W., Conneally, P. M., and Gusella, J. F. 1987. Genetic linkage of bilateral acoustic neurofibromatosis to a DNA marker on chromosome 22. *Nature* 329:246–248.

Rovasio, R. A., and Battiato, N. L. 1995. Role of early migratory neural crest cells in developmental anomalies induced by ethanol. *Int J Dev Biol* 39:421–422.

Rovasio, R. A., Delouvée, A., Yamada, K. M., Timpl, R., and Thiery, J.-P. 1983. Neural crest cell migration-requirements for endogenous fibronectin and high cell density. *J Cell Biol* 96:462–473.

Rovasio, R. A., and Thiery, J.-P. 1987. Mechanism of neural crest cell migration. *Microsc Elect Biol Cell* 11:81–100.

Rowe, A., and Brickell, P. M. 1995. Expression of the chicken retinoic X receptor-γ gene in migrating cranial neural crest cells. *Anat Embryol* 192:1–8.

Rowe, A., Eager, N. S. C., and Brickell, P. M. 1991. A member of the RXR nuclear-receptor family is expressed in neural-crest-derived cells of the developing chick peripheral nervous system. *Development* 111:771–778.

Rubenstein, A. E., Bunge, R.P., and Housman, D. E. (eds.) 1986. Neurofibromatosis. *Ann NY Acad Sci* 486:1–414.

Ruberte, E., Dolle, P., Krust, A., Zelent, A., Morriss-Kay, G. M., and Chambon, P. 1990. Specific spatial and temporal distribution of retinoic acid receptor g transcripts during mouse embryogenesis. *Development* 108:213–222.

Ruberte, E., Friederich, V., Morriss-Kay, G. M., and Chambon, P. 1992. Differential distribution patterns of CRABP-I and CRABP-II transcripts during mouse embryogenesis. *Development* 115:973–987.

Ruberte, E., Wood, H. B., and Morriss-Kay, G. M. 1997. Prorhombomeric subdivision of the mammalian embryonic hindbrain: Is it functionally meaningful? *Int J Dev Biol* 41:213–222.

Ruch, J. V., ed. 1995. Odontogenesis. *Int J Dev Biol* 39:1–297.

Ruiz, W. F., Mujwid, D. K., and Steffek, A. J. 1982. Scanning electron microscopy (SEM) of the extracellular matrix meshwork during cranial neural crest cell migration in chick embryos. In: Hawkes, S., and Wang, J. L., eds. *Extracellular Matrix.* New York: Academic Press. pp. 135–140.

Ruiz i Altaba, A., and Jessell, T. 1991. Retinoic acid modifies mesodermal patterning in early *Xenopus* embryos. *Genes Dev* 5:175–187.

Runyan, R. B., Maxwell, G. D., and Shur, B. D. 1986. Evidence for a novel enzymatic mechanism of neural crest cell migration on extracellular glycoconjugate matrices. *J Cell Biol* 102:432–441.

Russell, E. S. 1916. *Form and Function. A Contribution to the History of Animal Morphology.* London: John Murray. Reprinted with an introduction by G. V. Laude, 1982. Chicago: University of Chicago Press.

Sadaghiani, B., Crawford, B. J., and Vielkind, J. R. 1994. Changes in the distribution of extracellular matrix components during neural crest development in *Xiphophorus* spp. embryos. *Can J Zool* 72:1340–1353.

Sadaghiani, B., and Thiébaud, C. H. 1987. Neural crest development in the *Xenopus laevis* embryo, studied by interspecific transplantation and scanning electron microscopy. *Dev Biol* 124:91–110.

Sadaghiani, B., and Vielkind, J. R. 1989. Neural crest development in *Xiphophorus* fishes: Scanning electron and light microscopic studies. *Development* 105:487–504.

Sadaghiani, B., and Vielkind, J. R. 1990a. Explanted fish neural tubes give rise to differentiating neural crest cells. *Dev Growth Differ* 32:513–520.

Sadaghiani, B., and Vielkind, J. R. 1990b. Distribution and migration pathways of HNK-1-immunoreactive neural crest cells in teleost fish embryos. *Development* 110:197–209.

Sagemehl, M. 1882. *Untersuchungen über die Entwicklung der Spinalnerven.* Dorpat.

Saldivar, J. R., Krull, C. E., Krumlauf, R., Ariza-McNaughton, L., and Bronner-Fraser, M. 1996. Rhombomere of origin determines autonomous versus environmentally regulated expression of *Hoxa3* in the avian embryo. *Development* 122:895–904.

Saldivar, J. R., Sechrist, J. W., Krull, C. E., Ruffins, S., and Bronner-Fraser, M. 1997. Dorsal hindbrain ablation results in rerouting of neural crest migration and changes in gene expression, but normal hyoid development. *Development* 124:2729–2739.

Salvini-Plawen, L. 1980. Phylogenetischer Status und Bedeutung der mesenchymaten Bilateria. *Zool Jb Anat* 103:354–373.

Salzberg, A., and Bellen, H. J. 1996. Invertebrate versus vertebrate neurogenesis: Variations on the same theme? *Dev Genet* 18:1–10.

Salzgeber, B., and Guénet, J.-L. 1984. Studies on "Repeated Epilation" mouse mutant embryos: II. Development of limb, tail, and skin defects. *J Craniofac Genet Dev Biol* 4:95–114.

Sanders, E. J., and Cheung, E. 1988. Effects of HNK-1 monoclonal antibody on the substratum attachment and survival of neural crest and sclerotomal cells in culture. *J Cell Sci* 90:115–122.

Sanders, E. J., Prasad, S., and Cheung, E. 1988. Extracellular matrix synthesis is required for the movement of sclerotomal and neural crest cells on collagen. *Differ Ontog Neoplasia* 39:34–41.

Sansom, I. J. 1996. *Pseudoneotodus*—A histological study of an Ordovician to Devonian vertebrate lineage. *Zool J Linn Soc* 118:47–57.

Sansom, I. J., Smith, M. M., and Smith, M. P. 1996. Scales of thelodont and shark-like fishes from the Ordovician of Colorado. *Nature* 379:628–630.

Sansom, I. J., Smith, M. P., Armstrong, H. A., and Smith, M. M. 1992. Presence of the earliest vertebrate hard tissues in conodonts. *Science* 256:1308–1311.

Sansom, I. J., Smith, M. P., and Smith, M. M. 1994. Dentine in conodonts. *Nature* 368:591.

Sansom, I. J., Smith, M. P., Smith, M. M., and Turner, P. 1997. Astraspis—The anatomy and histology of an Ordovician fish. *Palaeontology* 40:625–643.

Sasai, Y., and de Robertis, E. M. 1997. Ectodermal patterning in vertebrate embryos. *Dev Biol* 182:5–20.

Sasai, Y., Lu, B., Steinbeisser, H., and de Robertis, E. M. 1995. Regulation of neural induction by the Chd and BMP-4 antagonistic patterning signals in *Xenopus*. *Nature* 376:333–336.

Satoh, N. 1994. *Developmental Biology of Ascidians.* Cambridge: Cambridge University Press.

Saxén, L., and Saxén, E. 1960. Malignant embryoma of the neural crest. *Cancer* 13:899–906.

Saxén, L., and Toivonen, S. 1962. *Primary Embryonic Induction.* London: Academic Press.

Schaeffer, B. 1977. The dermal skeleton in fishes. In: *Problems in Vertebrate Evolution.* Andrews, S. M., Miles, R. S., and Walker, A. D., eds. London: Academic Press. Linnean Society Symposium No. 4. pp. 25–52.

Schaeffer, B. 1987. Deuterostome monophyly and phylogeny. *Evol Biol* 21:179–235.

Schaeffer, B., and Thomson, K S. 1980. Reflections on agnathan-gnathosome relationships. In: Jacobs, L. L., ed. *Aspects of Vertebrate History. Essays in Honor of Edwin Harris Colber.* Flagstaff: Museum of Northern Arizona Press. pp. 19–33.

Schatterman, G. C., Morrison-Graham, K., van Koppen, A., Weston, J. A., and Bowen-Pope, D. F. 1992. Regulation and role of PDGF receptor a-subunit expression during embryogenesis. *Development* 115:123–131.

Scherson, T., Serbedzija, G., Fraser, S. E., and Bronner-Fraser, M. 1993. Regulative capacity of the cranial neural tube to form neural crest. *Development* 118:1049–1061.

Schilling, T. F. 1997. Genetic analysis of craniofacial development in the vertebrate embryo. *BioEssays* 19:459–468.

Schilling, T. F., and Kimmel, C. B. 1994. Segment and cell type lineage restrictions during pharyngeal arch development in the zebrafish embryo. *Development* 120:483–494.

Schilling, T. F., Piotrowski, T., Grandel, H., Brand, M., Heisenberg, C.-P., Jiang, Y.-J., Beuchle, D., Hammerschmidt, M., Kane, D. A., Mullins, M. C., van Eeden, F. J. M., Kelsh, R. N., Furutani-Seiki, M., Granato, M., Haffter, P., Odenthal, J., Warga, R. M., Trowe, T., and Nüsslein-Volhard, C. 1996. Jaw and branchial arch mutants in zebrafish 1: Branchial arches. *Development* 123:329–344.

Schilling, T. F., Walker, C., and Kimmel, C. B. 1996. The *chinless* mutation and neural crest cell interactions in zebrafish jaw development. *Development* 122:1417–1426.

Schmale, M. C., and Hensley, G. T. 1988. Transmissibility of a neurofibromatosis-like disease in bicolor damselfish. *Cancer Res* 48:3828–3833.

Schmale, M. C., Hensley, G. T., and Udey, L. R. 1983. Multiple Schwannomas in the bicolor damselfish, *Pomacentrus partitus*: A possible model of von Recklinghausen neurofibromatosis. *Am J Pathol* 112:238–241.

Schmitz, X., Papan, C., and Campos-Ortega, J. A. 1993. Neurulation in the anterior trunk region of the zebrafish, *Brachydanio rerio. Wilhelm Roux's Arch Dev Biol* 203:250–259.

Schlumberger, H. G. 1951. Limbus tumors as a manifestation of von Recklinghausen's neurofibromatosis in goldfish. *Am J Ophthalmol* 34:415–422.

Schoenwolf, G. C., ed. 1986. *Scanning Electron Microscopy Studies of Embryogenesis.* AMF O'Hare: Scanning Electron Microscopy, Inc.

Schoenwolf, G. C., and Alvarez, I. S. 1991. Specification of neuroepithelium and surface epithelium in avian transplantation chimeras. *Development* 112:713–722.

Schoenwolf, G. C., Chandler, N. B., and Smith, J. L. 1985. Analysis of the origins and early fates of neural crest cells in caudal regions of avian embryos. *Dev Biol* 110:467–479.

Schoenwolf, G. C., Everaert, S., Bortier, H., and Vakeet, L. 1989. Neural plate- and neural tube-forming potential of isolated epiblast areas in avian embryos. *Anat Embryol* 179:541–549.

Schoenwolf, G. C., and Nichols, D. H. 1984. Histological and ultrastructural studies on the origin of caudal neural crest cells in mouse embryos. *J Comp Neurol* 222:496–505.

Schoenwolf, G. C., and Sheard, P. 1990. Fate mapping the avian epiblast with focal injections of a fluorescent-histochemical marker: Ectodermal derivatives. *J Exp Zool* 255:323–339.

Schorle, H., Meier, P., Buchert, M., Jaenisch, R., and Mitchell, P. J. 1996. Transcription factor AP-2 essential for cranial closure and craniofacial development. *Nature* 381:235–238.

Schumacher, A., and Magnuson, T. 1997. Murine *Polycomb*- and *trithorax*-group genes regulate homeotic pathways and beyond. *Trends Genet* 13:167–170.

Scott, J. P. 1986. Critical periods in organizational processes. In: Falkner, F., and Tanner, J. M., eds. *Human Growth. A Comprehensive Treatise. Vol. 1. Developmental Biology, Prenatal Growth.* 2nd ed. New York and London: Plenum Press. pp. 181–198.

Searls, R. L., and Zwilling, E. 1964. Regeneration of the apical ectodermal ridge of the chick limb bud. *Dev Biol* 9:38–55.

Sechrist, J., Nieto, M. A., Zamanian, R. T., and Bronner-Fraser, M. 1995. Regulative response of the cranial neural tube after neural fold ablation: Spatiotemporal nature of neural crest regeneration and up-regulation of *Slug. Development* 121:4103–4115.

Sechrist, J., Scherson, T., and Bronner-Fraser, M. 1994. Rhombomere rotation reveals that multiple mechanisms contribute to the segmental pattern of hindbrain neural crest migration. *Development* 120:1777–1790.

Sechrist, J., Serbedzija, G. N., Scherson, T., Fraser, S. E., and Bronner-Fraser, M. 1993. Segmental migration of the hindbrain neural crest does not arise from its segmental generation. *Development* 118:691–703.

Sedgwick, A. 1894a. On the inadequacy of the cellular theory and on the early development of nerves, particularly of the third nerve and of the sympathetic in Elasmobranchii. *Q J Microsc Sci* 37:87–101. [1895] *Studies Morphol Lab Univ Camb* (A. Sedgwick, ed.) 6:93–107.

Sedgwick, A. 1894b. On the inadequacy of the cell theory and on the development of nerves. *Proc Camb Philos Soc* 8:(1895):248.

Sedgwick, A. 1895a. Remarks on the cell-theory. *Congr Int Zool C R* 1895:121–124.

Sedgwick, A. 1895b. Further remarks on the cell-theory with a reply to Mr. Bourne. *Q J Microsc Sci* 38:331–337.

Sedgwick, A. 1910. *Embryology. Encyclopaedia Britannica,* 11th ed. pp. 314–329.

Seifert, R., Jacob, M., and Jacob, H. J. 1993. The avian prechordal head region: A morphological study. *J Anat* 183:75–89.

Seizinger, B. R., Martuza, R. L., and Gusella, J. F. 1986. Loss of genes on chromosome 22 in tumorigenesis of human acoustic neuroma. *Nature* 322:644–647.

Selleck, M. A. J., and Bronner-Fraser, M. 1995. Origins of the avian neural crest: The role of neural plate-epidermal interactions. *Development* 121:525–538.

Selleck, M. A. J., Scherson, T. Y., and Bronner–Fraser, M. 1993. Origins of neural crest cell diversity. *Dev Biol* 159:1–11.

Selleck, M. A. J., and Stern, C. D. 1991. Fate mapping and cell lineage analysis of Hensen's node in the chick embryo. *Development* 112:615–626.

Sellman, S. 1940. Någro synpunkter på mekanismen vid Bildanet av Tandanlaget. *Odont Tidskr* 48:237–256.

Sellman, S. 1946. Some experiments on the determination of the larval tooth in *Amblystoma mexicanum. Odont Tidskr* 54:1–128.

Seno, T., and Nieuwkoop, P. D. 1958. The autonomous and dependent differentiations of the neural crest in amphibians. *Proc K Ned Akad Wet Ser C* 61:489–498.

Serbedzija, G. N., Bronner-Fraser, M., and Fraser, S. E. 1989. A vital dye analysis of the timing and pathways of avian trunk neural crest cell migration. *Development* 106:809–816.

Serbedzija, G. N., Bronner-Fraser, M., and Fraser, S. E. 1992. Vital dye analysis of cranial neural crest cell migration in the mouse embryo. *Development* 116:297–307.

Serbedzija, G. N., Bronner-Fraser, M., and Fraser, S. E. 1994. Developmental potential of trunk neural crest cells in the mouse. *Development* 120:1709–1718.

Serbedzija, G. N., Burgan, S., Fraser, S. E., and Bronner-Fraser, M. 1991. Vital dye labelling demonstrates a sacral neural crest contribution to the enteric nervous system of chick and mouse embryos. *Development* 111:857–866.

Serbedzija, G. N., Chen, J.-N., and Fishman, M. C. 1998. Regulation of the heart field of zebrafish. *Development* 125:1095–1101.

Serbedzija, G. N., Fraser, S. E., and Bronner-Fraser, M. 1990. Pathways of trunk neural crest cell migration in the mouse embryo as revealed by vital dye labelling. *Development* 108:605–612.

Serbedzija, G. N., and McMahon, A. P. 1997. Analysis of neural crest cell migration in *Splotch* mice using a neural crest-specific LacZ reporter. *Dev Biol* 185:139–147.

Servetnick, M., and Grainger, R. M. 1991. Homeogenetic neural induction in *Xenopus*. *Dev Biol* 147:73-82.

Seufert, D. W., and Hall, B. K. 1990. Tissue interactions involving cranial neural crest in cartilage formation in *Xenopus laevis* (Daudin). *Cell Differ Dev* 32:153–166.

Seufert, D. W., Hanken, J., and Klymkowsky, M. W. 1994. Type II collagen distribution during cranial development in *Xenopus laevis*. *Anat Embryol* 189:81–89.

Shah, N. M., Groves, A. K., and Anderson, D. J. 1996. Alternate neural crest cell fates are instructively promoted by TGFß superfamily members. *Cell* 85:331–343.

Shah, S. B., Skromme, I., Hume, C. R., Kessler, D., Lee, K. J., Stern, C. D., and Dodd, J. 1997. Misexpression of chick Vg1 in the marginal zone induces primitive streak formation. *Development* 124:5127–5138.

Sham, M. H., Vesque, C., Nonchev, S., Marshall, H., Frain, N., Das Gupta, R., Whiting, J., Wilkinson, D., Charnay, P., and Krumlauf, R. 1993. The zinc finger gene Krox 20 regulates *HoxB2* (Hox 2.8) during hindbrain segmentation. *Cell* 72:183–196.

Shankar, K. R., Chuong, C.-M., Jaskoll, T., and Melnick, M. 1994. Effect of in ovo retinoic acid exposure on forebrain neural crest: In vitro analysis reveals up-regulation of N-CAM and loss of mesenchymal phenotype. *Dev Dyn* 200:89–102.

Shankar, K. R., Jaskoll, T. F., and Melnick, M. 1992. Comparative 3-D views of normal and retinoic acid-treated neural crest in chick embryos. *Acta Stereol* 11(suppl 1):507–512.

Sharman, A. C., and Holland, P. W. H. 1996. Conservation, duplication, and divergence of developmental genes during chordate evolution. *Neth J Zool* 46:47–67.

Sharman, A. C., and Holland, P. W. H. 1998. Estimation of *Hox* gene cluster number in lampreys. *Int J Dev Biol* 42:617–620.

Sharpe, C. R. 1990. Regional neural induction in *Xenopus laevis*. *BioEssays* 12:591–596.

Sherman, L., Stocker, K. M., Morrison, R., and Ciment, G. 1993. Basic fibroblast growth factor (bFGF) acts intracellularly to cause the transdifferentiation of avian neural crest-derived Schwann cell precursors into melanocytes. *Development* 118:1313–1326.

Shigetani, Y., Aizawa, S., and Kuratani, S. C. 1995. Overlapping origins of pharyngeal arch crest cells on the postotic hind-brain. *Dev Growth Differ* 37:733–746.

Shimamura, K., Hartigan, D. J., Martinez, S., Puelles, L., and Rubenstein, J. L. R. 1995. Longitudinal organization of the anterior neural plate and neural tube. *Development* 121:3923–3933.

Shimamura, K., and Rubenstein, J. L. R. 1997. Inductive interactions direct early regionalization of the mouse forebrain. *Development* 124:2709–2718.

Shimeld, S. M., McKay, I. J., and Sharpe, P. T. 1996. The murine homeobox gene *Msx-3* shows highly restricted expression in the developing neural tube. *Mech Dev* 55:201–210.

Shu, D.-G., Conway Morris, S., and Zhang, X.-L. 1996. A Pikaia-like chordate from the Lower Cambrian of China. *Nature* 384:157–158.

Shu, D.-G., Zhang, X.-L., and Chen, L. 1996. Reinterpretation of *Yunnanozoon* as the earliest known hemichordate. *Nature* 380:428–430.

Shubin, N. H. 1991. The implications of "the Bauplan" for development and evolution of the tetrapod limb. In: Hinchliffe, J. R., Hurle, J. M., and Summerbell, D., eds. *Developmental Patterns of the Vertebrate Limb.* New York: Plenum Press. pp. 411–421.

Shubin, N. H. 1995. The evolution of paired fins and the origin of tetrapod limbs: Phylogenetic and transformational approaches. *Evol Biol* 28:39–86.

Shuey, D. L., Sadler, T. W., and Lauder, J. M. 1992. Serotonin as a regulator of craniofacial morphogenesis—site specific malformations following exposure to serotonin uptake inhibitors. *Teratology* 46:367–378.

Shur, B. D. 1982. Cell surface glycosyltransferase activities during normal and mutant (*T*/*T*) mesenchyme migration. *Dev Biol* 91:149–162.

Sieber-Blum, M. 1989a. Commitment of neural crest cells to the sensory neuron lineage. *Science* 243:1608–1611.

Sieber-Blum, M. 1989b. Inhibition of the adrenergic phenotype in cultured neural crest cells by norepinephrine uptake inhibitors. *Dev Biol* 136:372–380.

Sieber-Blum, M. 1990. Mechanisms of neural crest diversification. *Comments Dev Neurobiol* 4:225–249.

Sieber-Blum, M. 1991. Role of the neurotrophic factors BDNF and NGF in the commitment of pluripotent neural crest cells. *Neuron* 6:949–956.

Sieber-Blum, M., and Cohen, A. M. 1980. Clonal analysis of quail neural crest cells: They are pluripotent and differentiate in vitro in the absence of noncrest cells. *Dev Biol* 80:96–106.

Sieber-Blum, M., Kumar, S. R., and Riley, D. A. 1988. *In vitro* differentiation of quail neural crest cells into sensory-like neuroblasts. *Dev Brain Res* 39:69–83.

Sieber-Blum, M., and Sieber, F. 1981. Tumor-promoting phorbol-esters promote melanogenesis and prevent expression of the adrenergic phenotype in quail neural crest cells. *Differentiation* 20:117–123.

Sieber-Blum, M., and Sieber, F. 1985. *In vitro* analysis of quail neural crest cell differentiation. In: Bottenstein, J. E., and Sato, G., eds. *Cell Culture in the Neurosciences. Current Topics in Neurobiology.* New York: Plenum Publishing Co. pp. 193–222.

Sieber-Blum, M., and Zhang, J.-M. 1997. Growth factor action in neural crest cell diversification. *J Anat* 191:493–499.

Siebert, J. R., Graham, J. M. Jr., and MacDonald, C. 1985. Pathologic features of the CHARGE association: Support for involvement of the neural crest. *Teratology* 31:331–336.

Simeone, A., Acampora, D., Gulisano, M., Stornaiuolo, A., and Boncinelli, E. 1992. Nested expression domains of four homeobox genes in developing rostral brain. *Nature* 358:687–690.

Singer, C. 1959. *A History of Biology to About the Year 1900. A General Introduction to the Study of Living Things*. 3rd ed. London and New York: Abelard-Schuman.

Skreb, N., Svajger, A., and Levak-Svajger, B. 1976. *Developmental Potentialities of the Germ Layers in Mammals*. New York: Elsevier/North-Holland. CIBA Found. Symp. 40 (new series). pp. 27–45.

Slavkin, H. C., Sasano, Y., Kikunaga, S., Bessem, C., Bringas, P. Jr., Mayo, M., Luo, W., Mak, G., Rall, L., and Snead, M. L. 1990. Cartilage, bone and tooth induction during early embryonic mouse mandibular morphogenesis using serumless, chemically-defined medium. *Conn Tissue Res* 24:41–52.

Smith, A., Robinson, V., Patel, K., and Wilkinson, D. G. 1997. The EphA4 and EphB1 receptor tyrosine kinases and ephrin-B2 ligand regulate targeted migration of branchial neural crest cells. *Curr Biol* 7:561–570.

Smith, H. M. 1969. The Mexican axolotl: Some misconceptions and problems. *Bioscience* 19:593-597, 615.

Smith, J., and Fauquet, M. 1984. Glucocorticoids stimulate adrenergic differentiation in cultures of migrating and premigratory neural crest. *J Neurosci* 4:2160–2172.

Smith, K. K. 1997. Comparative patterns of craniofacial development in eutherian and metatherian mammals. *Evolution* 51:1663–1678.

Smith, K. K., and Schneider, R. A. 1998. Have gene knockouts caused evolutionary reversals in the mammalian first arch? *BioEssays* 20:245–255.

Smith, M. M. 1991. Putative skeletal neural crest cells in early Late Ordovician vertebrates from Colorado. *Science* 251:301–303.

Smith, M. M., and Hall, B. K. 1990. Developmental and evolutionary origins of vertebrate skeletogenic and odontogenic tissues. *Biol Rev Camb Philos Soc* 65:277–374.

Smith, M. M., and Hall, B. K. 1993. A developmental model for evolution of the vertebrate exoskeleton and teeth: The role of cranial and trunk neural crest. *Evol Biol* 27:387–448.

Smith, M. M., Hickman, A., Amanze, D., Lumsden, A., and Thorogood, P. 1994. Trunk neural crest origin of caudal fin mesenchyme in the zebrafish *Brachydanio rerio*. *Proc R Soc Lond B Biol Sci* 256:137–145.

Smith, R. W. 1849. *A Treatise on the Pathology: Diagnosis and Treatment of Neuroma*. Dublin: Hodges and Smith.

Smith, S. C. 1996. Pattern formation in the urodele mechanoreceptive lateral line: What features can be exploited for the study of development and evolution? *Int J Dev Biol* 40:727–733.

Smith, S. C., Graveson, A. C., and Hall, B. K. 1994. Evidence for a developmental and evolutionary link between placodal ectoderm and neural crest. *J Exp Zool* 270:292–301.

Smith, S. C., Lannoo, M. J., and Armstrong, J. B. 1988. Lateral-line neuromast development in *Ambystoma mexicanum* and a comparison with *Rana pipiens*. *J Morphol* 198:367–379.

Smith, S. C., Lannoo, M. J., and Armstrong, J. B. 1990. Development of the lateral-line system in the axolotl: Placode specification, guidance of migration, and the origin of polarity. *Anat Embryol* 182:171–180.

Smith Fernandez, A., Pieau, C., Repérant, J., Boncinelli, E., and Wassef, M. 1998. Expression of the *Emx-1* and *Dlx-1* homeobox genes define three molecularly distinct domains in the telencephalon of mouse, chick, turtle and frog embryos: Implications for the evolution of telencephalic subdivisions in amniotes. *Development* 125:2099–2111.

Smith-Thomas, L. C., Davis, J. P., and Epstein, M. L. 1986. The gut supports neurogenic differentiation of periocular mesenchyme, a chondrogenic neural crest-derived cell population. *Dev Biol* 115:293–300.

Smith-Thomas, L. C., and Fawcett, J. W. 1989. Expression of Schwann cell markers by mammalian neural crest cells *in vitro*. *Development* 105:251–262.

Smith-Thomas, L. C., Lott, I., and Bronner-Fraser, M. 1987. Effects of isotretinoin on the behavior of neural crest cells *in vitro*. *Dev Biol* 123:276–281.

Smits-van Prooije, A. E., Poelmann, R. E., Gesink, A. F., van Groeningen, M. J., and Vermeij-Keers, C. 1986. The cell surface coat in neurulating mouse and rat embryos, studied with lectins. *Anat Embryol* 175:111–117.

Smits-van Prooije, A. E., Vermeij-Keers, C., Dubbeldam, J. A., Mentink, M. M. T., and Poelmann, R. E. 1987. The formation of mesoderm and mesectoderm in presomite rat embryos cultured in vitro using WGA-Au as a marker. *Anat Embryol* 176:71–77.

Smits-van Prooije, A. E., Vermeijs-Keers, C., Poelmann, R. E., Mentink, M. M. T., and Dubbeldam, J. A. 1985. The neural crest in presomite to 40-somite murine embryos. *Acta Morphol Neerl-Scand* 23:99–114.

Smits-van Prooije, A. E., Vermeijs-Keers, C., Poelmann, R. E., Mentink, M. M. T., and Dubbeldam, J. A. 1988. The formation of mesoderm and mesectoderm in 5- to 40-somite rat embryos cultured in vitro, using WGA-Au as a marker. *Anat Embryol* 177:245–256.

Snow, M. H. L. 1981. Growth and its control in early mammalian development. *Br Med Bull* 37:221–226.

Snow, M. H. L., and Tam, P. P. L. 1979. Is compensatory growth a complicating factor in mouse teratology? *Nature* 279:555–557.

Sobotta, J. 1935. Beiträge zur Histogenese der sogenannten Ganglienleiste der Werbeltiere. Nach Untersuchungen an Selachiern und Amphibien. *Z Mikosk-Anat Forsch* 38:660–688.

Sohal, G. S., Bockman, D. E., Ali, M. M., and Tsai, N. T. 1996. DiI labeling and homeobox gene Islet-1 expression reveal the contribution of ventral neural tube cells to the formation of the avian trigeminal ganglion. *Int J Dev Neurosci* 14:419–427.

Song, Q., Mehler, M. F., and Kessler, J. A. 1998. Bone morphogenetic proteins induce apoptosis and growth factor dependence of cultured sympathoadrenal progenitor cells. *Dev Biol* 196:119–127.

Soriano, P. 1997. The PDGFa receptor is required for neural crest cell development and for normal patterning of the somites. *Development* 124:2691–2700.

Sordino, P., and Duboule, D. 1996. A molecular approach to the evolution of vertebrate paired appendages. *Trends Ecol Evol* 11:114–119.

Sordino, P., van der Hoeven, F., and Duboule, D. 1995. *Hox* gene expression in teleost fins and the origin of vertebrate digits. *Nature* 375:678–681.

Sower, S. A. 1998. Brain and pituitary hormones of lampreys, recent findings and their evolutionary significance. *Am Zool* 38:15–38.

Spence, S. G., and Poole, T. J. 1994. Developing blood vessels and associated extracellular matrix as substrates for neural crest migration in Japanese quail, *Coturnix coturnix japonica*. *Int J Dev Biol* 38:85–98.

Spranger, J., Benirschke, K., Hall, J. G., Lenz, W., Lowry, R. B., Opitz, J. M., Pinsky, L., Schwarzacher, H. G., and Smith, D. W. 1982. Errors of morphogenesis: Concepts and terms. *J Pediatr* 100:160–165.

Starck, D., and Siewing, R. 1980. Zue Diskussion der Begriffe Mesenchym und Mesoderm. *Zool Jb Anat* 103:374–388.

Stark, M. R., Sechrist, J., Bronner-Fraser, M., and Marcelle, C. 1997. Neural tube-ectoderm interactions are required for trigeminal placode formation. *Development* 124:4287–4295.

Stearner, S. P. 1946. Pigmentation studies in salamanders, with especial reference to the changes at metamorphosis. *Physiol Zool* 19:375–404.

Steele, C. E., Plenefisch, J. D., and Klein, N. W. 1982. Abnormal development of cultured rat embryos in rat and human sera prepared after vitamin A ingestion. *Experientia* 38:1237–1239.

Steen, T. P. 1968. Stability of chondrocyte differentiation and contribution of muscle to cartilage during limb regeneration in the axolotl (*Siredon mexicanum*). *J Exp Zool* 167:49–78.

Steen, T. P. 1970. Origin and differentiative capacities of cells in the blastema of the regenerating salamander limb. *Am Zool* 10:119–132.

Stefansson, K., Wollmann, R. L., Moore, B. W., and Arnason, B. G. W. 1982. S-100 protein in human chondrocytes. *Nature* 295:63–64.

Stejneger, L. 1907. Herpetology of Japan and adjacent territories. *Bull US Natl Mus* 58:1–577.

Stemple, D. L., and Anderson, D. J. 1992. Isolation of a stem cell for neurons and glia from the mammalian neural crest. *Cell* 71:973–985.

Stemple, D. L., and Anderson, D. J. 1993. Lineage diversification of the neural crest: *in vitro* investigations. *Dev Biol* 159:12–23.

Stennard, F., Ryan, K., and Gurdon, J. B. 1997. Markers of vertebrate mesoderm induction. *Curr Opin Genet Dev* 7:620–627.

Stern, C. D. 1990. The distinct mechanisms of segmentation? *Semin Dev Biol* 1:109–116.

Stern, C. D., Artinger, K. B., and Bronner-Fraser, M. 1991. Tissue interactions affecting the migration and differentiation of neural crest cells in the chick embryo. *Development* 113:207–216.

Stern, C. D., and Keynes, R. J. 1987. Interactions between somite cells: The formation and maintenance of segment boundaries in the chick embryo. *Development* 99:261–272.

Sternberg, J., and Kimber, S. J. 1986. The relationship between emerging neural crest cells and basement membranes in the trunk of the mouse embryo: A TEM and immunocytochemical study. *J Embryol Exp Morphol* 98:251–268.

Stewart, D. E., Kirby, M. L., and Sulik, K. K. 1986. Hemodynamic changes in chick embryos precede heart defects after cardiac neural crest ablation. *Circ Res* 59:545–550.

Stigson, M., and Kjellen, L. 1991. Large disulfide-stabilized proteoglycan complexes are synthesized by the epidermis of axolotl embryos. *Arch Biochem* 290:391–396.

Stigson, M., Löfberg, J., and Kjellen, L. 1997. Reduced epidermal expression of a Pg-M/versican-like proteoglycan in embryos of the white mutant axolotl. *Exp Cell Res* 236:57–65.

Stocker, K. M., Sherman, L., Rees, S., and Ciment, G. 1991. Basic FGF and TGF-ß1 influence commitment to melanogenesis in neural crest-derived cells of avian embryos. *Development* 111:635–645.

Stokes, M. D., and Holland, N. D. 1995. Embryo and larvae of a lancelet, *Branchiostoma floridae*, from hatching through metamorphosis: Growth in the laboratory and external morphology. *Acta Zool* 76:105–120.

Stone, L. S. 1922. Experiments on the development of the cranial ganglia and the lateral line sense organs in *Amblystoma punctatum*. *J Exp Zool* 35:421–496.

Stone, L. S. 1926. Further experiments on the extirpation and transplantation of mesectoderm in *Amblystoma punctatum*. *J Exp Zool* 44:95–131.

Stone, L. S. 1929. Experiments showing the role of migrating neural crest (mesectoderm) in the formation of head skeleton and loose connective tissue in *Rana palustris*. *Wilhelm Roux Arch EntwMech Org* 118:40–77.

Stone, L. S. 1932. Transplantation of the hyobranchial mesentoderm including the right lateral anlage of the second basibranchium in *Ambystoma punctatum*. *J Exp Zool* 62:109–123.

Storey, K. G., Crossley, J. M., de Robertis, E. M., Norris, W. E., and Stern, C. D. 1992. Neural induction and regionalization in the chick embryo. *Development* 114:729–741.

Strahan, R. 1958. Speculations on the evolution of the agnathan head. *Proc Centenary Bicentenary Cong Biol Singapore* 83–94.

Ströer, W. F. H. 1933. Experimentelle Untersuchungen über die Mundentwicklung bei den Urodelen. *Wilhelm Roux Arch EntwMech Org* 130:131–186.

Sulik, K. K. 1984. Craniofacial defects from genetic and teratogen-induced deficiencies in presomite embryos. *Birth Defects Orig Art Ser* 20 (3):79–98.

Sulik, K. K., Cook, C. S., and Webster, W. S. 1988. Teratogens and craniofacial malformations: Relationships to cell death. *Development* 103(suppl):213–231.

Sulik, K. K., and Johnston, M. C. 1983. Sequence of developmental alterations following acute ethanol exposure in mice: Craniofacial features of the fetal alcohol syndrome. *Am J Anat* 166:257–270.

Sulik, K. K., Johnston, M. C., and Webster, W. S. 1986. Retinoic acid embryopathy: Recently introduced drug causes predictable craniofacial malformations. *J Dental Res* 65:167.

Sumida, H., Akimoto, N., and Nakamura, H. 1989. Distribution of the neural crest cells in the heart of birds: A three-dimensional analysis. *Anat Embryol* 180:29–35.

Suzuki, A., Ueno, N., and Hemmati-Brivanlou, A. 1997. *Xenopus msx-1* mediates epidermal induction and neural inhibition by BMP4. *Development* 124:3037–3044.

Suzuki, H. R., and Kirby, M. L. 1997. Absence of neural crest cell regeneration from the postotic neural tube. *Dev Biol* 184:222–233

Suzuki, H. R., Padanilam, B. J., Vitale, E., Ramirez, F., and Solursh, M. 1991. Repeating developmental expression of G-Hox 7, a novel homeobox-containing gene in the chicken. *Dev Biol* 148:375–388.

Suzuki, N., Svensson, K., and Eriksson, V. J. 1996. High glucose concentration inhibits migration of rat cranial neural crest cells *in vitro*. *Diabetologia* 39:401–411.

Suzuki, T., Oohara, I., and Kurokawa, T. 1998. Hoxd-4 expression during pharyngeal arch development in flounder (*Paralichthys olivaceus*) embryos and effects of retinoic acid on expression. *Zool Sci* 15:57–67.

Svajger, A., Levak-Svajger, B., Kostovic-Knezevic, L., and Bradamante, Z. 1981. Morphogenetic behaviour of the rat embryonic ectoderm as a renal homograft. *J Embryol Exp Morphol* 65(suppl):243–267.

Swalla, B. J. 1992. The role of maternal factors in ascidian muscle development. *Semin Dev Biol* 3:287–295.

Swalla, B. J., and Jeffery, W. R. 1995. A maternal RNA localized in the yellow crescent is segregated to the larval muscle cells during ascidian development. *Dev Biol* 170:353–364.

Tacke, K., and Grunz, H. 1988. Close juxtaposition between inducing chordamesoderm and reacting neurectoderm is a prerequisite for neural induction in *Xenopus laevis*. *Cell Differ* 24:33–44.

Takahashi, K., Nuckolls, G. H., Tanaka, O., Semba, I., Takahashi, I., Dashner, R., Shum, L., and Slavkin, H. C. 1998. Adenovirus-mediated ectopic expression of *Msx2* in even-numbered rhombomeres induces apoptotic elimination of cranial neural crest cells *in ovo*. *Development* 125:1627–1635.

Takahashi, Y., Bontoux, M., and Le Douarin, N. M. 1991. Epithelio-mesenchymal interactions are critical for Quox-7 expression and membrane bone differentiation in the neural crest derived mandibular mesenchyme. *EMBO J.* 10:2387–2393.

Takahashi, Y., and Le Douarin, N. 1990. cDNA cloning of a quail homeobox gene and its expression in neural crest-derived mesenchyme and lateral plate mesoderm. *Proc Natl Acad Sci USA* 87:7482–7486.

Takamura, K., Okishima, T., Ohdo, S., and Hayakawa, K. 1990. Association of cephalic neural crest cells with cardiovascular development, particularly that of the semilunar valves. *Anat Embryol* 182:263–272.

Takiguchi-Hayashi, K., and Kitamura, K. 1993. The timing of appearance and pathway of migration of cranial neural crest-derived precursors of melanocytes in chick embryos. *Dev Growth Differ* 35:173–179.

Takihara, Y., Tomotsune, D., Shirai, M., Katoh-Fukui, Y., Nishii, K., Molaleb, M. A., Nomura, M., Tsuchiya, R., Fujita, Y., Shibata, Y., Higashinakagawa, T, and Shimada, K. 1997. Targeted disruption of the mouse homologue of the *Drosophila polyhomeotic* gene leads to altered anteroposterior patterning and neural crest defects. *Development* 124, 3673–3682.

Tam, P. P. L. 1989. Regionalization of the mouse embryonic ectoderm: Allocation of prospective ectodermal tissues during gastrulation. *Development* 107:55–68.

Tam, P. P. L., and Quinlan, G. A. 1996. Mapping vertebrate embryos. *Curr Biol* 6:104–106.

Tam, P. P. L., and Selwood, L. 1996. Development of lineages of primary germ layers, extra-embryonic membranes and fetus. *Reprod Fertil Dev* 8:803–805.

Tan, S.-S. 1991. Liver-specific and position-effect expression of a retinol-binding protein-*lacZ* fusion gene (*RBP-LacZ*) in transgenic mice. *Dev Biol* 146:24–37.

Tan, S.-S., and Morriss-Kay, G. M. 1984. Cephalic neural crest migration in the rat embryo. *J Embryol Exp Morphol* 82(suppl):123.

Tan, S.-S., and Morriss-Kay, G. M. 1985. The development and distribution of the cranial neural crest in the rat embryo. *Cell Tissue Res* 240:403–416.

Tan, S.-S., and Morriss-Kay, G. M. 1986. Analysis of cranial neural crest cell migration and early fates in postimplantation rat chimaeras. *J Embryol Exp Morphol* 98:21–58.

Tassabehji, M., Read, A. P., Newton, V. E., Harris, R., Balling, R., Gruss, P., and Strachan, T. 1992. Waardenburg's syndrome patients have mutations in the human homologue of the Pax-3 paired box gene. *Nature* 355:635–636.

Tassin, M. T., and Weill, R. 1981. Actions of excess vitamin A administered to pregnant mice during embryonic maxillae morphogenesis. *J Craniofac Genet Dev Biol* 1:299–314.

Taylor, L. E., Bennett, G. D., and Finnell, R. H. 1995. Altered gene expression in murine branchial arches following in utero exposure to retinoic acid. *J Craniofac Genet Dev Biol* 15:13–25.

Teichmann, H. 1955. Entwicklungsphysiologische Untersuchungen an der Nase des Alpen molches (*Triturus alpestris* Laur.). *Wilhelm Roux Arch EntwMech Org* 148:218–262.

Teichmann, H. 1959. Xenoplastischer Austausch der Nasenanlage zwischen Molch, Unke und Kröte. *Wilhelm Roux Arch EntwMech Org* 151:280–300.

Teichmann, H. 1961. Gestaltungsprinzipen der Nase von *Triturus*. *Embryologica* 6:110–118.

Teichmann, H. 1962. Experimente zur Analyse der Choanenentstehung und Formbildung der Nase bei *Triturus*. *Wilhelm Roux Arch EntwMech Org*. 153:455–485.

Teillet, M.-A., Kalcheim, C., and Le Douarin, N. M. 1987. Formation of the dorsal root ganglia in the avian embryo: Segmental origin and migratory behavior of neural crest progenitor cells. *Dev Biol* 120:329–347.

Teitelman, G. 1990. Insulin cells of pancreas extend neurites but do not arise from neuroectoderm. *Dev Biol* 142:368–379.

Terentiev, I. B. 1941. On the role played by the neural crest in the development of the dorsal fin in Urodela. *C R Acad Sci USSR* 31:91–94.

Thesleff, I. 1991. Tooth development. *Dental Update* November:382–397.

Thesleff, I. 1995. Homeobox genes and growth factors in regulation of craniofacial and tooth morphogenesis. *Acta Odontol Scand* 53:129–134.

Thesleff, I., and Sahlberg, C. 1996. Growth factors as inductive signals regulating tooth morphogenesis. *Semin Cell Dev Biol* 7:185–193.

Thesleff, I., and Sharpe, P. 1997. Signalling networks regulating dental development. *Mech Dev* 67:111–123.

Thesleff, I., Vaahtokari, A., and Partanen, A.-M. 1995. Regulation of organogenesis. Common molecular mechanisms regulating the development of teeth and other organs. *Int J Dev Biol* 39:35–50.

Thesleff, I., Vaahtokari, A., Kettungen, P., and Äberg, T. 1995. Epithelial-mesenchymal signalling during tooth development. *Conn Tissue Res* 32:9–15.

Thibaudeau, D. G., and Altig, R. 1988. Sequence of ontogenetic development and atrophy of the oral apparatus of six anuran tadpoles. *J Morphol* 197:63–70.

Thibaudeau, G., and Frost-Mason, S. K. 1992. Inhibition of neural crest cell differentiation by embryo ectodermal extract. *J Exp Zool* 261:431–440.

Thiery, J.-P., Duband, J. L., and Delouvée, A. 1982. Pathways and mechanisms of avian trunk neural crest cell migration and localization. *Dev Biol* 93:324–343.

Thisse, C., Thisse, B., and Postlethwait, J. H. 1995. Expression of *snail2*, a second member of the zebrafish *Snail* family, in cephalic mesendoderm and presumptive neural crest of wild-type and spadetail mutant embryos. *Dev Biol* 172:86–99.

Thomas, P., and Beddington, R. 1996. Anterior primitive endoderm may be responsible for patterning the anterior neural plate in the mouse embryo. *Curr Biol* 6:1487–1496.

Thomas, T., Kurihara, H., Yamagishi, H., Yazaki, Y., Olson, E. N., and Srivastava, D. 1998. A signaling cascade involving endothelin-1, dHAND and Msx1 regulates development of neural-crest-derived branchial arch mesenchyme. *Development* 125:3005–3014.

Thompson, D'A. W. 1942. On Growth and Form. Cambridge: Cambridge University Press.

Thomson, K. S. 1987. Speculations concerning the role of the neural crest in the morphogenesis and evolution of the vertebrate skeleton. In: Maderson, P. F. A., ed. *Developmental and Evolutionary Aspects of the Neural Crest*. New York: John Wiley & Sons. pp. 301–338.

Thomson, K. S. 1993. Segmentation, the adult skull and the problem of homology. In: Hanken, J., and Hall, B. K., eds. *The Vertebrate Skull. Vol. II. Patterns of Structural and Systematic Diversity*. Chicago: University of Chicago Press. pp. 36–68.

Thorndyke, M. C., and Probert, L. 1979. Calcitonin-like cells in the pharynx of the ascidian *Styela clava*. *Cell Tissue Res* 203:301–309.

Thorogood, P. V. 1987. Mechanisms of morphogenetic specification in skull development. In: Wolff, J. R., Sievers, J., and Berry, M., eds. *Mesenchymal/Epithelial Interactions in Neural Development*. Berlin: Springer-Verlag. NATO ASI Series, Cell Biology. pp. 141–152.

Thorogood, P. V. 1991. The development of the teleost fin and implications for our understanding of tetrapod limb evolution. In: Hinchliffe, J. R., Hurle, J. M., and Summerbell, D., eds. *Developmental Patterning of the Vertebrate Limb.* New York: Plenum Press. NATO ASI Series A: Life Sciences. pp. 347–354.

Thorogood, P. V. 1993a. Differentiation and morphogenesis of cranial skeletal tissues. In: Hanken, J., and Hall, B. K., eds. *The Skull. Vol. 1, Development.* Chicago: University of Chicago Press. pp. 112–152.

Thorogood, P. V. 1993b. Cranio-facial development: The problems of building a head. *Curr Biol* 3:705–708.

Thorogood, P. V. Bee, J., and Von der Mark, K. 1986. Transient expression of collagen type II at epithelio-mesenchymal interfaces during morphogenesis of the cartilaginous neurocranium. *Dev Biol* 116:497–509.

Thorogood, P. V., and Smith, L. 1984. Neural crest cells: The role of extracellular matrix in their differentiation and migration. In: Kemp, R. B., and Hinchliffe, J. R., eds. *Matrices and Cell Differentiation.* New York: Alan R. Liss, Inc. pp. 171–185.

Thorogood, P. V., Smith, L., Nicol, A., McGinty, R., and Garrod, D. 1982. Effects of vitamin A on the behaviour of migratory neural crest cells *in vitro. J Cell Sci* 57:331–350.

Tibbles, J. A. R., and Cohen, M. M. Jr. 1986. The Proteus syndrome: The elephant man diagnosed. *Br Med J* 293:683–685.

Tiedermann, H., Grunz, H., Loppnow-Blinde, B., and Tiedermann, H. 1994. Basic fibroblast growth factor can induce exclusively neural tissue in *Triturus* ectoderm explants. *Wilhelm Roux's Arch Dev Biol* 203:304–309.

Toerien, M. J. 1965a. Experimental studies on the columella-capsular interrelationships in the turtle, *Chelydra serpentina. J Embryol Exp Morphol* 14:265–272.

Toerien, M. J. 1965b. An experimental approach to the development of the ear capsule in the turtle, *Chelydra serpentina. J Embryol Exp Morphol* 13:141–149.

Tongiorgi, E., Bernhardt, R. R., Zinn, K., and Schachner, M. 1995. Tenascin-C mRNA is expressed in cranial neural crest cells, in some placodal derivatives, and in discrete domains of the embryonic zebrafish brain. *J Neurobiol* 28:391–407.

Toriello, H. V. 1988. New syndromes from old: The role of heterogeneity and variability in syndrome delineation. *Am J Med Genet* 1(suppl):50–70.

Tosney, K. W. 1982. The segregation and early migration of cranial neural crest cells in the avian embryo. *Dev Biol* 89:13–24.

Tosney, K. W., and Oakley, R. A. 1990. The perinotochordal mesenchyme acts as a barrier to axon advance in the chick embryo: Implications for a general mechanism of axonal guidance. *Exp Neurol* 109:75–89.

Trainor, P. A., and Tam, P. P. L. 1995. Cranial paraxial mesoderm and neural crest cells of the mouse embryo: Co-distribution in the craniofacial mesenchyme but distinct segregation in branchial arches. *Development* 121:2569–2582.

Trainor, T. A., Tan. S.-S., and Tam. P. P. L. 1994. Cranial paraxial mesoderm: Regionalization of cell fate and impact on craniofacial development in mouse embryos. *Development* 120:2397–2408.

Trampusch, H. A. L. 1941. On ear induction. *Acta Néerl Morphol* 4:195–213.

Tran, S., and Hall, B. K. 1989. Growth of the clavicle and development of clavicular secondary cartilage in the embryonic mouse. *Acta Anat* 135:200–207.

Tremblay, P., Kessel, M., and Gruss, P. 1995. A transgenic neuroanatomical marker identifies cranial neural crest deficiencies associated with the *Pax3* mutant *Splotch. Dev Biol* 171:317–329.

Treves, F. 1923. *The Elephant Man and Other Reminiscences.* Great Britain: Cassell & Co.

Trosko, J. E., Chang, C.-C., and Netzloff, M. 1982. The role of inhibited cell-cell communication in teratogenesis. *Teratog Carcinog Mutagen* 2:31–35.

Trueb, L., and Hanken, J. 1992. Skeletal development in *Xenopus laevis* (Anura: Pipidae). *J Morphol* 214:1–42.

Tschudi, J. J. 1838. *Classification der Batrachier.* Neuchatel.

Tsokos, M., Scarpa, S., Ross, R. A., and Triche, T. J. 1987. Differentiation of human neuroblastoma recapitulates neural crest development: Study of morphology, neurotransmitter enzymes and extracellular matrix proteins. *Am J Pathol* 128:484–496.

Tucker, A. S., Al Khamis, A., and Sharpe, P. T. 1998. Interactions between Bmp-4 and Msx-1 act to restrict gene expression to odontogenic mesenchyme. *Dev Dyn* 212:533–539.

Tucker, G. C., Aoyama, H., Lipinski, M., Turz, T., and Thiery, J. P. 1984. Identical reactivity of monoclonal antibodies HNK-1 and NC-1: Conservation in vertebrates on cells derived from neural primordium and on some leukocytes. *Cell Differ* 14:223–230.

Tucker, R. P., and Erickson, C. A. 1984. Morphology and behavior of quail neural crest cells in artificial three-dimensional extracellular matrices. *Dev Biol* 104:390–405.

Tucker, R. P., and McKay, S. E. 1991. The expression of tenascin by neural crest cells and glia. *Development* 112:1031–1039.

Tuckett, F., and Morriss-Kay, G. M. 1986. The distribution of fibronectin, laminin and entactin in the neurulating rat embryo studies by indirect immunofluorescence. *J Embryol Exp Morphol* 94:95–112.

Turley, E. A., Roth, S., and Weston, J. A. 1989. A model system that demonstrates interactions among extracellular matrix macromolecules. *Conn Tissue Res* 23:221–235.

Turner, D. L., and Weintraub, H. 1994. Expression of achaete-scute homolog 3 in *Xenopus* embryos converts ectodermal cells to a neural fate. *Genes Dev* 8:1434–1447.

Twitty, V. C. 1932. Influence of the eye on growth of its associated structures, studied by means of heteroplastic transplantation. *J Exp Zool* 61:333–374.

Twitty, V. C. 1936. Correlated genetic and embryological experiments on *Triturus.* I and II. *J Exp Zool* 74:239–302.

Twitty, V. C. 1944. Chromatophore migration as a response to mutual influences of the developing pigment cells. *J Exp Zool* 95:259–290.

Twitty, V. C. 1945. The developmental analysis of specific pigment patterns. *J Exp Zool* 100:141–178.

Twitty, V. C. 1949. Developmental analysis of amphibian pigmentation. *Growth Symp* 9:133–161.

Twitty, V. C., and Bodenstein, D. 1939. Correlated genetic and embryological experiments on *Triturus.* III. Further transplantation experiments on pigment development. IV: The study of pigment cell behavior *in vitro*. *J Exp Zool* 81:357–398.

Twitty, V. C., and Bodenstein, D. 1941. Experiments on the determination problem. I: The roles of ectoderm and neural crest in the development of the dorsal fin in Amphibia. II. Changes in ciliary polarity associated with the induction of fin epidermis. *J Exp Zool* 86:343–380.

Twitty, V. C., and Bodenstein, D. 1944. The effect of temporal and regional differentials on the development of grafted chromatophores. *J Exp Zool* 95:213–231.

Twitty, V. C., and Elliott, H. A. 1934. The relative growth of the amphibian eye, studied by means of transplantation. *J Exp Zool* 68:247–291.

Twitty, V. C., and Niu, M. C. 1948. Causal analysis of chromatophore migration. *J Exp Zool* 108:405–437.

Twitty, V. C., and Schwind, J. L. 1931. The growth of eyes and limbs transplanted heteroplastically between two species of *Amblystoma. J Exp Zool* 59:61–86.

Tyler, M. S., and Dewitt-Stott, R. A. 1986. Inhibition of membrane bone formation by vitamin A in the embryonic chick mandible. *Anat Rec* 214:193–197.

Urbánek, P., Fetka, I., Meisler, M. H., and Busslinger, M. 1997. Cooperation of *Pax2* and Pax5 in midbrain and cerebellum development. *Proc Natl Acad Sci USA* 94:5703–5708.

Vaessen, M.-J., Meikers, J. H. C., Bootsma, D., and van Kessel, A. G. 1990. The cellular retinoic-acid-binding protein is expressed in tissues associated with retinoic-acid-induced malformations. *Development* 110:371–378.

Vaglia, J. L., and Hall, B. K. 1997. Regulation of trunk neural crest during early development of teleost embryos. *J Morphol* 232:333A.

Vaglia, J. L., and Hall, B. K. 1999. Regulation of neural crest cell populations: occurrence, distribution and underlying mechanisms. *Int J Dev Biol* 43:95-110.

Valinsky, J. E., and Le Douarin, N. M. 1985. Production of plasminogen activator by migrating cephalic neural crest cells. *EMBO J* 4:1403–1406.

van Campenhout, E. 1930a. Historical survey of the development of the sympathetic nervous system. *Q Rev Biol* 5:23–50.

van Campenhout, E. 1930b. Contributions to the problem of the development of the sympathetic nervous system. *J Exp Zool* 56:295–320.

van Campenhout, E. 1931. Le développement du système nerveux sympathique chez le poulet. *Arch Biol Paris* 42:479–507.

van Campenhout, E. 1932. Further experiments on the origin of the enteric nervous system in the chick. *Physiol Zool* 5:333–353.

van Campenhout, E. 1941. Participation de l'épithélium à la constitution des plexus nerveux intrinsèques du dudénum de l'embryon de vache. *Arch Biol Paris* 52:473–508.

van Campenhout, E. 1946. The epithelioneural bodies. *Q Rev Biol* 21:327–347.

Vandersea, M. W., McCarthy, R. A., Fleming, P., and Smith, D. 1998. Exogenous retinoic acid during gastrulation induces cartilaginous and other craniofacial defects in *Fundulus heteroclitus. Biol Bull* 194:281–296.

Varley, J. E., Wehby, R. G., Rueger, D. C., and Maxwell, G. D. 1995. Number of adrenergic and islet-1 immunoreactive cells is increased in avian trunk neural crest cultures in the presence of human recombinant osteogenic protein-1. *Dev Dyn* 203:434–447.

Vermeij-Keers, C. 1972. Transformations in the facial region of the human embryo. *Adv Anat Embryol Cell Biol* 46(5):1–30.

Vermeij-Keers, C., and Poelman, R. E. 1980. The neural crest: A study on cell degeneration and the improbability of cell migration in the mouse embryo. *Neth J Zool* 30:74–81.

Verwoerd, C. D. A., van Oostrӧm, C. G., and Verwoerd-Verhoef, H. L. 1981. Otic placode and cephalic neural crest. *Acta Oto-Laryngol* 91:431–436.

Vielkind, J. R., Haasandela, H., Vielkind, U., and Anders, F. 1982. The induction of a specific pigment cell type by total genomic DNA injected into the neural crest region of fish embryos of the genus Xiphophorus. *Mol Gen Genetics* 185:379–389.

Vielkind, J. R., Tron, V. A., Schmidt, B. M., Dougherty, G. J., Ho, V. C., Woolcock, B. W., Sadaghiani, B., and Smith, C. J. 1993. A putative marker for human melanoma: A monoclonal antibody derived from the melanoma gene in the *Xiphophorus* melanoma model. *Am J Pathol* 143:656–662.

Vielle-Grosjean, I., Hunt, P., Gulisano, M., Boncinelli, E., and Thorogood, P. 1997. Branchial *Hox* gene expression and human craniofacial development. *Dev Biol* 183:49–60.

Vig, K. W. L., and Burdi, A. R., eds. 1988. *Craniofacial Morphogenesis and Dysmorphogenesis.* Monograph 21, Craniofacial Growth Series, Center for Human Growth and Development. Ann Arbor: University of Michigan.

Vincent, M., Duband, J.-L., and Thiery, J.-P. 1983. A cell surface determinant expressed early on migrating avian neural crest cells. *Dev Brain Res* 9:235–238.

Vincent, M., and Thiery, J.-P. 1984. A cell surface marker for neural crest and placodal cells: Further evolution in peripheral and central nervous system. *Dev Biol* 103:468–481.

Vischer, H. A. 1989a. The development of lateral-line receptors in *Eigenmannia* (Teleostei, Gymnotiformes). I. The mechanoreceptive lateral-line system. *Brain Behav Evol* 33:205–222.

Vischer, H. A. 1989b. The development of lateral-line receptors in *Eigenmannia* (Teleostei, Gymnotiformes). II. The electroreceptive lateral-line system. *Brain Behav Evol* 33:223–236.

Vogel, K. S., and Davies, A. M. 1993. Heterotopic transplantation of presumptive placodal ectoderm changes the fate of sensory neuron precursors. *Development* 119:263–276.

Vogt, W. 1925. Gestaltungsanalyse am Amphibienkeim mit örtlicher Vitalfärbung Vorwort über Wege und Ziele. I. Methodik und Wirkungsweise der örtlichen Vitalfärbung mit Agar als Farbträger. *Wilhelm Roux Arch EntwMech Org* 106:542–610.

von Baer, K. E.. 1828. *Über Entwicklungsgeschichte der Thiere. Beobactung und Reflexion.* Königsberg : Borntrager. Reprinted 1967 by Culture et Civilisation, Bruxelles.

Von der Mark, H., von der Mark, K., and Gat, S. 1976. Study of differential collagen synthesis during development of the chick embryo by immunofluorescence. I. Preparation of collagen type I and type II specific antibodies and their application to early stages of the chick embryo. *Dev Biol* 48:237–249.

von Kölliker, A., 1879. *Entwickelungsgeschichte des Menschen und der Höheren Thiere. Zweite ganz umgearbeitere Auflage.* Leipzig.

von Kölliker, A., 1884. Die Embryonalen Keimblätter und die Gewebe. *Zeit Wiss Zool* 40:179–213.

von Kölliker, A., 1889. *Handbuch der Gewebelehre des Menschen. 6. umgearbeitere Auflage. Erster Band: Die allgemeine Gewebelehre und die Systems der Haut.* Knochen und Muskeln. Leipzig.

von Kupffer, C. 1906. Die Morphogenie des Centralnervensystems [*Bdellostoma*]. In: Hertwig, O., ed. *Handbuch der Vergleichenden und Experimentellen Entwickelungslehre der Wirbeltiere.* Band II, Teil 3. Jena: Verlag von Gustav Fisher. pp. 24–38.

von Recklinghausen, F. D. 1882. *Über die multiplen fibromen der Haut und ihre Beziehung zu den multiplen Neuromen.* Berlin: A. Hirschwald.

von Schulte, H., and Tilney, F. 1915. Development of the neuraxis in the domestic cat to the stage of twenty-one somites. *Ann NY Acad Sci* 24:319–346.

Vukicevic, S., Kleinman, H. K., Luyten, F. P., Roberts, A. B., Roche, N. S., and Reddi, A. H. 1992. Identification of multiple active growth factors in basement membrane Matrigel suggests caution in interpretation of cellular activity related to extracellular matrix components. *Exp Cell Res* 202:1–8.

Vukicevic, S., Latin, V., Ping, C., Batorsky, R., Reddi, A. H., and Sampath, T. K. 1994. Localization of osteogenic protein-1 (bone morphogenetic protein-7) during human embryonic development: High affinity binding to basement membranes. *Biochem Biophys Res Commun* 198:693–700.

Wada, H., Holland, P. W. H., and Satoh, N. 1996. Origin of patterning in neural tubes. *Nature* 384:123.

Wada, H., Saiga, H., Satoh, N., and Holland, P. W. H. 1998. Tripartite organization of the ancestral chordate brain and the antiquity of placodes: Insights from ascidian *Pax-2/5/8, Hox* and *Otx* genes. *Development* 125:1113–1122.

Wagner, G. 1949. Die Bedeutung der Neuralleiste für die Kopfgestaltung der Amphibienlarven. Untersuchungen an Chimaeren von *Triton. Rev Suisse Zool* 56:519–620.

Wahlstrom, T., and Saxén, L. 1976. Malignant skin tumors of neural crest origin. *Cancer* 38:2022–2026.

Wake, D. B. 1976. On the correct scientific names of Urodeles. *Differentiation* 6:195.

Wake, D. B. 1993. Brainstem organization and branchiomeric nerves. *Acta Anat* 148:124–131.

Waldo, K. L., and Kirby, M. L. 1993. Cardiac neural crest contribution to the pulmonary artery and sixth aortic arch artery complex in chick embryos aged 6 to 18 days. *Anat Rec* 237:385–399.

Waldo, K. L., Kumiski, D., and Kirby, M. L. 1994. Association of the cardiac neural crest with development of the coronary arteries in the chick embryo. *Anat Rec* 239:315–331.

Waldo, K. L., Kumiski, D., and Kirby, M. L. 1996. Cardiac neural crest is essential for the persistence rather than the formation of an arch artery. *Dev Dyn* 205:281–292.

Waldo, K. L., Miyagawa-Tomita, S., Kumiski, D., and Kirby, M. L. 1998. Cardiac neural crest cells provide new insight into septation of the cardiac outflow tract: Aortic sac to ventricular septal closure. *Dev Biol* 196:129–144.

Wassersug, R. J., and Duellman, W. E. 1984. Oral structures and their development in egg-brooding hylid frog embryos and larvae: Evolutionary and ecological implications. *J Morphol* 182:1–37.

Watanabe Y., and Le Douarin, N. M. 1996. A role for BMP-4 in the development of subcutaneous cartilage. *Mech Dev* 57:69–78.

Waterman, R. E. 1975. SEM observations of surface alterations associated with neural tube closure in the mouse and hamster. *Anat Rec* 183:95–98.

Waterman, R. E. 1976. Topographical changes along the neural fold associated with neurulation in the hamster and mouse. *Am J Anat* 146:151–172.

Waterman, R. E. 1979. Embryonic and foetal tissues of vertebrates. In: Hodges, G. M., and Halowes, R. C., eds. *Biomedical Research Applications of Scanning Electron Microscopy.* New York: Academic Press. pp. 1–126.

Webb, J. F. 1990. Ontogeny and phylogeny of the trunk lateral line system in cichlid fishes. *J Zool (Lond)* 221:405–418.

Webb, J. F., and Noden, D. M. 1993. Ectodermal placodes: Contributions to the development of the vertebrate head. *Am Zool* 33:434–447.

Webb, J. F., and Northcutt, R. G. 1997. Morphology and distribution of pit organs and canal neuromasts in non-teleost bony fishes. *Brain Behav Evol* 50:139–151.

Webster, W. S. 1973. Embryogenesis of the enteric ganglia in normal mice and in mice that develop congenital aganglionic megacolon. *J Embryol Exp Morphol* 30:573–585.

Webster, W. S., Johnston, M. C., Lammer, E. J., and Sulik, K. K. 1986. Isotretinoin embryopathy and the cranial neural crest: An *in vivo* and *in vitro* study. *J Craniofac Genet Dev Biol* 6:211–222.

Wedden, S. E. 1987. Epithelial-mesenchymal interactions in the development of chick facial primordia and the target of retinoid action. *Development* 99:341–352.

Wedden, S. E., and Tickle, C. 1986. Quantitative analysis of the effects of retinoids on facial morphogenesis. *J Craniofac Genet Dev Biol* (suppl 2):169–178.

Wehrle-Haller, B., and Weston, J. A. 1995. Soluble and cell membrane forms of steel factor play distinct roles in melanocyte precursor dispersal and survival on the lateral neural crest migration pathway. *Development* 121:731–742.

Wehrle-Haller, B., and Weston, J. A. 1997. Receptor tyrosine kinase-dependent neural crest migration in response to differentially localized growth factors. *BioEssays* 19:337–345.

Weidenreich, F. 1912. Die Lokalization des Pigmentes und ihre Bedeutung in Ontogenie und Phylogenie der Wirbeltiere. *Z Morphol Anthrop* 2:59–140.

Weinstein, D. C., and Hemmati-Brivanlou, A. 1997. Neural induction in *Xenopus laevis*: Evidence for the default model. *Curr Opin Neurobiol* 7:7–12.

Weinstein, D. C., Honoré, E., and Hemmati-Brivanlou, A. 1997. Epidermal induction and inhibition of neural fate by translation initiation factor 4AIII. *Development* 124:4235–4242.

Weston, J. A. 1963. A radioautographic analysis of the migration and localization of trunk neural crest cells in the chick. *Dev Biol* 6:279–310.

Weston, J. A. 1970. The migration and differentiation of neural crest cells. *Adv Morphog* 8:41–114.

Weston, J. A. 1980. Role of the embryonic environment in neural crest morphogenesis. In: Pratt, R. M., and Christiansen, R. L., eds. *Current Research Trends in Prenatal Craniofacial Development*. Amsterdam: Elsevier/North Holland. pp. 27–45.

Weston, J. A. 1986. Phenotypic diversification in neural crest-derived cells: The time and stability of commitment during early development. *Curr Top Dev Biol* 20:195–210.

Weston, J. A. 1991. Sequential segregation and fate of developmentally restricted intermediate cell populations in the neural crest lineage. *Curr Topics Dev Biol* 25:133–153.

Whiteley, M., and Armstrong, J. B. 1991. Ectopic expression of a genomic fragment containing a homeobox causes neural defects in the axolotl. *Biochem Cell Biol* 69:366–374.

Whittaker, J. R. 1987. Cell lineages and determinants of cell fate in development. *Am Zool* 27:607–622.

Wicht, H., and Northcutt, R. G. 1995. Ontogeny of the head of the Pacific hagfish (*Eptatretus stouti,* Myxinoidea): Development of the lateral line system. *Philos Trans R Soc Lond B Biol Sci* 349:119–134.

Wilde, C. E. Jr. 1955. The urodele neuroepithelium. The differentiation *in vitro* of the cranial neural crest. *J Exp Zool* 130:573–591.

Wiley, M. J., Cauwenbergs, P., and Taylor, I. M. 1983. Effects of retinoic acid on the development of the facial skeleton in hamsters: Early changes involving cranial neural crest cells. *Acta Anat* 116:180–192.

Wilkinson, D. G. 1993. Molecular mechanisms of segmental patterning in the vertebrate hindbrain and neural crest. *BioEssays* 15:499–505.

Wilkinson, D. G. 1995. Genetic control of segmentation in the vertebrate hindbrain. *Perspect Dev Neurobiol* 3:29–38.

Willhite, C. C., Hill, R. M., and Irving, D. W. 1986. Isotretinoin-induced craniofacial malformations in humans and hamsters. *J Craniofac Genet Dev Biol* 2:193–209.

Williams, N. A., and Holland, P. W. H. 1996. Old head on young shoulders. *Nature* 383:490.

Williams, N. A., and Holland, P. W. H. 1998. Gene and domain duplication in the Chordate *Otx* gene family: Insights from Amphioxus *Otx*. *Mol Biol Evol* 15:600–607.

Williamson, D. A., Parrish, E. P., and Edelman, G. M. 1991a. Distribution and expression of two interactive extracellular matrix proteins, cytotactin and cytotactin-binding proteoglycan, during development of *Xenopus laevis.* I. Embryonic development. *J Morphol* 209:189–202.

Williamson, D. A., Parrish, E. P., and Edelman, G. M. 1991b. Distribution and expression of two interactive extracellular matrix proteins, cytotactin and cytotactin-binding proteoglycan, during development of *Xenopus laevis.* II. Metamorphosis. *J Morphol* 209:203–213.

Willis, R. A. 1962. *The Borderland of Embryology and Pathology.* 2nd ed. London: Butterworths & Co.

Wilson, D. B., and Wyatt, D. P. 1988. Closure of the posterior neuropore in the *vL* mutant mouse. *Anat Embryol* 178:559–563.

Wilson, D. B., and Wyatt, D. P. 1995. Alterations in cranial morphogenesis in the *Lp* mutant mouse. *J Craniofac Genet Dev Biol* 15:182–189.

Wilson, E. B. 1894. *The Embryological Criterion of Homology.* Wood's Hole Biological Lectures. Boston. pp. 101–124.

Wilson, H. V. 1899. The embryology of the sea bass (*Serranus atrarius*). *Bull US Fish Comm* 9:209–277.

Wilson, P. A., and Hemmati-Brivanlou, A. 1995. Induction of epidermis and inhibition of neural fate by BMP-4. *Nature* 376:331–333.

Wilson, T. J., Davidson, N. J., Boyd, R. L., and Gershwin, M. E. 1992. Phenotypic analysis of the chicken thymic microenvironment during ontogenic development. *Dev Immunol* 2:19–27.

Winograd, J., Reilly, M. P., Roe, R., Lutz, J., Laughner, E., Xu, X., Hu, L., Asakura, T., van der Kock, C., Strandberg, J. D., and Semenza, G. L. 1997. Perinatal lethality and multiple craniofacial malformations in *Msx2* transgenic mice. *Human Mol Genetics* 6:369–379.

Wittbrodt, J., Meyer, A., and Schartl, M. 1998. More genes in fish? *BioEssays* 20:511–515.

Woellwarth, C. V. 1961. Die rolle des Neuralleistenmaterials und der Temperatur bei der Determination der Augenlinse. *Embyologia* 6:219–242.

Woerdeman, M. W. 1945. On induction of a dorsal fin by the neural crest in axolotl embryos. *Acta Néerl Morphol* 5:378–384.

Wonsettler, A. L., and Webb, J. F. 1997. Morphology and development of the multiple lateral line canals on the trunk in two species of *Hexagrammis* (Scorpaeniformes, Hexagrammidae). *J Morphol* 233:195–214.

Woo, K., and Fraser, S. E. 1998. Specification of hindbrain fate in the zebrafish. *Dev Biol* 197:283–296.

Wood, A., Ashhurst, D. E., Corbett, A., and Thorogood, P. 1991. The transient expression of type II collagen at tissue interfaces during mammalian craniofacial development. *Development* 111:955–968.

Woods, A., and Couchman, J. R. 1998. Syndecans: Synergistic activators of cell adhesion. *Trends Cell Biol* 8:890–192.

Wright, G. M., Armstrong, L. A., Jacques, A. M., and Youson, J. H. 1988. Trabecular, nasal, branchial, and pericardial cartilages in the sea lamprey, *Petromyzon marinus*: Fine structure and immunohistochemical detection of elastin. *Am J Anat* 182:1–15.

Wright, G. M., Keeley, F. W., and Youson, J. H. 1983. Lamprin: A new vertebrate protein comprising the major structural protein of adult lamprey cartilage. *Experientia* 39:495–496.

Wright, G. M., Keeley, F. W., Youson, J. H., and Babineau, D. L. 1984. Cartilage in the Atlantic hagfish, *Myxine glutinosa. Am J Anat* 169:407–424.

Wright, G. M., and Youson, J. H. 1982. Ultrastructure of mucocartilage in the larval sea lamprey, *Petromyzon marinus* L. *Am J Anat* 165:39–51.

Wright, G. M., and Youson, J. H. 1983. Ultrastructure of cartilage from young adult sea lamprey, *Petromyzon marinus* L.: A new type of vertebrate cartilage. *Am J Anat* 167:59–70.

Xu, R.-H., Jaebong, K., Taira, M., Shuning, Z., Sredni, D., and Hsiang-Fu, K. 1995. A dominant negative bone morphogenetic protein 4 receptor causes neuralization of *Xenopus* ectoderm. *Biochem Biophys Res Commun* 212:212–219.

Yamada, T. 1990. Regulations in the induction of the organized neural system in amphibian embryos. *Development* 110:653–659.

Yan, Y.-L., Hatta, K., Riggleman, B., and Postlethwait, J. H. 1995. Expression of a type II collagen gene in the zebrafish embryonic axis. *Dev Dyn* 203:363–376.

Yavarone, M. S., Shuey, D. L., Tamir, H., Sadler, T. W., and Lauder, J. M. 1993. Serotonin and cardiac morphogenesis in the mouse embryo. *Teratology* 47:573–584.

Yip, J. W. 1986. Migratory patterns of sympathetic ganglioblasts and other neural crest derivatives in chick embryos. *J Neurosci* 6:3465–473.

Yip, J. E., Kokich, V. H., and Shepard, T. H. 1980. The effect of high doses of retinoic acid on prenatal craniofacial development of *Macaca nemestrina. Teratology* 21:29–38.

Yntema, C. L. 1943. An experimental study of the origin of the sensory neurons and sheath cells of the IXth and Xth cranial nerves in *Amblystoma punctatum. J Exp Zool* 92:93–120.

Yntema, C. L. 1955. Ear and nose. In: Willier, B. H., Weiss, P. A., and Hamburger, V., eds. *Analysis of Development.* Philadelphia: Saunders. pp. 415–428.

Yntema, C. L., and Hammond, W. S. 1945. Depletions and abnormalities in the cervical sympathetic system of the chick following extirpation of the neural crest. *J Exp Zool* 100:237–263.

Yntema, C. L., and Hammond, W. S. 1947. The development of the autonomic nervous system. *Biol Rev Camb Philos Soc* 22:344–357.

Yu, B. D., Hanson, R. D., Hess, J. L., Horning, S. E., and Korsmeyer, S. J. 1998. MLL, a mammalian *trithorax*-group gene, functions as a transcriptional maintenance factor in morphogenesis. *Proc Natl Acad Sci USA* 95:10632–10636.

Yuan, S., and Schoenwolf, G. C. 1998. *De novo* induction of the organizer and formation of the primitive streak in an experimental model of notochord reconstitution in avian species. *Development* 125:201–213.

Zagris, N., and Chung, A. E. 1990. Distribution and functional role of laminin during induction of the embryonic axis in the chick embryo. *Differentiation* 43:81–86.

Zelditch, M. L. 1988. Ontogenetic variation in patterns of phenotypic integration in the laboratory rat. *Evolution* 42:28–41.

Zelditch, M. L., and Carmichael, A. C. 1989. Ontogenetic variation in patterns of developmental and functional integration in skulls of *Sigmodon fulviventer. Evolution* 43:814–824.

Zelkowitz, M. 1981. Neurofibromatosis fibroblasts: Abnormal growth and binding to epidermal growth factor. In: Riccardi, V. M., and Mulvihill, J. J., eds. *Neurofibromatosis (von Recklinghausen Disease).* New York: Raven Press. pp. 173–189.

Zhang, J., Hagopian-Donaldson, S., Serbedzija, G., Elsemore, J., Plehn-Dujowich, D., McMahon, A. P., Flavell, R. A., and Williams, T. 1996. Neural tube, skeletal and body wall defects in mice lacking transcription factor AP-2. *Nature* 381:238–241.

Ziller, C., Dupin, E., Brazeau, P., Paulin, D., and Le Douarin, N. M. 1983. Early segregation of a neuronal precursor cell line in the neural crest as revealed by culture in a chemically defined medium. *Cell* 32:627–638.

Zottoli, S. J., and Seyfarth, E.-A. 1994. Julia B. Platt (1857–1935): Pioneer comparative embryologist and neuroscientist. *Brain Behav Evol* 43:91–106

Index

Figures in the color insert are indexed as I-1 to I-4 to indicate insert pages 1-4.